# Advances in Experimental Medicine and Biology

Volume 1203

More information about this series at http://www.springer.com/series/5584

Marlene Oeffinger • Daniel Zenklusen
Editors

# The Biology of mRNA: Structure and Function

Springer

*Editors*
Marlene Oeffinger
Systems biology (IRCM), Biochemistry and Molecular Medicine (UdeM), Experimental Medicine (McGill)
Institut de recherches cliniques de Montréal (IRCM), Université de Montréal, McGill University
Montreal, QC, Canada

Daniel Zenklusen
Department of Biochemistry and Molecular Medicine
Université de Montréal
Montréal, QC, Canada

ISSN 0065-2598 ISSN 2214-8019 (electronic)
Advances in Experimental Medicine and Biology
ISBN 978-3-030-31433-0 ISBN 978-3-030-31434-7 (eBook)
https://doi.org/10.1007/978-3-030-31434-7



This Springer imprint is published by the registered company Springer Nature Switzerland AG.
The registered company address is: Gewerbestrasse 11, 6330 Cham, Switzerland

# Preface

Regulating gene expression is an essential task for every cellular system in which mRNA serves as the central messenger molecule of the gene expression pathway bridging the information stored in DNA and providing the template for protein synthesis. Yet, mRNA is far from a simple copy of genetic information, and the path from its making to serving in proteins synthesis is one of the most complex cellular processes, involving hundreds of factors acting at different stages. Moreover, mRNA does not simply exist in cells in isolation, but rather associates with RNA binding proteins (RBPs) that are crucial in determining its fate. The formation of mRNA–protein complexes (mRNPs) starts as early as transcription initiation, and the continued dynamic interaction with proteins defines all steps in the complex life cycle of an mRNA molecule—all the way to its degradation. Not surprisingly, defects in any of these steps, including mutations in proteins implicated in the different stages of mRNA biogenesis, are associated with many disease phenotypes and have been increasingly linked to neurological diseases. This emphasizes the need for a better mechanistic understanding of mRNA metabolism and a more detailed understanding of the processes mediating the different steps of the gene regulation pathway, be it in the context of healthy cells or disease models. As it stands, the field is well on its way to tackle this challenge, applying ever new experimental approaches to expand our knowledge and insight into mRNA structure and function.

In this book, we aim to offer an overview on the many aspects of mRNA regulation and the approaches used to study different facets of this complex process. In Chap. 1, Wende, Friedhoff, and Strässer describe the interdependent nature of RNA transcription and mRNP assembly and how this determines the ability of mRNPs to be exported to the cytoplasm, the latter of which can be achieved by many different pathways as detailed by Scott, Aguilar, Kramar, and Oeffinger in Chap. 2. In Chap. 3, Wegener and Müller-McNicoll outline the role of one of the main classes of mRNA binding proteins, the serine and arginine-rich protein family (SR proteins), as key determinants of mRNP formation, identity, and fate. mRNPs are subjected to different quality control mechanisms, and in Chap. 4, Schmid and

Jensen illustrates the structural and functional features of one of the cellular RNA turnover machineries, the RNA exosome. Many of the regulatory processes acting on mRNAs, including turnover, are mediated by proteins binding to their 5′ and 3′ untranslated regions (UTRs). In Chap. 5, Beilharz, See, and Boag outlines further how studies in the nematode *Caenorhabditis elegans* have contributed to our understanding of how 3′ UTR sequences and their binding proteins participate in controlling protein expression in space and time. In Chap. 6, Fakim and Fabian follow with a chapter that describes how communication between 3′ and 5′ end of an mRNA regulates the cytoplasmic fate of mRNAs as well as of different viral RNAs, modulating translation and turnover, while in Chap. 7, Bouvrette, Blanchette, and Lécuyer describe the use of computational approaches that define RNA subcellular localization to define RNA sequence motifs for RBP binding to 3′ UTRs. Many RNA binding proteins have now been shown to assemble into membrane-less organelles and these assemblies have been associated with a number of neurodegenerative diseases, as outlined in Chap. 8, Sidibé and Vande Velde. As microscopy has been a central approach to study RNA metabolism, in Chap. 9, Adivarahan and Zenklusen provide a perspective on the role of RNA imaging in shaping our current view of all different aspects of mRNA life. And in Chap. 10, Sauvageau looks at lncRNAs, molecules with many mRNA-like features, and how the same approaches used to study mRNPs allow us to dissect their binding partners and divergent functions. We are grateful to the authors who have contributed their time, thought, and words and have made this book possible.

Montreal, Canada

Marlene Oeffinger
Daniel Zenklusen

# Contents

# Chapter 1
# Mechanism and Regulation of Co-transcriptional mRNP Assembly and Nuclear mRNA Export

**Wolfgang Wende, Peter Friedhoff, and Katja Sträßer**

**Abstract** mRNA is the "hermes" of gene expression as it carries the information of a protein-coding gene to the ribosome. Already during its synthesis, the mRNA is bound by mRNA-binding proteins that package the mRNA into a messenger ribonucleoprotein particle (mRNP). This mRNP assembly is important for mRNA stability and nuclear mRNA export. It also often regulates later steps in the mRNA lifetime such as translation and mRNA degradation in the cytoplasm. Thus, mRNP composition and accordingly the assembly of nuclear mRNA-binding proteins onto the mRNA are of crucial importance for correct gene expression. Here, we review our current knowledge of the mechanism of co-transcriptional mRNP assembly and nuclear mRNA export. We introduce the proteins involved and elaborate on what is known about their functions so far. In addition, we discuss the importance of regulated mRNP assembly in changing environmental conditions, especially during stress. Furthermore, we examine how defects in mRNP assembly cause diseases and how viruses exploit the host's nuclear mRNA export pathway. Finally, we summarize the questions that need to be answered in the future.

**Keywords** mRNA · RNA-binding protein · RBP · mRNA assembly · mRNP · Nuclear mRNA export

## 1.1 Introduction

Messenger RNAs (mRNAs) are couriers that bring the genome-encoded information to synthesize proteins to ribosomes. In eukaryotes, the genomic information is stored in the nucleus, whereas the ribosomes reside in the cytoplasm. Thus, all

---

W. Wende · P. Friedhoff · K. Sträßer (✉)
Department of Biology and Chemistry, Institute of Biochemistry, Justus Liebig University Giessen, Giessen, Germany
e-mail: wolfgang.wende@chemie.bio.uni-giessen.de; peter.friedhoff@chemie.bio.uni-giessen.de; katja.straesser@chemie.bio.uni-giessen.de

 
M. Oeffinger, D. Zenklusen (eds.), *The Biology of mRNA: Structure and Function*, Advances in Experimental Medicine and Biology 1203,
https://doi.org/10.1007/978-3-030-31434-7_1

mRNAs have to be transported from the nucleus through the nuclear pores to the cytoplasm.

RNA polymerase II (RNAPII) synthesizes mRNAs by transcribing protein-coding genes (Fig. 1.1). Transcription occurs in three distinct steps, initiation, elongation, and termination. This transcription cycle is regulated by various post-transcriptional modifications, including modifications of the C-terminal domain (CTD) of the largest subunit of RNAPII (Buratowski 2009; Harlen and Churchman 2017; Jeronimo et al. 2016; Zaborowska et al. 2016). The CTD is composed of heptapeptide repeats with the consensus sequence YSPTSPS, and the differential posttranslational modification of the CTD controls the recruitment of many regulatory factors and mRNA processing factors to RNAPII (Jeronimo et al. 2013). Phosphorylation of the CTD is probably the best understood regulatory process of the transcription cycle (Heidemann et al. 2012). When RNAPII first binds to the promoter, the CTD has a low phosphorylation status. During initiation, the CTD becomes phosphorylated on serine 5 (S5P) and serine 7 (S7P), whereas during elongation, phosphorylation of S5P strongly decreases and phosphorylation of Y1, S2, and T4 increases in the yeast *S. cerevisiae*. At the end of a transcription unit, Y1P decreases upstream of the polyadenylation site, whereas S2P and T4P decrease downstream of the polyadenylation site in yeast. In humans, this pattern of CTD phosphorylation is highly conserved, except that Y1P is high at the transcription start site and decreases shortly downstream (Harlen and Churchman 2017; Heidemann et al. 2012; Jeronimo et al. 2016; Zaborowska et al. 2016). Analysis of the nascent RNA associated with the differently phosphorylated forms of RNAPII showed that S2P and S5P increase just downstream of the transcription start site, slowly decrease during elongation, and drop shortly before the transcription termination site (Nojima et al. 2015). Furthermore, S5P peaks over exonic sequences, and S2P increases at the cleavage and polyadenylation site (Nojima et al. 2015).

During and after synthesis, the (pre-)mRNA is processed: It is capped at its 5′ end, introns are removed by splicing, and, after release of the transcript by cleavage at its 3′ end, the mRNA becomes polyadenylated. In addition to the control of the transcription cycle, specific CTD modifications coordinate transcription with mRNA processing events by binding to the appropriate mRNA processing factors (Harlen and Churchman 2017; Hsin and Manley 2012; Jeronimo et al. 2016). Importantly, the CTD is crucial for the assembly of nuclear mRNA-binding proteins (mRBPs) onto the mRNA, i.e., the assembly of messenger ribonucleoprotein particles (mRNPs) (Meinel and Strasser 2015). Here, the CTD also functions as a landing platform for proteins that package the mRNA into an mRNP. The function of the CTD in mRNP assembly will be discussed in detail in Sect. 1.2: co-transcriptional mRNP assembly.

Besides the CTD, the chromatin state influences mRNA processing events, such as splice site recognition and choice, by affecting the elongation rate of RNAPII and/or recruitment of the spliceosome and splicing factors via specific epigenetic marks [for reviews, see Brown et al. (2012), Dargemont and Babour (2017), and Luco et al. (2010)]. Furthermore, recent studies indicate that chromatin structure also influences 3′ end processing (Dargemont and Babour 2017). Important in this

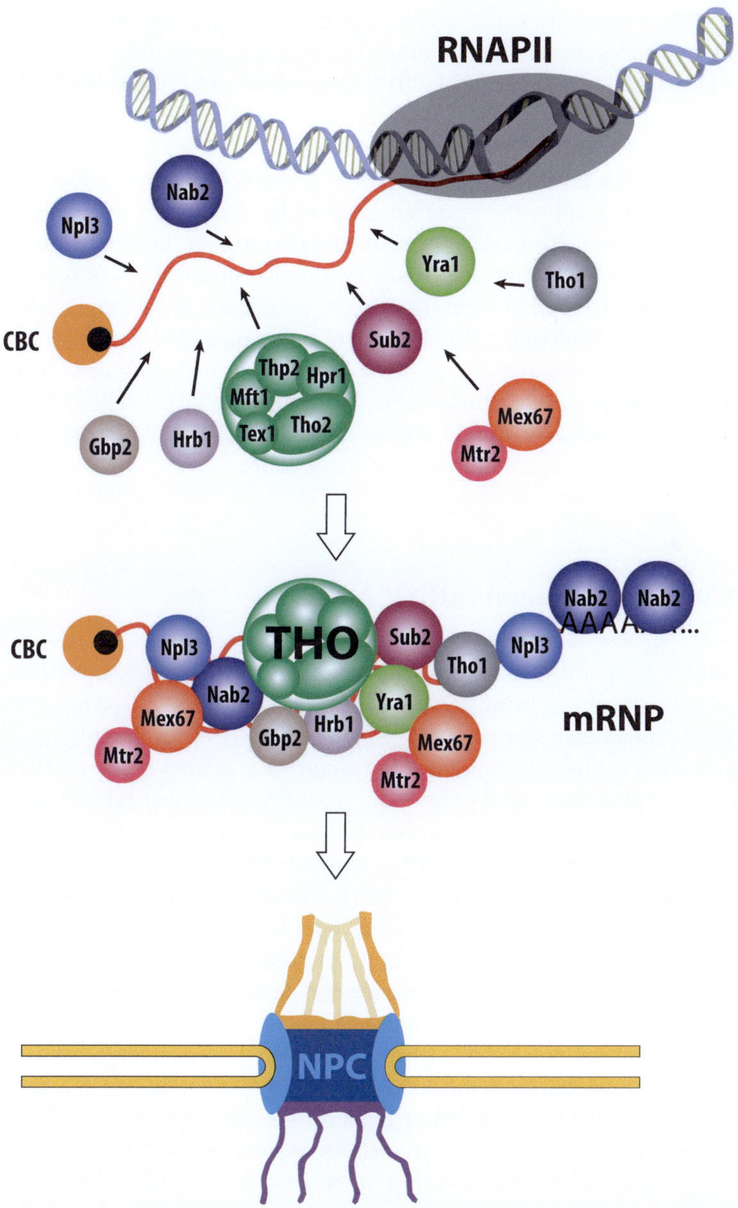


**Fig. 1.1** Model of mRNP assembly. RNA polymerase II (RNAPII, gray) transcribes protein-coding genes and synthesizes the corresponding mRNA. Already during transcription, the mRNA is processed (not shown). In addition, nuclear mRNA-binding proteins (colored) bind to the mRNA leading to the formation of an mRNP. Binding of nuclear mRNA-binding proteins to the mRNA is required for mRNA stability and to mediate nuclear mRNA export through the nuclear pore complexes (NPCs). The composition of an mRNP often controls cytoplasmic events such as localization, translation rate, and stability of the mRNA

context, mRNP assembly is also influenced by chromatin state as chromatin modifications or chromatin remodelers affect the recruitment of different mRNP components to the mRNA (Brown et al. 2012; Dargemont and Babour 2017; Luco et al. 2010). Lastly, chromatin is also involved in the quality control of nuclear mRNP assembly.

Notably, mRNA never exists by itself but is always bound by proteins. mRNA processing factors bind—often co-transcriptionally—to the mRNA and carry out capping, splicing, and cleavage/polyadenylation. Moreover, nuclear mRBPs bind to the mRNA and package it into an mRNP. This protects the mRNA from degradation and is necessary for export of the mRNA to the cytoplasm. Interestingly, the function and destination of numerous mRNAs are not only embedded within its nucleotide sequence but also determined by proteins bound to it, including their cytoplasmic localization, translation rate, and half-life (Gehring et al. 2017). Thus, the composition of the mRNP and, accordingly, the assembly of nuclear mRBPs onto the mRNA are of crucial importance for correct gene expression.

## 1.2 Co-transcriptional mRNP Assembly

The assembly of an mRNP begins co-transcriptionally when nuclear mRBPs bind to the mRNA as it emerges from RNAPII (Fig. 1.1) (Bjork and Wieslander 2017; Lei et al. 2001; Meinel and Strasser 2015; Singh et al. 2015). As the different steps of gene expression are highly coordinated, the correct assembly of an mRNP depends on many other processes. In *S. cerevisiae*, probably all nuclear mRNP components are known, but their functions have remained largely enigmatic, except that most of them have to assemble into the mRNP to ensure mRNA stability and nuclear export (Table 1.1). Thus, many questions remain about the function of the different mRNP components in mRNP assembly as well as later steps in the lifetime of an mRNA (see Sect. 1.6). Homologs of most of these proteins have been identified in higher eukaryotes, underscoring their importance for nuclear mRNP assembly and thus gene expression (Table 1.1). The recruitment mechanism for many of these proteins to the mRNP has been described at least in some detail. By and large, nuclear mRNP components are recruited to an mRNP by binding directly to the mRNA and/or after initially binding to the transcription machinery. The recruitment and function of the different proteins that associate with mRNAs to form an mRNP will be discussed below by focusing on mRNP assembly in the yeast *S. cerevisiae*. The orchestrated assembly of mRNP component leads to a mature and export-competent mRNP, ready to be exported to the cytoplasm Nuclear mRNA export is mediated by the mRNA export receptor Mex67-Mtr2/NXF1-NXT1 that directly interacts with components of the nuclear pore complex.

**Table 1.1** Proteins involved in nuclear mRNP formation

| *S. cerevisiae* | *H. sapiens* | Protein complex | Function(s) |
|---|---|---|---|
| Cbc1/Cbp80/Sto1 | Cbp80/NCBP1 | CBC | Large subunit of the CBC |
| Cbc2/Cbp20 | Cbp20/NCBP2 | CBC | Small subunit of the CBC |
| | NCBP3 | | Alternative form of NCBP2 important under stress conditions |
| Hpr1 | THOC1 | TREX/THO | Transcription elongation, prevention of hyper-recombination, transcription-coupled DNA repair (TCR), nuclear mRNA export |
| Tho2 | THOC2 | TREX/THO | |
| Tex1 | THOC3 | TREX/THO | |
| | THOC5 | TREX/THO | |
| | THOC6 | TREX/THO | |
| | THOC7 | TREX/THO | |
| Mft1 | | TREX/THO | |
| Thp2 | | TREX/THO | |
| Gbp2 | | TREX | |
| Hrb1 | | TREX | |
| Sub2 | UAP56/DDX39B | TREX | |
| | DDX39A | TREX | |
| Yra1 | ALYREF/THOC4 | TREX | |
| | UIF | | Additional human TREX subunits |
| | LUZP4 | | |
| | CHTOP | | |
| | POLDIP3 | | |
| | ZC3H11A | | |
| Tho1 | CIP29/SARNP | | Nuclear mRNA export; in humans classified as TREX subunit |
| Npl3 | | | Transcription elongation, 3′ end formation, nuclear mRNA export |
| Nab2 | ZC3H14 | | Poly(A) binding, nuclear mRNA export |
| Sac3 | GANP | THSC | Nuclear mRNA export, chromatin modification, THSC was also named TREX-2 |
| Thp1 | PCID2 | THSC | |
| Cdc31 | Centrin/CENP | THSC | |
| Sus1 | ENY1 | THSC | |
| Sem1 | DSS1 | THSC | |
| Pab1 | PABPN | | Poly(A) binding |
| Mex67 | NXF1/TAP | mRNA exporter | mRNA exporter |
| Mtr2 | NXT1/p15 | mRNA exporter | |

This table summarizes the proteins involved in nuclear mRNP formation in *S. cerevisiae* and their human homologs. In addition, the protein complex they belong to (if any) and known function (s) are listed.

### 1.2.1 Cap-Binding Complex

The cap-binding complex (CBC) binds to the $m^7G$-cap structure at the 5′ end of the mRNA and consists of a large and a small subunit: Cbp80 and Cbp20 in *S. cerevisiae* and NCBP1 and NCBP2 in human cells. CBC binds to the 5′-$m^7G$-cap via its Cbp20 subunit, while Cbp80 is needed for high-affinity binding of the CBC to the cap as well as the interaction with other proteins. Thus, the CBC is recruited to the mRNA by a direct interaction with the cap structure. Functionally, the CBC is important to protect the mRNA from degradation as well as for transcription elongation, splicing, nuclear mRNA export, and translation [reviewed in Gonatopoulos-Pournatzis and Cowling (2014)].

### 1.2.2 The TREX Complex

The TREX complex is one of the first complexes associating with mRNAs and couples *tr*anscription to mRNA *ex*port (Strasser et al. 2002). It promotes transcription elongation and binds to the mRNA in a co-transcriptional manner contributing to mRNP assembly (Heath et al. 2016). TREX consists of the heteropentameric THO complex (consisting of Tho1, Hpr1, Mft1, Thp2, and Tex1), the nuclear mRNA export factors Sub2 and Yra1, and the SR-like proteins Gbp2 and Hrb1 (Hurt et al. 2004; Strasser et al. 2002). TREX recruitment to the site of transcription is complex and mediated by interaction of THO with the transcription machinery, in particular the S2 and S2-S5 phosphorylated CTD, as well as interactions of several of its subunits with the nascent RNA (Abruzzi et al. 2004; Meinel et al. 2013). Furthermore, in *S. cerevisiae*, TREX occupancy at intron-containing and also intronless genes requires the Prp19 complex (Prp19C), a complex with a well-established function in splicing (Chanarat et al. 2011). Thus, Prp19C functions in TREX recruitment in addition to its function in splicing (Chanarat et al. 2011, 2012). The complexity of TREX recruitment is best illustrated by its subunit Yra1, which is recruited to the mRNP by its interaction with the TREX component Sub2, a helicase of the DEAD-box family (Strasser and Hurt 2001), as well as Pcf11, a component of the cleavage and polyadenylation complex (Johnson et al. 2009). Moreover, Dbp2, another DEAD-box family helicase, is also required to recruit Yra1 to the mRNA (Ma et al. 2013). In addition, Yra1 is recruited by the RNA itself (Meinel et al. 2013). Furthermore, ubiquitylation of both the histone H2B and Swd2, a component of the H3K4 methyltransferase complex and the cleavage and polyadenylation complex, is necessary for the recruitment of Yra1 as well as Nab2 (see below) to the mRNA (Vitaliano-Prunier et al. 2012). Yra1 in turn directly interacts with and thereby recruits the mRNA exporter Mex67-Mtr2 to the mRNP and thus contributes directly to nuclear mRNA export (Strasser and Hurt 2000) (also see below).

Although some TREX subunits are specific for *S. cerevisiae* and others for *H. sapiens* (Table 1.1), TREX and its functions are generally well conserved in many organisms, suggesting their physiological relevance (Heath et al. 2016). However, the mechanism of recruitment seems to differ from lower to higher eukaryotes. Consistent with a higher occurrence of introns and an important role of splicing in mRNP assembly in human cells, TREX recruitment is thought to occur by the spliceosome during splicing and involves the interaction of ALYREF, the human homolog of Yra1, with the exon junction complex (EJC) component eIF4AIII (Gromadzka et al. 2016; Masuda et al. 2005). The EJC is deposited on mRNAs just upstream of each exon-exon junction during splicing (Boehm and Gehring 2016). This connection between splicing and TREX most likely explains the early finding that splicing enhances mRNA export (Valencia et al. 2008). TREX also interacts with PRP19C in human cells, suggesting that PRP19C might also be involved in TREX recruitment to the mRNP in higher eukaryotes, as it is in yeast—however, this might occur during splicing rather than transcription, although the two processes are highly coupled in human cells and might be difficult to separate (Chanarat and Strasser 2013; Dufu et al. 2010). In addition, TREX is recruited to the 5′ end of the mRNA by the interaction of its component ALYREF with the CBC (Cheng et al. 2006; Nojima et al. 2007). Recently, a transcriptome-wide study revealed that ALYREF is present not only at the 5′ but also at the 3′ ends of mRNAs in a CBP80- and PABPN1 (polyadenylate-binding nuclear protein 1)-dependent manner, respectively (Shi et al. 2017). Furthermore, TREX is recruited to naturally intronless genes independently of splicing by recognizing specific RNA sequence elements (Lei et al. 2011, 2013; Shi et al. 2017). In addition, as in *S. cerevisiae*, TREX may have a role in transcription since ALYREF functions in transcription of at least a subset of genes (Stubbs and Conrad 2015). Also similar to yeast, ALYREF interacts with UAP56/DDX39B, the human homolog of Sub2, which recruits ALYREF to the mRNP. ALYREF in turn, recruits the export receptor NXF1-NXT1, the human homologs of Mex67-Mtr2, to the mRNP (Luo et al. 2001; Taniguchi and Ohno 2008). Furthermore, several other adaptors exist for the recruitment of NXF1-NXT1 to the mRNA (Luo et al. 2001; Taniguchi and Ohno 2008), such as UIF (UAP56-interacting factor) (Hautbergue et al. 2009). CHTOP (chromatin target of PRMT1 protein) is another mRNA export adaptor that likely functions similarly to ALYREF and UIF and, like ALYREF, requires UAP56 for its loading onto the mRNA (Chang et al. 2013). In mammalian cells, two other proteins, POLDIP3 (polymerase delta-interacting protein 3) and ZC3H11A (zinc finger CCCH domain-containing protein 11A), associate with the TREX complex in an ATP-dependent manner and function in nuclear mRNA export (Folco et al. 2012). Furthermore, UAP56 has a paralog, DDX39A (URH49), which probably serves overlapping functions to UAP56 (Pryor et al. 2004; Yamazaki et al. 2010). However, the functions of POLDIP3, ZC3H11A, and DDX39A remain to be determined. Taken together, several adaptors are likely to work together to load the mRNA export receptor Mex67-Mtr2/NXF1-NXT1 onto the mRNA.

A connection between chromatin and mRNP assembly, and here especially the TREX complex, also exists in human cells. ALYREF interacts with IWS1 (interacts with SUPT6H), a chromatin remodeler that in turn interacts with the transcription elongation factor SPT6, which is recruited to the transcription machinery by binding to the S2 phosphorylated CTD (Yoh et al. 2007). As depletion of IWS1 leads to a decrease of ALYREF at genes and a nuclear mRNA export defect (Yoh et al. 2007), IWS1 and SPT6 are probably needed for the co-transcriptional recruitment of ALYREF to the mRNA and thus correct mRNP assembly. UIF interacts not only with UAP56 but also with the histone chaperone FACT (facilitates chromatin transcription), an interaction that is required for the recruitment of UIF to the mRNA (Hautbergue et al. 2009).

### 1.2.3 Tho1 and CIP29

Tho1 is a conserved nuclear mRBP that was proposed to function complementarily to Sub2 as both, overexpression of *THO1* and overexpression of *SUB2*, suppress the defects of a *Δhpr1* strain (Jimeno et al. 2006). Furthermore, Tho1 is recruited to transcribed genes in a THO- and RNA-dependent manner (Jimeno et al. 2006). Interestingly, CIP29, the human homolog of Tho1, interacts with human TREX (hTREX) and is recruited to the mRNA in a splicing- and cap-dependent manner (Dufu et al. 2010). In addition, CIP29 interacts with UAP56, the homolog of Sub2, in an ATP-dependent manner (Dufu et al. 2010). In *Arabidopsis*, mutants in MOS11, the plant homolog of Tho1/CIP29, exhibit nuclear accumulation of poly(A) RNA (Germain et al. 2010). Thus, Tho1/CIP29/MOS11 is probably recruited to the nuclear mRNP during transcription, and its recruitment to mRNPs is required for nuclear mRNA export. However, the mechanism of its recruitment as well as its function as part of the nuclear mRNP remains to be elucidated.

### 1.2.4 Npl3, Nab2, and Mammalian SR-Proteins

Several SR (serine, arginine)- and SR-like mRBPs are involved in packaging of the mRNA into an mRNP. Npl3 is an SR-like protein with roles in transcription elongation, splicing, 3′ end processing, as well as nuclear mRNA export (Bucheli and Buratowski 2005; Dermody et al. 2008; Kress et al. 2008; Lee et al. 1996). Npl3 is recruited co-transcriptionally by direct interaction with the mRNA and the S2 phosphorylated CTD (Dermody et al. 2008; Meinel et al. 2013) and is therefore another mRNP component that binds to the mRNA at an early step of mRNP assembly. Association and dissociation of Npl3 with and from the mRNA are regulated by a phosphorylation cycle; the nuclear phosphatase Glc7 dephosphorylates Npl3, and only in this dephosphorylated form Npl3 binds to mRNA and recruits the mRNA exporter Mex67-Mtr2 (Gilbert and Guthrie 2004). In the cytoplasm, Sky1 phosphorylates Npl3 at one of the eight SR motifs (S411), which mediates its release from mRNA (Gilbert et al. 2001). Thus, Npl3 is a component of nuclear mRNPs that recruits the mRNA exporter Mex67-Mtr2 to the mRNA and accompanies the mRNA to the cytoplasm.

Nab2 is a serine-rich nuclear poly(A)-binding protein that functions in poly (A) tail length control, nuclear mRNP assembly, and nuclear mRNA export (Batisse et al. 2009; Green et al. 2002; Hector et al. 2002). Nab2 binds to RNA, and, consistently, RNA is needed for its recruitment to the site of transcription (Anderson et al. 1993; Meinel et al. 2013). Furthermore, as for Yra1, ubiquitylation of H2B and Swd2 as well as the RNA helicase Dbp2 are necessary for Nab2 recruitment (Ma et al. 2013; Vitaliano-Prunier et al. 2012). Interestingly, Nab2 dimerizes upon binding to the RNA (Aibara et al. 2017). Like Npl3, Nab2 packages the mRNA into an mRNP and recruits the mRNA exporter Mex67-Mtr2 to the mRNA. However, the exact function of Nab2 and its molecular mode of action are still unclear.

In human cells, several SR-proteins serve as adaptor proteins for the main export receptor NXF1-NXT1 (Huang et al. 2003; Lai and Tarn 2004; Muller-McNicoll et al. 2016). The SR-proteins 9G8 and SRp20/SRSF3 interact with NXF1 in a manner competitive to ALYREF (Huang et al. 2003). SRSF1–7 bind mRNA adjacent to NXF1 with SRSF3 being the most potent NXF1 adaptor (Muller-McNicoll et al. 2016). SRSF3 and SRSF7 regulate 3′ UTR length in an opposing manner to each other, but both recruit NXF1 to the mRNA indicating a role in controlling the expression of transcripts with alternative 3′ ends (Muller-McNicoll et al. 2016). Interestingly, SR-proteins interact with NXF1 in their nonphosphorylated form, similar to Npl3 in yeast, one example being the SR-protein ASF/SF2 that binds to nuclear mRNPs in its hypophosphorylated form (Lai and Tarn 2004). The functions of ZC3H14, the human homolog of Nab2, seem to be conserved as, for example, poly(A) binding was shown in *H. sapiens*, *M. musculus*, *R. norvegicus*, and *D. melanogaster* (Kelly et al. 2014). In contrast, it is not known whether ZC3H14 also functions in mRNP assembly and nuclear mRNA export. Nevertheless, several SR- and SR-like proteins are components of nuclear mRNPs and important for nuclear mRNA export (see also Chap. 3).

### 1.2.5 THSC Complex

The THSC complex, also named TREX-2, is composed of Thp1, Sac3, Sus1, Cdc31, and Sem1. THSC is required for nuclear mRNA export and interacts with NXF1-NXT1/Mex67-Mtr2 as well as the nuclear pore complex (NPC) (Fischer et al. 2002; Umlauf et al. 2013; Wickramasinghe et al. 2010). Interestingly, in yeast THSC also interacts with two complexes involved in transcription, the SAGA histone acetylase complex and the promoter-bound mediator complex (Rodriguez-Navarro et al. 2004; Schneider et al. 2015). This suggests that, if these physical interactions are functionally linked, THSC could facilitate nuclear mRNA export by coupling transcription to the transport through the nuclear pore complex. However, function(s) and possible mechanism of THSC remain to be explored.

### 1.2.6 The mRNA Exporter Mex67-Mtr2

The mRNA exporter Mex67-Mtr2 binds directly to the mRNA as well as nuclear pore proteins and thus exports the mRNP out of the nucleus and to the cytoplasm (Segref et al. 1997; Strasser et al. 2000). Their human homologs were first named TAP-p15 and later renamed NXF1-NXT1. As all proteins involved in the mRNP assembly, Mex67-Mtr2 is recruited to the mRNA during transcription through interactions with multiple proteins as described above. The proteins recruiting Mex67-Mtr2/NXF1-NXT1 to the mRNA are therefore often called export adaptors and include the TREX complex, specifically its components Hpr1 and Yra1, as well as Npl3 and Nab2 (Gilbert and Guthrie 2004; Gwizdek et al. 2006; Iglesias et al. 2010; Strasser and Hurt 2000). In mammalian cells, TREX (via its components ALYREF and THOC5), CHTOP, several SR-proteins, and ZC3H3 (zinc finger CCCH domain-containing protein 3) recruit NXF1-NXT1 to the mRNA (Chang et al. 2013; Huang et al. 2003; Hurt et al. 2009; Viphakone et al. 2012). Most likely, these export adaptors function to increase the weak intrinsic affinity of Mex67-Mtr2/NXF1-NXT1 for RNA as, for example, NXF1 binds to RNA weakly and its affinity is increased by the adaptor proteins ALYREF and THOC5 (Viphakone et al. 2012). These interactions are partially regulated by posttranslational modifications such as phosphorylation (see above) and ubiquitylation (Nino et al. 2013). The specific—and maybe transcript-specific—functions of each of these export adaptors largely remain unexplored as well as their mechanistic function in mRNP assembly.

### 1.2.7 General Aspects of mRNP Assembly

The correct assembly of an mRNP is important for many later steps of gene expression such as mRNA stability and nuclear mRNA export as well as various cytoplasmic processes such as mRNA localization, translation rate, and mRNA degradation. Thus, the composition of each mRNP is crucial for physiologically correct gene expression patterns. During mRNP assembly, the recruitment of individual nuclear mRBPs to the mRNA can be achieved in various ways: the RNA itself, other mRBPs, chromatin, and/or the CTD of RNAPII. Binding to the RNA is thought to be mostly sequence-unspecific: only the SR-like TREX components Gbp2 and Hrb1 show a preference for degenerate sequence motifs (Baejen et al. 2014; Riordan et al. 2011; Tuck and Tollervey 2013), and Nab2 prefers degenerate GUA-rich motifs in addition to the poly(A) tail (Baejen et al. 2014; Riordan et al. 2011; Tuck and Tollervey 2013). Thus, the question arises how these mRNA export adaptors, especially all the ones that do not show any RNA sequence specificity, find their "correct place" on the mRNA. The lack of sequence specificity suggests that most mRNA export adaptors are recruited to the mRNA by protein-protein interaction(s) and are subsequently loaded onto the mRNA. The absence of highly specific RNA-binding motifs within these mRNP components might be necessary to bind to all the different mRNAs that share limited sequence identity, as their primary sequences have evolved for protein coding. In addition, the interplay of all the proteins that assemble the mRNA into an mRNP is not known (see also Sect. 1.6).

## 1.3 RNA Helicases Involved in mRNP Assembly

RNA helicases form a large group among RNA-binding proteins: There are roughly 40 RNA helicases in *S. cerevisiae* and about 70 in humans (Jankowsky and Harris 2015). They are involved in almost all aspects of RNA life such as mRNP assembly, splicing, export, transport and storage, translation activation and inhibition, as well as mRNA degradation. Frequently, a particular RNA helicase is involved in multiple pathways (Bourgeois et al. 2016). RNA helicases are members of the P-loop nucleoside-triphosphatase (NTPase) superfamily, which use nucleoside triphosphate (NTP) binding and NTP hydrolysis to exert their function and unwind or remodel RNA structures as well as mRNPs (Linder and Jankowsky 2011). Based on sequence motifs and structural features, helicases were grouped into six superfamilies (SF 1–6). Their RecA-like domains, which lie within one polypeptide chain (RecA1 and RecA2) or within different subunits, form the core of the enzymes (Gorbalenya and Koonin 1993; Singleton et al. 2007). Variable N- and C-terminal extensions and, less frequently, insertion in the RecA-like domains determine specificity and function of the RNA helicase (Linder and Jankowsky 2011). Systematic sequence analysis of RNA helicases of bacterial, eukaryotic, viral, and archaeal origin revealed that most RNA helicases belong to the SF2 family (Jankowsky et al. 2011; Moukhtar et al. 2017).

RNA helicases can switch between open, half-open, and closed conformations depending on the bound nucleotide state (ATP, ADP + $P_i$, ADP, or empty), association with RNA, and/or regulatory proteins (Sloan and Bohnsack 2018). Both, RNAs and protein cofactors, can affect the affinity for nucleotides, the rate of hydrolysis, or binding to other factors (Putnam and Jankowsky 2013; Rudolph and Klostermeier 2015; Sloan and Bohnsack 2018). Many of the RNA helicase protein cofactors share common interaction domains, e.g., the MIF4G (middle of eIF4G) domain that binds to eIF4A-like DEAD-box RNA helicases and the G-patch domain that binds the OB (oligosaccharide-binding) fold of DEAH-box helicases (Aravind and Koonin 1999; Sloan and Bohnsack 2018). In contrast to the highly processive DNA and RNA helicases involved in replication, RNA helicases involved in mRNP assembly often unwind only few base pairs (10–15 bp) and rather function in the clamping and remodeling of RNA and mRNP structures (Linder and Jankowsky 2011; Sloan and Bohnsack 2018). Due to their ability to modulate RNP structure and composition, helicases are likely key players in the orchestrated assembly and remodeling of mRNPs at different stages of the gene expression pathway. Most RNA helicases involved in mRNP assembly and export (excluding splicing) belong to the DEAD-box family (SF2 superfamily), which is the largest and best-studied family (Bourgeois et al. 2016; Linder and Jankowsky 2011; Linder et al. 1989).

In the following sections, we will discuss three helicases involved in nuclear mRNP packaging in more detail: Yeast Dbp2/human p68 (DDX5) and the two eIF4A-like RNA helicases yeast Sub2/human UAP56 (DDX39B) and yeast Dbp5/ human DBP5 (DDX19B). RNA helicases primarily involved in splicing are covered in a recent review by Ficner et al. (2017).

### *1.3.1 Dbp2/p68 (DDX5)*

Yeast Dbp2/human p68 (DDX5) participates in multiple processes within mRNP metabolism including transcription, splicing, RNA export, RNA decay, microRNA processing, rRNA processing, as well as storage/transport and shuttles between nucleus and cytoplasm [reviewed in Bourgeois et al. (2016) and Xing et al. (2018)]. Human p68 (DDX5) is one of the first proteins for which an RNA helicase activity was demonstrated (Hirling et al. 1989). Dbp2 is involved in co-transcriptional mRNP assembly as well as nuclear mRNA export (Cloutier et al. 2012). Genetic and physical interaction of Dbp2 with Yra1 (human ALYREF) and its role in loading Nab2 (human ZC3H14) and Mex67 (human NXF1) onto the mRNA demonstrated its role in mRNP assembly and export (Ma et al. 2013). The interaction between Yra1 and Dpb2 is an example for the regulation of an RNA helicase by a non-MIF4G or G-patch domain protein (Sloan and Bohnsack 2018). Yra1 inhibits the single-stranded RNA (ssRNA) binding and unwinding activity of Dbp2, thereby terminating mRNP rearrangements by Dbp2 (Ma et al. 2016). Loss of Yra1–Dbp2 interaction leads to accumulation of Dbp2 on mRNA (Ma et al. 2016). Crystal structures are only available for the recA1 domains of human DDX5 in the absence of (residues 52–304 PDB-ID 4A4D) and in complex with ADP (residues 71–304; PDB-ID 3FE2) (Dutta et al. 2012; Schutz et al. 2010). Thus, little is known about how interaction partners bind to and regulate the function of Dbp2 (DDX5).

### *1.3.2 Yeast Sub2/Human UAP56 (DDX39B)*

Human UAP56 was first uncovered as an *U*2AF65 (U2 small nuclear RNA auxiliary factor 2) *a*ssociated *p*rotein required for the U2 snRNP-branchpoint interaction during mRNA splicing (Fleckner et al. 1997). Its yeast ortholog Sub2—initially identified as a *su*ppressor of *brr1-1*—was also first known for its function in mRNA splicing (Noble and Guthrie 1996). However, Sub2 was later discovered to also function in nuclear mRNP assembly as part of the TREX complex, where it interacts directly with Yra1 (Strasser and Hurt 2001; Strasser et al. 2002). Likewise, the THO complex assembles in an ATP-dependent manner with UAP56, ALYREF, and CIP29 (yeast Tho1) to form the human TREX complex (Chi et al. 2013; Kota et al. 2008) that links mRNA processing to export (Zhou et al. 2000). Sub2's function in nuclear mRNA export is well documented for its orthologs from several species: *S. cerevisiae* (Jensen et al. 2001; Strasser et al. 2002), human (Luo et al. 2001), *D. melanogaster* (Gatfield et al. 2001; Ma et al. 2013), and *C. elegans* (MacMorris et al. 2003) [reviewed in Linder and Stutz (2001)]. In addition, roles of Sub2 in RNA transport and storage (Meignin and Davis 2008) and R-loop prevention (Gaillard et al. 2007; Gomez-Gonzalez et al. 2011) have been demonstrated.

In vitro, both human UAP56 (Shen et al. 2007) and yeast Sub2 (Ma et al. 2013; Saguez et al. 2013) are bona fide ATPases that bind ssRNA and unwind DNA or RNA from partial duplex substrates containing a 3′-overhang. In its ATP-bound state, human UAP56 forms a complex with ssRNA and ALYREF or CHTOP or CIP29 in vitro. ALYREF stimulates the ATPase and CIP29 the helicase activity of UAP56 leading to dissociation of UAP56 from the complex (Chang et al. 2013; Dufu et al. 2010; Taniguchi and Ohno 2008). Similarly, the ATPase activity of yeast Sub2 is stimulated by a C-terminal fragment of Yra1 (Ren et al. 2017). Upon UAP56 release, NXF1 binds to ALYREF (Hautbergue et al. 2008; Taniguchi and Ohno 2008). In addition to unwinding, yeast Sub2 can displace proteins such as Mud2, the potential homolog of human U2AF65, from mRNA during splicing (Kistler and Guthrie 2001; Linder and Jankowsky 2011).

Structures have been determined of human UAP56 in complex with different adenine nucleotides (residues 46–426; PDB-ID 1XTI, 1XTJ, 1XTK, 1T6N, 1T5I) (Shi et al. 2004; Zhao et al. 2004). Recently, yeast Sub2 was crystallized in complex with components of the THO complex (residues 62–270; 280–444 PDB-ID 5SUQ) at low resolution (6 Å) and at 2.6 Å resolution in complex with RNA and a C-terminal fragment of Yra1 (residues 64–444; PDB-ID 5SUP), which contains the C-box also found in UIF, CHTOP, Luzp4, and ALYREF (Ren et al. 2017). Notably, although the resolution of the THO-Sub2 complex was low, the interaction of a MIF4G-like domain of one of the THO proteins with Sub2 was observed similar to those of Gle1 with Dbp5, eIF4G with eIF4A, and CNOT1 with DDX6 (Ren et al. 2017). Despite these advances, it remains elusive which protein of the THO complex harbors a MIF4G-like domain and how RNA binding and helicase activity of Sub2 are controlled.

### *1.3.3 Yeast Dbp5/Human DBP5 (DDX19B)*

Another well-studied helicase implicated in different steps of mRNP metabolism is Dbp5. Dbp5 localizes mainly to the cytoplasmic side of nuclear rim, but was shown to associate with transcribed genes, suggesting an early role in the mRNA life cycle. Its role is best understood during nuclear mRNA export (Tieg and Krebber 2013). Directional export of larger complexes such as mRNPs out of the nucleus needs energy, and at least part of this energy requirement is provided by an ATP-helicase cycle involving Gle1 [tethered to the NPC via the nucleoporin hCG1 (Strahm et al.

1999)], Dbp5/DDX19, and inositol hexakisphosphate (IP6) (Alcazar-Roman et al. 2006; Delaleau and Borden 2015; Hodge et al. 2011; Noble et al. 2011). In this cycle, Dbp5 plays a key role in mRNP remodeling by removing Mex67 and Nab2 from the mRNP at the cytoplasmic site of the nuclear pore complex (Kohler and Hurt 2007; Weirich et al. 2006).

There are several structures of Dbp5 (DDX19) in complex with various interaction partners (all of which lack the first 50–90 residues) that provide insight into how Dpb5 is regulated by RNA, nucleotide, Gle1, Nup159, or Nup214 (Kubitscheck and Siebrasse 2017). The interaction between Dbp5 and Gle1 is mediated via the MIF4G domain of Gle1 and binding of IP6, which simulates the helicase activity of Dbp5 by releasing the unwound RNA (Kubitscheck and Siebrasse 2017; Montpetit et al. 2011). The details of this remodeling are still unclear as is the number of nuclear transport receptors removed by a single Dbp5 (Kubitscheck and Siebrasse 2017). In contrast to Dbp5, the human DDX19B functions independently of IP6 and is activated by Gle1, which induces a conformational change to remove auto-inhibition by the N-terminal helix of DDX19B (Lin et al. 2018).

## 1.4 mRNP Assembly and mRNA Export Under Stress

Adapting to changing environmental conditions, in particular an appropriate response to different stresses, is challenging for an organism. Cells have to switch quickly from one physiological state to an alternative program to ensure survival. Indispensable for this switch are the preferential synthesis and nuclear export of stress response transcripts, while general mRNAs are sequestered in the nucleus [reviewed in Zander and Krebber (2017)].

In *S. cerevisiae*, heat stress triggers the phosphorylation of Nab2 by the kinase Slt2 (Carmody et al. 2010) as well as the dissociation of Nab2, Npl3, Gbp2, Hrb1, and the export receptor Mex67-Mtr2 from non-heat shock transcripts, consequently blocking their nuclear export (Zander et al. 2016). The sequestration of Nab2, Yra1, and Mlp1 in nuclear foci and the aggregation of Gbp2 prevent their rebinding to nascent mRNA (Fig. 1.2a) (Carmody et al. 2010; Wallace et al. 2015). However, the mechanism of this process is still enigmatic. Simultaneously, heat shock transcripts are able to bypass mRNA quality control by binding to Mex67 without the help of any of the conventional adaptor proteins (Zander et al. 2016). Instead, Mex67 is recruited to the site of transcription by its interaction with Hsf1, the heat shock transcription factor 1, which is bound to the heat-shock-promoter element (Fig. 1.2b) (Zander et al. 2016). The export block for general mRNAs and the facilitated export of heat shock transcripts ensure a fast response upon stress (For further information regarding nuclear degradation and quality control of mRNA, see Chap. 4).

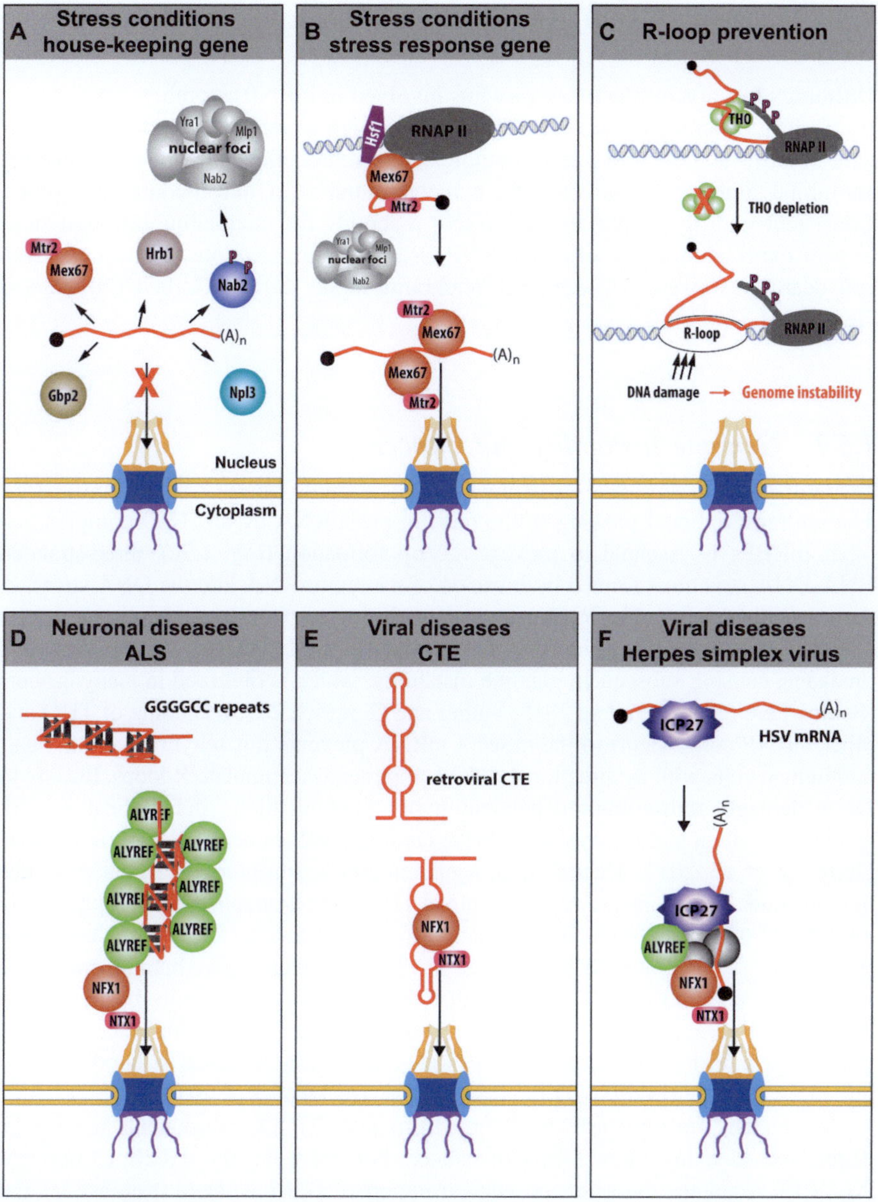


**Fig. 1.2** mRNP assembly in stress and disease conditions. (**a**) Stress conditions lead to the modification of export factors (like Nab2), their dissociation from housekeeping mRNAs, and their aggregation in nuclear foci, preventing normal mRNA export. (**b**) During heat stress, the transcription factor Hsf1 induces the transcription of stress response genes and recruits Mex67-Mtr2 to their transcripts mediating their selective nuclear export in *S. cerevisiae*. (**c**) Binding of the THO complex to the nascent mRNA prevents R-loop formation and contributes to genome stability. (**d**) In amyotrophic lateral sclerosis (ALS), binding of ALYREF to the GGGGCC repeats enriches the

## 1.5 Misregulation of mRNP Assembly Is Linked to Diseases

Orthologs for most of the many proteins involved in mRNP assembly can be found from yeast to human (Table 1.1). These have a high degree of evolutionary conservation and a tendency to higher complexity in multicellular organisms, reflecting additional biological functions and a higher number of intron-containing genes. Consistent with the importance of mRNP assembly for modulating different steps of gene expression, dysfunctional mRNP assembly causes various diseases and even cell death [reviewed in Carey and Wickramasinghe (2018), Corbett (2018), and Heath et al. (2016)].

### *1.5.1 Genome Instability and Cancer*

The co-transcriptional packing of the nascent pre-mRNA by the THO complex and other mRBPs is essential to prevent R-loop formation (Fig. 1.2c), three-stranded RNA-DNA structures formed by invasion of nascent mRNA into the DNA strand of the transcribed locus. The RNA reanneals with the template strand leaving a single-stranded coding strand, which thus becomes susceptible to DNA damage or strand breakage, causing subsequent genome instability, which is observed in many cancers (Santos-Pereira and Aguilera 2015; Sollier and Cimprich 2015). Binding of THO and other mRNP components to the nascent mRNA prevents this rehybridization. Yeast and human cells with dysfunctional THO components accumulate R loops, leading to DNA damage, transcription-dependent hyper-recombination, and replication fork stalling (Aguilera and Garcia-Muse 2012; Dominguez-Sanchez et al. 2011a; Gomez-Gonzalez et al. 2011; Huertas and Aguilera 2003). Intriguingly, recent evidence demonstrates a crosstalk between the human THO and chromatin modifiers promoting a local and transient chromatin compaction to prevent R-loop formation (Castellano-Pozo et al. 2013; Salas-Armenteros et al. 2017). Interestingly, human oncogenic viruses can also induce R-loop accumulation and genome instability by binding to TREX. ORF57, a protein of Kaposi's sarcoma-associated herpesvirus, can sequester the complete host TREX complex after infection, causing increased R-loop formation and DNA damage contributing to tumorigenesis (Jackson et al. 2014).

Apart from the influence on R-loop formation, TREX components have been implicated in many other forms of cancer. For example, the TREX component ALYREF regulates the selective nuclear export of DNA damage response-related

---

**Fig. 1.2** (continued) local concentration of RNA export factors and overrules the normal nuclear retention of these abnormal RNAs. (**e**) Simple retroviruses contain a constitutive transport element (CTE). The CTE binds with high affinity to NXF1 promoting nuclear export of the viral RNA. (**f**) The HSV protein ICP27 binds selectively to intronless virus transcripts and recruits ALYREF. This interaction allows the nuclear export of the viral mRNA via the NXF1 pathway

transcripts in response to phosphatidylinositol-trisphosphate production, regulated by the inositol polyphosphate multikinase (Wickramasinghe et al. 2013). This finding raises the question which other factors involved in mRNP biogenesis can modulate the export efficiency of specific classes of mRNA and may provide an explanation why dysregulation of the mRNA export machinery is found in numerous cancers (Adams et al. 2017; Vohhodina et al. 2017; Wickramasinghe and Laskey 2015). THOC1 expression is upregulated in many cancer cells, preferentially in ovarian, colon, and lung tumors, reflecting the fact that fast- growing tumors require efficient mRNP biogenesis (Chinnam et al. 2014; Dominguez-Sanchez et al. 2011b; Guo et al. 2005, 2012; Lapek et al. 2017; Li et al. 2007; Liu et al. 2015). However, downregulation of THOC1 expression was observed in testicular and skin cancer (Dominguez-Sanchez et al. 2011b). Dysregulation of ALYREF was also observed in a wide variety of tumors (Dominguez-Sanchez et al. 2011b; Saito et al. 2013). In oral squamous cell carcinoma cells, ALYREF is upregulated, thereby sequestering the metastasis modulators RRP1B (ribosomal RNA processing 1B) and CD82 (Cluster of Differentiation 82) leading to the promotion of metastasis (Saito et al. 2013). Phosphorylation of THOC5 (on T225) by the leukemogenic protein tyrosine kinase was observed in chronic myeloid leukemia (Griaud et al. 2013), and the expression of LUZP4, an mRNA export adaptor complementing ALYREF that is normally restricted to testis, is frequently upregulated in a range of tumors (Viphakone et al. 2015). LUZP4 (Leucine zipper protein 4) is preferentially expressed in melanoma cells where it is required for growth (Viphakone et al. 2015). Furthermore, depletion of DDX39B perturbs BRCA1 expression (Yamazaki et al. 2010), SARNP/CIP29 is upregulated in leukemia (Fukuda et al. 2002), and SARNP/CIP29 protein fusions with the MLL (mixed lineage leukemia) protein occur in acute myelomonocytic leukemia (AMMoL) (Hashii et al. 2004). Importantly, depletion of ALYREF (Saito et al. 2013), LUZP4 (Viphakone et al. 2015), or THOC1 (Guo et al. 2005; Li et al. 2005) inhibits the proliferation of tumor cells and reduces their metastatic capacity.

In addition to the TREX complex, the nuclear pore-associated THSC/TREX-2 complex promotes mRNA export of selected transcripts (Schubert and Köhler 2016; Schneider et al. 2015). Therefore, similar to TREX, the THSC component GANP is upregulated in many cancers (Bhatia et al. 2014; Fujimura et al. 2005; Sakaguchi and Maeda 2016; Wickramasinghe et al. 2014; Wickramasinghe and Laskey 2015). Overall, these data suggest that modulating the activity of TREX components and other mRNP components by drugs could provide novel therapeutic strategies to target cancer.

### 1.5.2 Misregulation of mRNP Assembly Linked to Neurodevelopmental Disorders

In addition to cancer, dysfunction of the mRNP assembly machinery is linked to numerous neurological diseases (Boehringer and Bowser 2018). THOC2 missense mutations leading to a partial loss of function cause X-linked syndromic intellectual disability (Kumar et al. 2015, 2018). Furthermore, the importance of THOC2 function in neuronal development was shown in a child with a nonprogressive form of congenital ataxia, cognitive impairment, and cerebellar hypoplasia (Di Gregorio et al. 2013). Here, a chromosomal translocation created a PTK2-THOC2 gene fusion with THOC2 expression knockdown. The protein tyrosine kinase 2 (PKT2) is known to be involved in axonal guidance and neurite growth. However, inactivation of PKT2 alone does not cause the phenotype indicating a specific role of THOC2 knockdown leading to cognitive impairment (Di Gregorio et al. 2013). Knockout of THOC5 in mice dopaminergic neurons causes a nuclear export defect of synaptic transcripts and finally the degeneration of the neurons (Maeder et al. 2018). THOC6 mutations resulting in mislocalization of the protein to the cytoplasm were reported in patients with intellectual disability (Amos et al. 2017; Beaulieu et al. 2013). Recently, mutations in the ZC3H14 gene, the yeast Nab2 ortholog, have been linked to a nonsyndromic form of autosomal recessive intellectual disability (Fasken and Corbett 2016). In *Drosophila*, dNab2 interacts with the Fragile X Protein ortholog and is needed for normal neuronal function (Bienkowski et al. 2017; Kelly et al. 2015). Interestingly, Nab2 also functions in RNA polymerase III (RNAPIII) transcription in *S. cerevisiae* (Reuter et al. 2015). It remains to be shown whether the function of Nab2 in RNAPIII transcription is conserved in higher eukaryotes and, if so, which function of Nab2 is causative for its disease phenotype.

A prominent example for TREX-associated neuronal diseases is amyotrophic lateral sclerosis (ALS), a progressive neurodegenerative disorder that results in the loss of motor neurons, muscle atrophy, and progressive paralysis. GGGGCC repeat expansions of C9orf72 fold into a G-quadruplex secondary structure and represent the most common genetic variant of ALS. The abnormal repeat pre-mRNA should be retained in the nucleus but is instead exported to the cytoplasm. In the cytoplasm, RNA foci form and a repeat-associated non-AUG (RAN) translation leads to the production of toxic dipeptide repeat proteins promoting progressive disease (Walsh et al. 2015). The nuclear export adaptor ALYREF was found to be sequestered by the GGGGCC RNA repeat in the nucleus causing the aberrant export of C9orf72 pre-mRNA from the nucleus (Fig. 1.2d) (Cooper-Knock et al. 2014; Hautbergue et al. 2017). In a *Drosophila* C9orf72 model system, mutations of ALYREF act as a potential suppressor of neurodegeneration (Freibaum et al. 2015). Furthermore, Matrin 3, a protein linked to ALS, interacts directly with TREX proteins highlighting the role mRNP biogenesis and nuclear export in the pathogenesis of ALS (Boehringer et al. 2017). Recently, it was shown that targeting the C9orf72 GGGGCC RNA repeats by binding of a small molecule is a potential treatment strategy for ALS (Simone et al. 2018). For further information regarding RNA granules and their role in neurodegenerative disease, see Chap. 8.

### *1.5.3 Nuclear mRNP Export Highjacked by Viruses*

Viruses utilize the host gene expression machinery and interfere with cellular processes to maximize their own replication and production of infectious virions. In general, replication and gene transcription of DNA viruses occur in the host nucleus. To export their transcripts, viruses exploit the host export receptor NXF1 or CRM1 (chromosome region maintenance 1), an export receptor of the importin β family, and some viruses are even capable of blocking the export of host mRNPs [reviewed in Kuss et al. (2013) and Yarbrough et al. (2014)].

Viral transcripts of simple retroviruses, such as Mason-Pfizer monkey virus, bind directly to the RNA export receptor NXF1. These viral transcripts contain a structured RNA element, the constitutive transport element (CTE), which binds with high affinity to NXF1 and thus mediates its own nuclear export (Fig. 1.2e) (Bachi et al. 2000; Gruter et al. 1998). RNAs of the hepatitis B virus, a DNA virus, contain a posttranscriptional element (SEP1) that recruits TREX via cellular factors like ZC3H18, ensuring efficient viral mRNA nuclear export (Chi et al. 2014). The herpes simplex virus ICP27 protein acts as a viral-specific export adaptor. ICP27 binds selectively to intronless viral transcripts and recruits the nuclear TREX complex, in particular ALYREF (Tunnicliffe et al. 2011, 2018). This interaction introduces the viral mRNA to the NXF1 pathway, subsequently directing it to the nuclear pore for export to the cytoplasm (Fig. 1.2). This strategy is also employed by other viruses with functionally homologous proteins: EB2 in Epstein-Barr virus (Hiriart et al. 2003), UL69 in human cytomegalovirus binding to DDX39B (Lischka et al. 2006), IE4 in varicella-zoster virus (Ote et al. 2009), the trimeric nucleoprotein in influenza virus interacting with UAP56 (Balasubramaniam et al. 2013; Hu et al. 2017; Chiba et al. 2018), and ORF57 in *Herpesvirus saimiri* and human Kaposi's sarcoma-associated herpesvirus (Jackson et al. 2014; Schumann et al. 2016a; Tunnicliffe et al. 2014). Furthermore, knockout of the stress-induced ZC3H11A protein, which associates with the TREX complex, showed that efficient growth of several viruses is dependent on this nuclear zinc finger protein (Younis et al. 2018). Small-molecule inhibitors that selectively inhibit the ATPase activity of the TREX component UAP56 result in effective inhibition of viral ribonucleoprotein (vRNP) formation, viral lytic replication, and infectious virion production of the oncogenic herpesvirus Kaposi's sarcoma-associated herpesvirus (KSHV) (Schumann et al. 2016b). Importantly, as all human herpesviruses use conserved mRNA processing pathways involving hTREX components, this approach could be used for pan-herpesvirus inhibition (Schumann et al. 2016b; Schumann and Whitehouse 2017). In summary, many viruses rely on the cellular machinery for nuclear mRNP assembly and export for their life cycle. Thus, components of this machinery are targets for novel antiviral drugs.

## 1.6 Perspectives

Studies in many model systems have allowed to build a comprehensive picture of the assembly and composition of nuclear mRNPs. It is believed that most—if not all—proteins involved in this process have been characterized in the yeast *S. cerevisiae*, and that many have been identified in other eukaryotes, including humans. In addition, the recruitment mechanisms of many mRNP components have been, at least partially, determined. Furthermore, for many nuclear mRNP components it is known which process or processes they function in.

Nevertheless, major questions remain. Even as we know for many mRNP components when and where they bind to the (pre-) mRNA, we still know little about the molecular function of many of these proteins. Moreover, the overall mechanism of the coordinated assembly of all these proteins with the mRNA to form an mRNP is still unknown. Also, it is still unclear how many of the nuclear mRBPs find "their" specific place on the mRNA to form a well-proportioned mRNP, especially since most of them bind in a sequence-unspecific manner. Furthermore, deletions of proteins or of specific domains within proteins that abrogate protein-protein interactions were used to determine the interactions and functions of each protein. Thus, the specific role of their RNA-binding activity has often remained enigmatic. In addition, how the mRNP might undergo changes during its biogenesis, often termed remodeling, is still unclear. Furthermore, whether such rearrangements result in major changes in the overall organization of mRNPs or rather small rearrangements of local secondary structure is equally enigmatic. It is also still mostly ambiguous how the cell recognizes an mRNP that is not correctly formed or assembled, e.g., due to the absence of an mRNP component, and thus degrades the mRNA. Moreover, in the past few years, a plethora of new RNA-binding proteins has been discovered (Beckmann et al. 2016; Hentze et al. 2018), and at least some of these might function in nuclear mRNP assembly, which needs to be elucidated.

The exact composition and stoichiometry of all components of an mRNP are also still unknown. The cellular pool of mRNPs is heterogeneous due to the many different mRNAs of various lengths, and purification of specific mRNPs, e.g., mRNPs containing one kind of mRNA, is biochemically very challenging. Accordingly, little is known about the structure of an mRNP and its mRNA. mRNPs have been purified from two well-studied systems, *Chironomus tentans* (Balbiani ring genes 1 and 2; BR1 and BR2) and *S. cerevisiae*, and visualized by electron microscopy. BR1/2 mRNPs contain an about 40-kb-long mRNA, first fold in 7–10 nm fibers that then reorganize into 26 nm ribbons that in turn form ring-shaped structures and further compact structures of about 50 nm in diameter (Bjork and Wieslander 2015, 2017; Skoglund et al. 1986). Analysis of mRNPs purified from *S. cerevisiae* by electron microscopy revealed elongated, rod-like structures with a diameter of 5–7 nm and increasing length depending on the length of the mRNA (Batisse et al. 2009). Recently, such a linear organization and rod-like structure have also been suggested for nuclear mRNPs in mammalian cells using a proximity RNA ligation approach or single-molecule fluorescence coupled to super-resolution microscopy experiments (Adivarahan et al. 2018; Metkar et al. 2018). However, structures of nuclear mRNPs at a higher resolution still await their elucidation.

Last but not least, it remains to be determined how the composition of an mRNP determines the fate of its mRNA (Gehring et al. 2017). So far, only few examples, such as the control of mRNA stability by the promoter (Trcek et al. 2011), are understood at the molecular level. We need to further unravel how a differential assembly occurs on different mRNAs and how this is influenced by different conditions. Thus, an exciting time lies ahead of us.

## References

Abruzzi KC, Lacadie S, Rosbash M (2004) Biochemical analysis of TREX complex recruitment to intronless and intron-containing yeast genes. EMBO J 23:2620–2631

Adams RL, Mason AC, Glass L, Aditi, Wente SR (2017) Nup42 and IP6 coordinate Gle1 stimulation of Dbp5/DDX19B for mRNA export in yeast and human cells. Traffic 18:776–790

Adivarahan S, Livingston N, Nicholson B, Rahman S, Wu B, Rissland OS, Zenklusen D (2018) Spatial organization of single mRNPs at different stages of the gene expression pathway. Mol Cell 72:727–738 e725

Aguilera A, Garcia-Muse T (2012) R loops: from transcription byproducts to threats to genome stability. Mol Cell 46:115–124

Aibara S, Gordon JM, Riesterer AS, McLaughlin SH, Stewart M (2017) Structural basis for the dimerization of Nab2 generated by RNA binding provides insight into its contribution to both poly(A) tail length determination and transcript compaction in Saccharomyces cerevisiae. Nucleic Acids Res 45:1529–1538

Alcazar-Roman AR, Tran EJ, Guo SL, Wente SR (2006) Inositol hexakisphosphate and Gle1 activate the DEAD-box protein Dbp5 for nuclear mRNA export. Nat Cell Biol 8:711–U131

Amos JS, Huang L, Thevenon J, Kariminedjad A, Beaulieu CL, Masurel-Paulet A, Najmabadi H, Fattahi Z, Beheshtian M, Tonekaboni SH et al (2017) Autosomal recessive mutations in THOC6 cause intellectual disability: syndrome delineation requiring forward and reverse phenotyping. Clin Genet 91:92–99

Anderson JT, Wilson SM, Datar KV, Swanson MS (1993) NAB2: a yeast nuclear polyadenylated RNA-binding protein essential for cell viability. Mol Cell Biol 13:2730–2741

Aravind L, Koonin EV (1999) G-patch: a new conserved domain in eukaryotic RNA-processing proteins and type D retroviral polyproteins. Trends Biochem Sci 24:342–344

Bachi A, Braun IC, Rodrigues JP, Pante N, Ribbeck K, von Kobbe C, Kutay U, Wilm M, Gorlich D, Carmo-Fonseca M et al (2000) The C-terminal domain of TAP interacts with the nuclear pore complex and promotes export of specific CTE-bearing RNA substrates. RNA 6:136–158

Baejen C, Torkler P, Gressel S, Essig K, Soding J, Cramer P (2014) Transcriptome maps of mRNP biogenesis factors define pre-mRNA recognition. Mol Cell 55:745–757

Balasubramaniam VR, Hong Wai T, Ario Tejo B, Omar AR, Syed Hassan S (2013) Highly pathogenic avian influenza virus nucleoprotein interacts with TREX complex adaptor protein Aly/REF. PLoS One 8:e72429

Batisse J, Batisse C, Budd A, Bottcher B, Hurt E (2009) Purification of nuclear poly(A)-binding protein Nab2 reveals association with the yeast transcriptome and a messenger ribonucleoprotein core structure. J Biol Chem 284:34911–34917

Beaulieu CL, Huang L, Innes AM, Akimenko MA, Puffenberger EG, Schwartz C, Jerry P, Ober C, Hegele RA, McLeod DR et al (2013) Intellectual disability associated with a homozygous missense mutation in THOC6. Orphanet J Rare Dis 8:62

Beckmann BM, Castello A, Medenbach J (2016) The expanding universe of ribonucleoproteins: of novel RNA-binding proteins and unconventional interactions. Pflugers Arch: Eur J Physiol 468:1029–1040

Bhatia V, Barroso SI, Garcia-Rubio ML, Tumini E, Herrera-Moyano E, Aguilera A (2014) BRCA2 prevents R-loop accumulation and associates with TREX-2 mRNA export factor PCID2. Nature 511:362–365

Bienkowski RS, Banerjee A, Rounds JC, Rha J, Omotade OF, Gross C, Morris KJ, Leung SW, Pak C, Jones SK et al (2017) The conserved, disease-associated RNA binding protein dNab2 interacts with the fragile X protein ortholog in drosophila neurons. Cell Rep 20:1372–1384

Bjork P, Wieslander L (2015) The Balbiani ring story: synthesis, assembly, processing, and transport of specific messenger RNA-protein complexes. Annu Rev Biochem 84:65–92

Bjork P, Wieslander L (2017) Integration of mRNP formation and export. Cell Mol Life Sci 74:2875–2897

Boehm V, Gehring NH (2016) Exon junction complexes: supervising the gene expression assembly line. Trends Genet 32:724–735

Boehringer A, Bowser R (2018) RNA nucleocytoplasmic transport defects in neurodegenerative diseases. Adv Neurobiol 20:85–101

Boehringer A, Garcia-Mansfield K, Singh G, Bakkar N, Pirrotte P, Bowser R (2017) ALS associated mutations in Matrin 3 alter protein-protein interactions and impede mRNA nuclear export. Sci Rep 7:14529

Bourgeois CF, Mortreux F, Auboeuf D (2016) The multiple functions of RNA helicases as drivers and regulators of gene expression. Nat Rev Mol Cell Biol 17:426–438

Brown SJ, Stoilov P, Xing Y (2012) Chromatin and epigenetic regulation of pre-mRNA processing. Hum Mol Genet 21:R90–R96

Bucheli ME, Buratowski S (2005) Npl3 is an antagonist of mRNA 3′ end formation by RNA polymerase II. EMBO J 24:2150–2160

Buratowski S (2009) Progression through the RNA polymerase II CTD cycle. Mol Cell 36:541–546

Carey KT, Wickramasinghe VO (2018) Regulatory potential of the RNA processing machinery: implications for human disease. Trends Genet 34:279–290

Carmody SR, Tran EJ, Apponi LH, Corbett AH, Wente SR (2010) The mitogen-activated protein kinase Slt2 regulates nuclear retention of non-heat shock mRNAs during heat shock-induced stress. Mol Cell Biol 30:5168–5179

Castellano-Pozo M, Santos-Pereira JM, Rondon AG, Barroso S, Andujar E, Perez-Alegre M, Garcia-Muse T, Aguilera A (2013) R loops are linked to histone H3 S10 phosphorylation and chromatin condensation. Mol Cell 52:583–590

Chanarat S, Strasser K (2013) Splicing and beyond: the many faces of the Prp19 complex. Biochim Biophys Acta 1833:2126–2134

Chanarat S, Seizl M, Strasser K (2011) The Prp19 complex is a novel transcription elongation factor required for TREX occupancy at transcribed genes. Genes Dev 25:1147–1158

Chanarat S, Burkert-Kautzsch C, Meinel DM, Strasser K (2012) Prp19C and TREX: interacting to promote transcription elongation and mRNA export. Transcription 3:8–12

Chang CT, Hautbergue GM, Walsh MJ, Viphakone N, van Dijk TB, Philipsen S, Wilson SA (2013) Chtop is a component of the dynamic TREX mRNA export complex. EMBO J 32:473–486

Cheng H, Dufu K, Lee CS, Hsu JL, Dias A, Reed R (2006) Human mRNA export machinery recruited to the 5′ end of mRNA. Cell 127:1389–1400

Chi B, Wang Q, Wu G, Tan M, Wang L, Shi M, Chang X, Cheng H (2013) Aly and THO are required for assembly of the human TREX complex and association of TREX components with the spliced mRNA. Nucleic Acids Res 41:1294–1306

Chi BK, Wang K, Du YH, Gui B, Chang XY, Wang LT, Fan J, Chen S, Wu XD, Li GH et al (2014) A sub-element in PRE enhances nuclear export of intronless mRNAs by recruiting the TREX complex via ZC3H18. Nucleic Acids Res 42:7305–7318

Chiba S, Hill-Batorski L, Neumann G, Kawaoka Y (2018) The cellular DExD/H-Box RNA helicase UAP56 Co-localizes with the influenza A virus NS1 protein. Front Microbiol 9:2192

Chinnam M, Wang Y, Zhang X, Gold DL, Khoury T, Nikitin AY, Foster BA, Li Y, Bshara W, Morrison CD et al (2014) The Thoc1 ribonucleoprotein and prostate cancer progression. J Natl Cancer Inst 106

Cloutier SC, Ma WK, Nguyen LT, Tran EJ (2012) The DEAD-box RNA helicase Dbp2 connects RNA quality control with repression of aberrant transcription. J Biol Chem 287:26155–26166

Cooper-Knock J, Walsh MJ, Higginbottom A, Robin Highley J, Dickman MJ, Edbauer D, Ince PG, Wharton SB, Wilson SA, Kirby J et al (2014) Sequestration of multiple RNA recognition motif-containing proteins by C9orf72 repeat expansions. Brain 137:2040–2051

Corbett AH (2018) Post-transcriptional regulation of gene expression and human disease. Curr Opin Cell Biol 52:96–104

Dargemont C, Babour A (2017) Novel functions for chromatin dynamics in mRNA biogenesis beyond transcription. Nucleus (Austin, Tex.) 8:482–488

Delaleau M, Borden KL (2015) Multiple export mechanisms for mRNAs. Cell 4:452–473

Dermody JL, Dreyfuss JM, Villen J, Ogundipe B, Gygi SP, Park PJ, Ponticelli AS, Moore CL, Buratowski S, Bucheli ME (2008) Unphosphorylated SR-like protein Npl3 stimulates RNA polymerase II elongation. PLoS One 3:e3273

Di Gregorio E, Bianchi FT, Schiavi A, Chiotto AM, Rolando M, Verdun di Cantogno L, Grosso E, Cavalieri S, Calcia A, Lacerenza D et al (2013) A de novo X;8 translocation creates a PTK2-THOC2 gene fusion with THOC2 expression knockdown in a patient with psychomotor retardation and congenital cerebellar hypoplasia. J Med Genet 50:543–551

Dominguez-Sanchez MS, Barroso S, Gomez-Gonzalez B, Luna R, Aguilera A (2011a) Genome instability and transcription elongation impairment in human cells depleted of THO/TREX. PLoS Genet 7:e1002386

Dominguez-Sanchez MS, Saez C, Japon MA, Aguilera A, Luna R (2011b) Differential expression of THOC1 and ALY mRNP biogenesis/export factors in human cancers. BMC Cancer 11:77

Dufu K, Livingstone MJ, Seebacher J, Gygi SP, Wilson SA, Reed R (2010) ATP is required for interactions between UAP56 and two conserved mRNA export proteins, Aly and CIP29, to assemble the TREX complex. Genes Dev 24:2043–2053

Dutta S, Gupta G, Choi YW, Kotaka M, Fielding BC, Song J, Tan YJ (2012) The variable N-terminal region of DDX5 contains structural elements and auto-inhibits its interaction with NS5B of hepatitis C virus. Biochem J 446:37–46

Fasken MB, Corbett AH (2016) Links between mRNA splicing, mRNA quality control, and intellectual disability. RNA Dis 3:e1448

Ficner R, Dickmanns A, Neumann P (2017) Studying structure and function of spliceosomal helicases. Methods 125:63–69

Fischer T, Strasser K, Racz A, Rodriguez-Navarro S, Oppizzi M, Ihrig P, Lechner J, Hurt E (2002) The mRNA export machinery requires the novel Sac3p-Thp1p complex to dock at the nucleoplasmic entrance of the nuclear pores. EMBO J 21:5843–5852

Fleckner J, Zhang M, Valcarcel J, Green MR (1997) U2AF(65) recruits a novel human DEAD box protein required for the U2 snRNP-branchpoint interaction. Genes Dev 11:1864–1872

Folco EG, Lee CS, Dufu K, Yamazaki T, Reed R (2012) The proteins PDIP3 and ZC11A associate with the human TREX complex in an ATP-dependent manner and function in mRNA export. PLoS One 7:e43804

Freibaum BD, Lu Y, Lopez-Gonzalez R, Kim NC, Almeida S, Lee KH, Badders N, Valentine M, Miller BL, Wong PC et al (2015) GGGGCC repeat expansion in C9orf72 compromises nucleocytoplasmic transport. Nature 525:129–133

Fujimura S, Xing Y, Takeya M, Yamashita Y, Ohshima K, Kuwahara K, Sakaguchi N (2005) Increased expression of germinal center-associated nuclear protein RNA-primase is associated with lymphomagenesis. Cancer Res 65:5925–5934

Fukuda S, Wu DW, Stark K, Pelus LM (2002) Cloning and characterization of a proliferation-associated cytokine-inducible protein, CIP29. Biochem Biophys Res Commun 292:593–600

Gaillard H, Wellinger RE, Aguilera A (2007) A new connection of mRNP biogenesis and export with transcription-coupled repair. Nucleic Acids Res 35:3893–3906

Gatfield D, Le Hir H, Schmitt C, Braun IC, Kocher T, Wilm M, Izaurralde E (2001) The DExH/D box protein HEL/UAP56 is essential for mRNA nuclear export in Drosophila. Curr Biol 11:1716–1721

Gehring NH, Wahle E, Fischer U (2017) Deciphering the mRNP code: RNA-bound determinants of post-transcriptional gene regulation. Trends Biochem Sci 42:369–382

Germain H, Qu N, Cheng YT, Lee E, Huang Y, Dong OX, Gannon P, Huang S, Ding P, Li Y et al (2010) MOS11: a new component in the mRNA export pathway. PLoS Genet 6:e1001250

Gilbert W, Guthrie C (2004) The Glc7p nuclear phosphatase promotes mRNA export by facilitating association of Mex67p with mRNA. Mol Cell 13:201–212

Gilbert W, Siebel CW, Guthrie C (2001) Phosphorylation by Sky1p promotes Npl3p shuttling and mRNA dissociation. RNA 7:302–313

Gomez-Gonzalez B, Garcia-Rubio M, Bermejo R, Gaillard H, Shirahige K, Marin A, Foiani M, Aguilera A (2011) Genome-wide function of THO/TREX in active genes prevents R-loop-dependent replication obstacles. EMBO J 30:3106–3119

Gonatopoulos-Pournatzis T, Cowling VH (2014) Cap-binding complex (CBC). Biochem J 457:231–242

Gorbalenya AE, Koonin EV (1993) Helicases - amino-acid-sequence comparisons and structure-function-relationships. Curr Opin Struct Biol 3:419–429

Green DM, Marfatia KA, Crafton EB, Zhang X, Cheng X, Corbett AH (2002) Nab2p is required for poly(A) RNA export in *Saccharomyces cerevisiae* and is regulated by arginine methylation via Hmt1p. J Biol Chem 277:7752–7760

Griaud F, Pierce A, Gonzalez Sanchez MB, Scott M, Abraham SA, Holyoake TL, Tran DD, Tamura T, Whetton AD (2013) A pathway from leukemogenic oncogenes and stem cell chemokines to RNA processing via THOC5. Leukemia 27:932–940

Gromadzka AM, Steckelberg AL, Singh KK, Hofmann K, Gehring NH (2016) A short conserved motif in ALYREF directs cap- and EJC-dependent assembly of export complexes on spliced mRNAs. Nucleic Acids Res 44:2348–2361

Gruter P, Tabernero C, von Kobbe C, Schmitt C, Saavedra C, Bachi A, Wilm M, Felber BK, Izaurralde E (1998) TAP, the human homolog of Mex67p, mediates CTE-dependent RNA export from the nucleus. Mol Cell 1:649–659

Guo S, Hakimi MA, Baillat D, Chen X, Farber MJ, Klein-Szanto AJ, Cooch NS, Godwin AK, Shiekhattar R (2005) Linking transcriptional elongation and messenger RNA export to metastatic breast cancers. Cancer Res 65:3011–3016

Guo S, Liu M, Godwin AK (2012) Transcriptional regulation of hTREX84 in human cancer cells. PLoS One 7:e43610

Gwizdek C, Iglesias N, Rodriguez MS, Ossareh-Nazari B, Hobeika M, Divita G, Stutz F, Dargemont C (2006) Ubiquitin-associated domain of Mex67 synchronizes recruitment of the mRNA export machinery with transcription. Proc Natl Acad Sci U S A 103:16376–16381

Harlen KM, Churchman LS (2017) The code and beyond: transcription regulation by the RNA polymerase II carboxy-terminal domain. Nat Rev Mol Cell Biol 18:263–273

Hashii Y, Kim JY, Sawada A, Tokimasa S, Hiroyuki F, Ohta H, Makiko K, Takihara Y, Ozono K, Hara J (2004) A novel partner gene CIP29 containing a SAP domain with MLL identified in infantile myelomonocytic leukemia. Leukemia 18:1546–1548

Hautbergue GM, Hung ML, Golovanov AP, Lian LY, Wilson SA (2008) Mutually exclusive interactions drive handover of mRNA from export adaptors to TAP. Proc Natl Acad Sci U S A 105:5154–5159

Hautbergue GM, Hung ML, Walsh MJ, Snijders AP, Chang CT, Jones R, Ponting CP, Dickman MJ, Wilson SA (2009) UIF, a new mRNA export adaptor that works together with REF/ALY, requires FACT for recruitment to mRNA. Curr Biol 19:1918–1924

Hautbergue GM, Castelli LM, Ferraiuolo L, Sanchez-Martinez A, Cooper-Knock J, Higginbottom A, Lin Y-H, Bauer CS, Dodd JE, Myszczynska MA et al (2017) SRSF1-dependent nuclear export inhibition of C9ORF72 repeat transcripts prevents neurodegeneration and associated motor deficits. Nat Commun 8:16063

Heath CG, Viphakone N, Wilson SA (2016) The role of TREX in gene expression and disease. Biochem J 473:2911–2935

Hector RE, Nykamp KR, Dheur S, Anderson JT, Non PJ, Urbinati CR, Wilson SM, Minvielle-Sebastia L, Swanson MS (2002) Dual requirement for yeast hnRNP Nab2p in mRNA poly (A) tail length control and nuclear export. EMBO J 21:1800–1810

Heidemann M, Hintermair C, Voss K, Eick D (2012) Dynamic phosphorylation patterns of RNA polymerase II CTD during transcription. Biochim Biophys Acta 1829:55–62

Hentze MW, Castello A, Schwarzl T, Preiss T (2018) A brave new world of RNA-binding proteins. Nat Rev Mol Cell Biol 19:327–341

Hiriart E, Bardouillet L, Manet E, Gruffat H, Penin F, Montserret R, Farjot G, Sergeant A (2003) A region of the Epstein-Barr virus (EBV) mRNA export factor EB2 containing an arginine-rich motif mediates direct binding to RNA. J Biol Chem 278:37790–37798

Hirling H, Scheffner M, Restle T, Stahl H (1989) RNA helicase activity associated with the human p68 protein. Nature 339:562–564

Hodge CA, Tran EJ, Noble KN, Alcazar-Roman AR, Ben-Yishay R, Scarcelli JJ, Folkmann AW, Shav-Tal Y, Wente SR, Cole CN (2011) The Dbp5 cycle at the nuclear pore complex during mRNA export I: dbp5 mutants with defects in RNA binding and ATP hydrolysis define key steps for Nup159 and Gle1. Genes Dev 25:1052–1064

Hsin JP, Manley JL (2012) The RNA polymerase II CTD coordinates transcription and RNA processing. Genes Dev 26:2119–2137

Hu Y, Gor V, Morikawa K, Nagata K, Kawaguchi A (2017) Cellular splicing factor UAP56 stimulates trimeric NP formation for assembly of functional influenza viral ribonucleoprotein complexes. Sci Rep 7:14053

Huang Y, Gattoni R, Stevenin J, Steitz JA (2003) SR splicing factors serve as adapter proteins for TAP-dependent mRNA export. Mol Cell 11:837–843

Huertas P, Aguilera A (2003) Cotranscriptionally formed DNA:RNA hybrids mediate transcription elongation impairment and transcription-associated recombination. Mol Cell 12:711–721

Hurt E, Luo MJ, Rother S, Reed R, Strasser K (2004) Cotranscriptional recruitment of the serine-arginine-rich (SR)-like proteins Gbp2 and Hrb1 to nascent mRNA via the TREX complex. Proc Natl Acad Sci U S A 101:1858–1862

Hurt JA, Obar RA, Zhai B, Farny NG, Gygi SP, Silver PA (2009) A conserved CCCH-type zinc finger protein regulates mRNA nuclear adenylation and export. J Cell Biol 185:265–277

Iglesias N, Tutucci E, Gwizdek C, Vinciguerra P, Von Dach E, Corbett AH, Dargemont C, Stutz F (2010) Ubiquitin-mediated mRNP dynamics and surveillance prior to budding yeast mRNA export. Genes Dev 24:1927–1938

Jackson BR, Noerenberg M, Whitehouse A (2014) A novel mechanism inducing genome instability in Kaposi's sarcoma-associated herpesvirus infected cells. PLoS Pathog 10:e1004098

Jankowsky E, Harris ME (2015) Specificity and nonspecificity in RNA-protein interactions. Nat Rev Mol Cell Biol 16:533–544

Jankowsky A, Guenther U-P, Jankowsky E (2011) The RNA helicase database. Nucleic Acids Res 39:D338–D341

Jensen TH, Boulay J, Rosbash M, Libri D (2001) The DECD box putative ATPase Sub2p is an early mRNA export factor. Curr Biol 11:1711–1715

Jeronimo C, Bataille AR, Robert F (2013) The writers, readers, and functions of the RNA polymerase II C-terminal domain code. Chem Rev 113:8491–8522

Jeronimo C, Collin P, Robert F (2016) The RNA polymerase II CTD: the increasing complexity of a low-complexity protein domain. J Mol Biol 428:2607–2622

Jimeno S, Luna R, Garcia-Rubio M, Aguilera A (2006) Tho1, a novel hnRNP, and Sub2 provide alternative pathways for mRNP biogenesis in yeast THO mutants. Mol Cell Biol 26:4387–4398

Johnson SA, Cubberley G, Bentley DL (2009) Cotranscriptional recruitment of the mRNA export factor Yra1 by direct interaction with the 3′ end processing factor Pcf11. Mol Cell 33:215–226

Kelly SM, Leung SW, Pak C, Banerjee A, Moberg KH, Corbett AH (2014) A conserved role for the zinc finger polyadenosine RNA binding protein, ZC3H14, in control of poly(A) tail length. RNA 20:681–688

Kelly SM, Bienkowski R, Banerjee A, Melicharek DJ, Brewer ZA, Marenda DR, Corbett AH, Moberg KH (2015) The Drosophila ortholog of the Zc3h14 RNA binding protein acts within neurons to pattern axon projection in the developing brain. Dev Neurobiol. 76:93–106

Kistler AL, Guthrie C (2001) Deletion of MUD2, the yeast homolog of U2AF65, can bypass the requirement for sub2, an essential spliceosomal ATPase. Genes Dev 15:42–49

Kohler A, Hurt E (2007) Exporting RNA from the nucleus to the cytoplasm. Nat Rev Mol Cell Biol 8:761–773

Kota KP, Wagner SR, Huerta E, Underwood JM, Nickerson JA (2008) Binding of ATP to UAP56 is necessary for mRNA export. J Cell Sci 121:1526–1537

Kress TL, Krogan NJ, Guthrie C (2008) A single SR-like protein, Npl3, promotes pre-mRNA splicing in budding yeast. Mol Cell 32:727–734

Kubitscheck U, Siebrasse JP (2017) Kinetics of transport through the nuclear pore complex. Semin Cell Dev Biol 68:18–26

Kumar R, Corbett MA, van Bon BW, Woenig JA, Weir L, Douglas E, Friend KL, Gardner A, Shaw M, Jolly LA et al (2015) THOC2 mutations implicate mRNA-export pathway in X-linked intellectual disability. Am J Hum Genet 97:302–310

Kumar R, Gardner A, Homan CC, Douglas E, Mefford H, Wieczorek D, Ludecke HJ, Stark Z, Sadedin S, Broad CMG et al (2018) Severe neurocognitive and growth disorders due to variation in THOC2, an essential component of nuclear mRNA export machinery. Hum Mutat 39:1126–1138

Kuss SK, Mata MA, Zhang L, Fontoura BM (2013) Nuclear imprisonment: viral strategies to arrest host mRNA nuclear export. Viruses 5:1824–1849

Lai MC, Tarn WY (2004) Hypophosphorylated ASF/SF2 binds TAP and is present in messenger ribonucleoproteins. J Biol Chem 279:31745–31749

Lapek JD Jr, Greninger P, Morris R, Amzallag A, Pruteanu-Malinici I, Benes CH, Haas W (2017) Detection of dysregulated protein-association networks by high-throughput proteomics predicts cancer vulnerabilities. Nat Biotechnol 35:983

Lee MS, Henry M, Silver PA (1996) A protein that shuttles between the nucleus and the cytoplasm is an important mediator of RNA export. Genes Dev 10:1233–1246

Lei EP, Krebber H, Silver PA (2001) Messenger RNAs are recruited for nuclear export during transcription. Genes Dev 15:1771–1782

Lei H, Dias AP, Reed R (2011) Export and stability of naturally intronless mRNAs require specific coding region sequences and the TREX mRNA export complex. Proc Natl Acad Sci U S A 108:17985–17990

Lei H, Zhai B, Yin S, Gygi S, Reed R (2013) Evidence that a consensus element found in naturally intronless mRNAs promotes mRNA export. Nucleic Acids Res 41:2517–2525

Li Y, Wang X, Zhang X, Goodrich DW (2005) Human hHpr1/p84/Thoc1 regulates transcriptional elongation and physically links RNA polymerase II and RNA processing factors. Mol Cell Biol 25:4023–4033

Li Y, Lin AW, Zhang X, Wang Y, Wang X, Goodrich DW (2007) Cancer cells and normal cells differ in their requirements for Thoc1. Cancer Res 67:6657–6664

Lin DH, Correia AR, Cai SW, Huber FM, Jette CA, Hoelz A (2018) Structural and functional analysis of mRNA export regulation by the nuclear pore complex. Nat Commun 9:2319

Linder P, Jankowsky E (2011) From unwinding to clamping – the DEAD box RNA helicase family. Nat Rev Mol Cell Biol 12:505–516

Linder P, Stutz F (2001) mRNA export: travelling with DEAD box proteins. Curr Biol 11:R961–R963

Linder P, Lasko PF, Ashburner M, Leroy P, Nielsen PJ, Nishi K, Schnier J, Slonimski PP (1989) Birth of the D-E-A-D box. Nature 337:121–122

Lischka P, Toth Z, Thomas M, Mueller R, Stamminger T (2006) The UL69 transactivator protein of human cytomegalovirus interacts with DEXD/H-Box RNA helicase UAP56 to promote cytoplasmic accumulation of unspliced RNA. Mol Cell Biol 26:1631–1643

Liu C, Yue B, Yuan C, Zhao S, Fang C, Yu Y, Yan D (2015) Elevated expression of Thoc1 is associated with aggressive phenotype and poor prognosis in colorectal cancer. Biochem Biophys Res Commun 468:53–58

Luco RF, Pan Q, Tominaga K, Blencowe BJ, Pereira-Smith OM, Misteli T (2010) Regulation of alternative splicing by histone modifications. Science 327:996–1000

Luo MJ, Zhou ZL, Magni K, Christoforides C, Rappsilber J, Mann M, Reed R (2001) Pre-mRNA splicing and mRNA export linked by direct interactions between UAP56 and Aly. Nature 413:644–647

Ma WK, Cloutier SC, Tran EJ (2013) The DEAD-box protein Dbp2 functions with the RNA-binding protein Yra1 to promote mRNP assembly. J Mol Biol 425:3824–3838

Ma WK, Paudel BP, Xing Z, Sabath IG, Rueda D, Tran EJ (2016) Recruitment, duplex unwinding and protein-mediated inhibition of the dead-box RNA helicase Dbp2 at actively transcribed chromatin. J Mol Biol 428:1091–1106

MacMorris M, Brocker C, Blumenthal T (2003) UAP56 levels affect viability and mRNA export in *Caenorhabditis elegans*. RNA 9:847–857

Maeder CI, Kim JI, Liang X, Kaganovsky K, Shen A, Li Q, Li Z, Wang S, Xu XZS, Li JB et al (2018) The THO complex coordinates transcripts for synapse development and dopamine neuron survival. Cell 174:1436–1449.e1420

Masuda S, Das R, Cheng H, Hurt E, Dorman N, Reed R (2005) Recruitment of the human TREX complex to mRNA during splicing. Genes Dev 19:1512–1517

Meignin C, Davis I (2008) UAP56 RNA helicase is required for axis specification and cytoplasmic mRNA localization in Drosophila. Dev Biol 315:89–98

Meinel DM, Strasser K (2015) Co-transcriptional mRNP formation is coordinated within a molecular mRNP packaging station in *S. cerevisiae*. Bioessays 37:666–677

Meinel DM, Burkert-Kautzsch C, Kieser A, O'Duibhir E, Siebert M, Mayer A, Cramer P, Soding J, Holstege FC, Strasser K (2013) Recruitment of TREX to the transcription machinery by its direct binding to the phospho-CTD of RNA polymerase II. PLoS Genet 9:e1003914

Metkar M, Ozadam H, Lajoie BR, Imakaev M, Mirny LA, Dekker J, Moore MJ (2018) Higher-order organization principles of pre-translational mRNPs. Mol Cell 72:715–726.e713

Montpetit B, Thomsen ND, Helmke KJ, Seeliger MA, Berger JM, Weis K (2011) A conserved mechanism of DEAD-box ATPase activation by nucleoporins and InsP6 in mRNA export. Nature 472:238–242

Moukhtar M, Chaar W, Abdel-Razzak Z, Khalil M, Taha S, Chamieh H (2017) ARCPHdb: a comprehensive protein database for SF1 and SF2 helicase from archaea. Comput Biol Med 80:185–189

Muller-McNicoll M, Botti V, de Jesus Domingues AM, Brandl H, Schwich OD, Steiner MC, Curk T, Poser I, Zarnack K, Neugebauer KM (2016) SR proteins are NXF1 adaptors that link alternative RNA processing to mRNA export. Genes Dev 30:553–566

Nino CA, Herissant L, Babour A, Dargemont C (2013) mRNA nuclear export in yeast. Chem Rev 113:8523–8545

Noble SM, Guthrie C (1996) Transcriptional pulse-chase analysis reveals a role for a novel snRNP-associated protein in the manufacture of spliceosomal snRNPs. EMBO J 15:4368–4379

Noble KN, Tran EJ, Alcazar-Roman AR, Hodge CA, Cole CN, Wente SR (2011) The Dbp5 cycle at the nuclear pore complex during mRNA export II: nucleotide cycling and mRNP remodeling by Dbp5 are controlled by Nup159 and Gle1. Genes Dev 25:1065–1077

Nojima T, Hirose T, Kimura H, Hagiwara M (2007) The interaction between cap-binding complex and RNA export factor is required for intronless mRNA export. J Biol Chem 282:15645–15651

Nojima T, Gomes T, Grosso ARF, Kimura H, Dye MJ, Dhir S, Carmo-Fonseca M, Proudfoot NJ (2015) Mammalian NET-seq reveals genome-wide nascent transcription coupled to RNA processing. Cell 161:526–540

Ote I, Lebrun M, Vandevenne P, Bontems S, Medina-Palazon C, Manet E, Piette J, Sadzot-Delvaux C (2009) Varicella-zoster virus IE4 protein interacts with SR proteins and exports mRNAs through the TAP/NXF1 pathway. PLoS One 4:e7882

Pryor A, Tung L, Yang Z, Kapadia F, Chang TH, Johnson LF (2004) Growth-regulated expression and G0-specific turnover of the mRNA that encodes URH49, a mammalian DExH/D box protein that is highly related to the mRNA export protein UAP56. Nucleic Acids Res 32:1857–1865

Putnam AA, Jankowsky E (2013) DEAD-box helicases as integrators of RNA, nucleotide and protein binding. BBA-Gene Regul Mech 1829:884–893

Ren Y, Schmiege P, Blobel G (2017) Structural and biochemical analyses of the DEAD-box ATPase Sub2 in association with THO or Yra1. Elife 6

Reuter LM, Meinel DM, Strasser K (2015) The poly(A)-binding protein Nab2 functions in RNA polymerase III transcription. Genes Dev 29:1565–1575

Riordan DP, Herschlag D, Brown PO (2011) Identification of RNA recognition elements in the Saccharomyces cerevisiae transcriptome. Nucleic Acids Res 39:1501–1509

Rodriguez-Navarro S, Fischer T, Luo MJ, Antunez O, Brettschneider S, Lechner J, Perez-Ortin JE, Reed R, Hurt E (2004) Sus1, a functional component of the SAGA histone acetylase complex and the nuclear pore-associated mRNA export machinery. Cell 116:75–86

Rudolph MG, Klostermeier D (2015) When core competence is not enough: functional interplay of the DEAD-box helicase core with ancillary domains and auxiliary factors in RNA binding and unwinding. Biol Chem 396:849–865

Saguez C, Gonzales FA, Schmid M, Boggild A, Latrick CM, Malagon F, Putnam A, Sanderson L, Jankowsky E, Brodersen DE et al (2013) Mutational analysis of the yeast RNA helicase Sub2p reveals conserved domains required for growth, mRNA export, and genomic stability. RNA 19:1363–1371

Saito Y, Kasamatsu A, Yamamoto A, Shimizu T, Yokoe H, Sakamoto Y, Ogawara K, Shiiba M, Tanzawa H, Uzawa K (2013) ALY as a potential contributor to metastasis in human oral squamous cell carcinoma. J Cancer Res Clin Oncol 139:585–594

Sakaguchi N, Maeda K (2016) Germinal center B-cell-associated nuclear protein (GANP) involved in RNA metabolism for B cell maturation. In: Alt FW (ed) Advances in immunology. Academic, New York, pp 135–186

Salas-Armenteros I, Pérez-Calero C, Bayona-Feliu A, Tumini E, Luna R, Aguilera A (2017) Human THO–Sin3A interaction reveals new mechanisms to prevent R-loops that cause genome instability. EMBO J 36:3532–3547

Santos-Pereira JM, Aguilera A (2015) R loops: new modulators of genome dynamics and function. Nat Rev Genet 16:583–597

Schneider M, Hellerschmied D, Schubert T, Amlacher S, Vinayachandran V, Reja R, Pugh BF, Clausen T, Kohler A (2015) The nuclear pore-associated TREX-2 complex employs mediator to regulate gene expression. Cell 162:1016–1028

Schubert T, Köhler A (2016) Mediator and TREX-2: emerging links between transcription initiation and mRNA export. Nucleus 7(2):126–131

Schumann S, Whitehouse A (2017) Targeting the human TREX complex to prevent herpesvirus replication: what is new? Future Virol 12:81–83

Schumann S, Baquero-Perez B, Whitehouse A (2016a) Interactions between KSHV ORF57 and the novel human TREX proteins, CHTOP and CIP29. J Gen Virol 97:1904–1910

Schumann S, Jackson BR, Yule I, Whitehead SK, Revill C, Foster R, Whitehouse A (2016b) Targeting the ATP-dependent formation of herpesvirus ribonucleoprotein particle assembly as an antiviral approach. Nat Microbiol 2:16201

Schutz P, Karlberg T, van den Berg S, Collins R, Lehtio L, Hogbom M, Holmberg-Schiavone L, Tempel W, Park HW, Hammarstrom M et al (2010) Comparative structural analysis of human DEAD-box RNA helicases. PLoS One 5:e12791

Segref A, Sharma K, Doye V, Hellwig A, Huber J, Luhrmann R, Hurt E (1997) Mex67p, a novel factor for nuclear mRNA export, binds to both poly(A)+ RNA and nuclear pores. EMBO J 16:3256–3271

Shen JP, Zhang LD, Zhao R (2007) Biochemical characterization of the ATPase and helicase activity of UAP56, an essential pre-mRNA splicing and mRNA export factor. J Biol Chem 282:22544–22550

Shi H, Cordin O, Minder CM, Linder P, Xu RM (2004) Crystal structure of the human ATP-dependent splicing and export factor UAP56. Proc Natl Acad Sci U S A 101:17628–17633

Shi M, Zhang H, Wu X, He Z, Wang L, Yin S, Tian B, Li G, Cheng H (2017) ALYREF mainly binds to the 5′ and the 3′ regions of the mRNA in vivo. Nucleic Acids Res 45:9640–9653

Simone R, Balendra R, Moens TG, Preza E, Wilson KM, Heslegrave A, Woodling NS, Niccoli T, Gilbert-Jaramillo J, Abdelkarim S et al (2018) G-quadruplex-binding small molecules ameliorate *C9orf72* FTD/ALS pathology *in vitro* and *in vivo*. EMBO Mol Med 10:22–31

Singh G, Pratt G, Yeo GW, Moore MJ (2015) The clothes make the mRNA: past and present trends in mRNP fashion. Annu Rev Biochem 84:325–354

Singleton MR, Dillingham MS, Wigley DB (2007) Structure and mechanism of helicases and nucleic acid translocases. Annu Rev Biochem 76:23–50

Skoglund U, Andersson K, Strandberg B, Daneholt B (1986) Three-dimensional structure of a specific pre-messenger RNP particle established by electron microscope tomography. Nature 319:560–564

Sloan KE, Bohnsack MT (2018) Unravelling the mechanisms of RNA helicase regulation. Trends Biochem Sci 43:237–250

Sollier J, Cimprich KA (2015) Breaking bad: R-loops and genome integrity. Trends Cell Biol 25:514–522

Strahm Y, Fahrenkrog B, Zenklusen D, Rychner E, Kantor J, Rosbash M, Stutz F (1999) The RNA export factor Gle1p is located on the cytoplasmic fibrils of the NPC and physically interacts with the FG-nucleoporin Rip1p, the DEAD-box protein Rat8p/Dbp5p and a new protein Ymr255p. EMBO J 18:5761–5777

Strasser K, Hurt E (2000) Yra1p, a conserved nuclear RNA-binding protein, interacts directly with Mex67p and is required for mRNA export. EMBO J 19:410–420

Strasser K, Hurt E (2001) Splicing factor Sub2p is required for nuclear mRNA export through its interaction with Yra1p. Nature 413:648–652

Strasser K, Bassler J, Hurt E (2000) Binding of the Mex67p/Mtr2p heterodimer to FXFG, GLFG, and FG repeat nucleoporins is essential for nuclear mRNA export. J Cell Biol 150:695–706

Strasser K, Masuda S, Mason P, Pfannstiel J, Oppizzi M, Rodriguez-Navarro S, Rondon AG, Aguilera A, Struhl K, Reed R et al (2002) TREX is a conserved complex coupling transcription with messenger RNA export. Nature 417:304–308

Stubbs SH, Conrad NK (2015) Depletion of REF/Aly alters gene expression and reduces RNA polymerase II occupancy. Nucleic Acids Res 43:504–519

Taniguchi I, Ohno M (2008) ATP-dependent recruitment of export factor Aly/REF onto intronless mRNAs by RNA helicase UAP56. Mol Cell Biol 28:601–608

Tieg B, Krebber H (2013) Dbp5 – from nuclear export to translation. Biochim Biophys Acta 1829:791–798

Trcek T, Larson DR, Moldon A, Query CC, Singer RH (2011) Single-molecule mRNA decay measurements reveal promoter-regulated mRNA stability in yeast. Cell 147:1484–1497

Tuck AC, Tollervey D (2013) A transcriptome-wide atlas of RNP composition reveals diverse classes of mRNAs and lncRNAs. Cell 154:996–1009

Tunnicliffe RB, Hautbergue GM, Kalra P, Jackson BR, Whitehouse A, Wilson SA, Golovanov AP (2011) Structural basis for the recognition of cellular mRNA export factor REF by herpes viral proteins HSV-1 ICP27 and HVS ORF57. PLoS Pathog 7:e1001244

Tunnicliffe RB, Hautbergue GM, Wilson SA, Kalra P, Golovanov AP (2014) Competitive and cooperative interactions mediate RNA transfer from herpesvirus saimiri ORF57 to the mammalian export adaptor ALYREF. PLoS Pathog 10:e1003907

Tunnicliffe RB, Tian X, Storer J, Sandri-Goldin RM, Golovanov AP (2018) Overlapping motifs on the herpes viral proteins ICP27 and ORF57 mediate interactions with the mRNA export adaptors ALYREF and UIF. Sci Rep 8:15005

Umlauf D, Bonnet J, Waharte F, Fournier M, Stierle M, Fischer B, Brino L, Devys D, Tora L (2013) The human TREX-2 complex is stably associated with the nuclear pore basket. J Cell Sci 126:2656–2667

Valencia P, Dias AP, Reed R (2008) Splicing promotes rapid and efficient mRNA export in mammalian cells. Proc Natl Acad Sci U S A 105:3386–3391

Viphakone N, Hautbergue GM, Walsh M, Chang CT, Holland A, Folco EG, Reed R, Wilson SA (2012) TREX exposes the RNA-binding domain of Nxf1 to enable mRNA export. Nat Commun 3:1006

Viphakone N, Cumberbatch MG, Livingstone MJ, Heath PR, Dickman MJ, Catto JW, Wilson SA (2015) Luzp4 defines a new mRNA export pathway in cancer cells. Nucleic Acids Res 43:2353–2366

Vitaliano-Prunier A, Babour A, Herissant L, Apponi L, Margaritis T, Holstege FC, Corbett AH, Gwizdek C, Dargemont C (2012) H2B ubiquitylation controls the formation of export-competent mRNP. Mol Cell 45:132–139

Vohhodina J, Barros EM, Savage AL, Liberante FG, Manti L, Bankhead P, Cosgrove N, Madden AF, Harkin DP, Savage KI (2017) The RNA processing factors THRAP3 and BCLAF1 promote the DNA damage response through selective mRNA splicing and nuclear export. Nucleic Acids Res 45:12816–12833

Wallace EW, Kear-Scott JL, Pilipenko EV, Schwartz MH, Laskowski PR, Rojek AE, Katanski CD, Riback JA, Dion MF, Franks AM et al (2015) Reversible, specific, active aggregates of endogenous proteins assemble upon heat stress. Cell 162:1286–1298

Walsh MJ, Cooper-Knock J, Dodd JE, Stopford MJ, Mihaylov SR, Kirby J, Shaw PJ, Hautbergue GM (2015) Invited review: decoding the pathophysiological mechanisms that underlie RNA dysregulation in neurodegenerative disorders: a review of the current state of the art. Neuropathol Appl Neurobiol 41:109–134

Weirich CS, Erzberger JP, Flick JS, Berger JM, Thorner J, Weis K (2006) Activation of the DExD/H-box protein Dbp5 by the nuclear-pore protein Gle1 and its coactivator InsP6 is required for mRNA export. Nat Cell Biol 8:668–676

Wickramasinghe VO, Laskey RA (2015) Control of mammalian gene expression by selective mRNA export. Nat Rev Mol Cell Biol 16:431–442

Wickramasinghe VO, McMurtrie PI, Mills AD, Takei Y, Penrhyn-Lowe S, Amagase Y, Main S, Marr J, Stewart M, Laskey RA (2010) mRNA export from mammalian cell nuclei is dependent on GANP. Curr Biol 20:25–31

Wickramasinghe VO, Savill JM, Chavali S, Jonsdottir AB, Rajendra E, Gruner T, Laskey RA, Babu MM, Venkitaraman AR (2013) Human inositol polyphosphate multikinase regulates transcript-selective nuclear mRNA export to preserve genome integrity. Mol Cell 51:737–750

Wickramasinghe VO, Andrews R, Ellis P, Langford C, Gurdon JB, Stewart M, Venkitaraman AR, Laskey RA (2014) Selective nuclear export of specific classes of mRNA from mammalian nuclei is promoted by GANP. Nucleic Acids Res 42:5059–5071

Xing Z, Ma WK, Tran EJ (2018) The DDX5/Dbp2 subfamily of DEAD-box RNA helicases. Wiley interdisciplinary reviews. RNA 10:e1519

Yamazaki T, Fujiwara N, Yukinaga H, Ebisuya M, Shiki T, Kurihara T, Kioka N, Kambe T, Nagao M, Nishida E et al (2010) The closely related RNA helicases, UAP56 and URH49, preferentially form distinct mRNA export machineries and coordinately regulate mitotic progression. Mol Biol Cell 21:2953–2965

Yarbrough ML, Mata MA, Sakthivel R, Fontoura BMA (2014) Viral subversion of nucleocytoplasmic trafficking. Traffic 15:127–140

Yoh SM, Cho H, Pickle L, Evans RM, Jones KA (2007) The Spt6 SH2 domain binds Ser2-P RNAPII to direct Iws1-dependent mRNA splicing and export. Genes Dev 21:160–174

Younis S, Kamel W, Falkeborn T, Wang H, Yu D, Daniels R, Essand M, Hinkula J, Akusjarvi G, Andersson L (2018) Multiple nuclear-replicating viruses require the stress-induced protein ZC3H11A for efficient growth. Proc Natl Acad Sci U S A 115:E3808–E3816

Zaborowska J, Egloff S, Murphy S (2016) The pol II CTD: new twists in the tail. Nat Struct Mol Biol 23:771–777

Zander G, Krebber H (2017) Quick or quality? How mRNA escapes nuclear quality control during stress. RNA Biol 14:1–7

Zander G, Hackmann A, Bender L, Becker D, Lingner T, Salinas G, Krebber H (2016) mRNA quality control is bypassed for immediate export of stress-responsive transcripts. Nature 540:593

Zhao R, Shen J, Green MR, MacMorris M, Blumenthal T (2004) Crystal structure of UAP56, a DExD/H-box protein involved in pre-mRNA splicing and mRNA export. Structure 12:1373–1381

Zhou ZL, Luo M, Straesser K, Katahira J, Hurt E, Reed R (2000) The protein Aly links pre-messenger-RNA splicing to nuclear export in metazoans. Nature 407:401–405

# Chapter 2
# It's Not the Destination, It's the Journey: Heterogeneity in mRNA Export Mechanisms

**Daniel D. Scott, L. Carolina Aguilar, Mathew Kramar, and Marlene Oeffinger**

**Abstract** The process of creating a translation-competent mRNA is highly complex and involves numerous steps including transcription, splicing, addition of modifications, and, finally, export to the cytoplasm. Historically, much of the research on regulation of gene expression at the level of the mRNA has been focused on either the regulation of mRNA synthesis (transcription and splicing) or metabolism (translation and degradation). However, in recent years, the advent of new experimental techniques has revealed the export of mRNA to be a major node in the regulation of gene expression, and numerous large-scale and specific mRNA export pathways have been defined. In this chapter, we will begin by outlining the mechanism by which most mRNAs are homeostatically exported ("bulk mRNA export"), involving the recruitment of the NXF1/TAP export receptor by the Aly/REF and THOC5 components of the TREX complex. We will then examine various mechanisms by which this pathway may be controlled, modified, or bypassed in order to promote the export of subset(s) of cellular mRNAs, which include the use of metazoan-specific

D. D. Scott · M. Kramar
Institut de recherches cliniques de Montréal, Montréal, QC, Canada

Faculty of Medicine, Division of Experimental Medicine, McGill University, Montréal, QC, Canada
e-mail: daniel.scott@ircm.qc.ca; mathew.kramar@mail.mcgill.ca

L. C. Aguilar
Institut de recherches cliniques de Montréal, Montréal, QC, Canada
e-mail: LisbethCarolina.Aguilar@ircm.qc.ca

M. Oeffinger (✉)
Institut de recherches cliniques de Montréal, Montréal, QC, Canada

Faculty of Medicine, Division of Experimental Medicine, McGill University, Montréal, QC, Canada

Faculté de Médecine, Département de Biochimie, Université de Montréal, Montréal, QC, Canada
e-mail: marlene.oeffinger@ircm.qc.ca



M. Oeffinger, D. Zenklusen (eds.), *The Biology of mRNA: Structure and Function*, Advances in Experimental Medicine and Biology 1203,
https://doi.org/10.1007/978-3-030-31434-7_2

orthologs of bulk mRNA export factors, specific *cis* RNA motifs which recruit mRNA export machinery via specific *trans*-acting-binding factors, posttranscriptional mRNA modifications that act as "inducible" export *cis* elements, the use of the atypical mRNA export receptor, CRM1, and the manipulation or bypass of the nuclear pore itself. Finally, we will discuss major outstanding questions in the field of mRNA export heterogeneity and outline how cutting-edge experimental techniques are providing new insights into and tools for investigating the intriguing field of mRNA export heterogeneity.

**Keywords** mRNA export · NXF1 · CRM1 · Sequence elements · Nuclear pore complex

## 2.1 Introduction

The defining feature of eukaryotic organisms is the presence of the nucleus, which compartmentalizes the vast majority of the cell's DNA. While this compartmentalization has numerous advantages for the cell including reducing the DNA mutation rate and allowing more efficient regulation and replication of eukaryotes' large genomes, it has resulted in the physical separation of the cell's mechanisms for mRNA production (transcription, processing, maturation, and packaging into mRNPs) and mRNA metabolism (translation, localization, and degradation) into two compartments, physically separated by the nuclear membrane.

Material entering and departing the nucleus passes through the nuclear pore complex (NPC), a multi-megadalton protein complex which spans the inner and outer nuclear membranes and creates a pore through which molecules can pass between the cytoplasm and the nucleus (Beck and Hurt 2017). While molecules of less than ~30 kDa are capable of diffusing freely through the nuclear pore, nuclear mRNPs require an active and energy-dependent process to facilitate their translocation. This process, termed "mRNA export," involves the sequential loading and remodeling of a series of mRNA export factors onto the mRNP which, collectively, identify mature and correctly processed mRNAs, recruit them to the NPC, and facilitate their translocation through the nuclear pore before releasing them into the cytoplasm (Björk and Wieslander 2017; Folkmann et al. 2011; Oeffinger and Zenklusen 2012; Okamura et al. 2015).

The first investigations into the mechanistic basis of mRNA export occurred in the late 1980s through the observation that nuclear mRNA splicing in *S. cerevisiae* was able to control the subsequent availability of mRNAs to the translation machinery in the cytoplasm (Legrain and Rosbash 1989). However, it was not until the mid-to-late 1990s that work on retroviral RNA dynamics revealed the existence of unique mRNA export pathways and their particular *cis*- and *trans*-acting factors (Cullen 1998; Jarmolowski et al. 1994). Since these early observations, it has become clear that mRNA export, far from being a passive mechanism of bulk transport, is a highly complex system in which numerous overlapping and competing pathways act to control the translocation across the nuclear pore in response to competing spatial, temporal, and environmental demands (Delaleau and Borden 2015; Wickramasinghe et al. 2014). New mRNA export pathways, regulatory systems, and interfaces with other cellular processes are being reported on a regular basis, with each new discovery adding to our comprehension of how the regulation of mRNA export contributes to cellular homeostasis and disease.

Despite these recent advances, however, much still remains to be discovered in the field, and numerous fundamental questions regarding the machinery and regulation of mRNA export remain unanswered (Okamura et al. 2015). The considerable complexity has made it difficult to dissect the precise molecular biology of mRNA export (Delaleau and Borden 2015; Wickramasinghe et al. 2014), exacerbated by a lack of experimental techniques capable of addressing the complex hypotheses that have been proposed. Historically, the field of mRNA export research has relied heavily on the use of model systems such as *Xenopus* oocytes, in which numerous steps that intimately couple transcription and mRNA processing to mRNA export factor loading and maturation are bypassed by the microinjection of mature, in vitro-transcribed mRNAs. Similarly, a reliance on bulk mRNA-focused quantification techniques, particularly oligo(dT) fluorescence in situ hybridization (FISH), prevents the quantification of mRNA export defects on a transcript-specific basis, resulting in either over- or underreporting of phenotypic severity (Guria et al. 2011; Katahira et al. 2009; Rehwinkel et al. 2004). In recent years, the emergence of new technologies and approaches in microscopy, transcriptomics, and proteomics has finally enabled us to address these limitations; in so doing, these techniques—and others still to come—are altering the paradigms of research in the field, allowing researchers to address new and fundamental questions regarding the mRNA export machinery and its regulation.

In this chapter, we will examine in detail the numerous overlapping and complementary mRNA export pathways that have been described thus far in higher eukaryotes. We will discuss the key unifying principles and unique features of these pathways and will outline important outstanding questions in the field. In so doing, we will take a particular look at how cutting-edge techniques are allowing researchers to dissect these pathways in unprecedented detail and how novel techniques are expected to provide new paradigms for the investigation of mRNA export research. This chapter will focus primarily on work conducted in metazoans, with examples from other eukaryotes drawn upon where relevant.

## 2.2 The Ground State: Bulk Export Pathways for mRNAs

The most thoroughly explored and described pathway of mRNA export is the "bulk mRNA export" pathway which exports the majority of cellular mRNAs during homeostatic growth conditions (Fig. 2.1). Components of this pathway are highly conserved throughout the eukaryotes and are often—though not always—essential proteins for the growth of the cell (Björk and Wieslander 2017; Okamura et al.

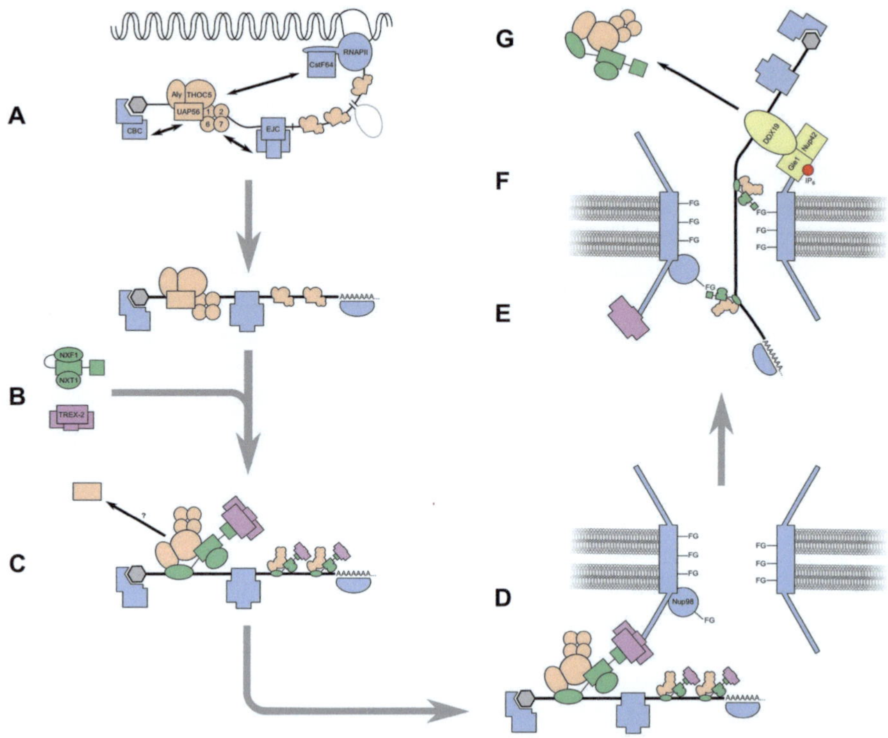


**Fig. 2.1** Schematic of the bulk mRNA export pathway in metazoans. Nascent mRNA (**a**) is co-transcriptionally assembled with maturation factors including the cap-binding complex (CBC), exon-exon junction complex (EJC), and the 3′ processing machinery, including the component CstF64 (blue). Each of these components acts to recruit one or more copies of the TREX complex (orange)—including the mRNA export adaptor Aly/REF (*Aly*) and the coadaptor THOC5—which are deposited along the length of the mRNA, ultimately forming a TREX-coated, transcribed mRNA (**b**). Aly/REF and THOC5 bind the export receptor heterodimer NXF1:NXT1 (green), causing a structural rearrangement that exposes NXF1's RNA-binding domain and allows hand-off of RNA from Aly/REF:THOC5 to NXF1:NXT1 (**c**). Upon association with the chaperone complex TREX-2 (purple), the NXF1:NXT1-loaded mRNA is translocated to the nuclear pore via an interaction between TREX-2 and the nuclear basket (**d**). After quality control and/or structural rearrangements, the FG-Nup-binding C-terminal domain of NXF1 is released by TREX-2 and can interact with the proximal FG-Nup NUP98 (**e**) and then other FG-Nups deeper within the nuclear pore channel (**f**). Upon reaching the cytoplasmic surface of the nuclear pore, the NXF1:NXT1:TREX complex is unloaded from the cargo mRNA by the actions of the DDX19:Gle1:Nup42 complex (yellow) bound by $IP_6$ (red), releasing the mRNA for further cytoplasmic maturation and/or metabolism (**g**). See Sect. 2.2 for a detailed description of this pathway. See text for more details

2015). Instead of relying on specific sequence or structural elements of the mRNA transcript to identify mRNA cargoes, the bulk mRNA export pathway is instead closely interfaced with the co-transcriptional processes of mRNA maturation, thereby (a) preventing the preferential export of particular mRNA species and

(b) ensuring exclusive export of mature, translation-competent mRNAs to the cytoplasm.

The key initiating factor in the bulk mRNA export pathway is the TREX complex, a conserved multiprotein complex composed in humans of the mRNA export adaptor Aly/REF (Yra1 in *S. cerevisiae*), the helicase UAP56/DDX39B (Sub2), CIP29 (Tho1), and the THO subcomplex, composed of the conserved subunits hHpr1/THOC1 (Hpr1), hTho2/THOC2 (Tho2), hTex1 (Tex1), THOC7 (Mft1), and the metazoan-specific THOC5 and THOC6 (Katahira 2012). TREX assembles in a highly cooperative fashion which requires ATP binding of the scaffolding UAP56 helicase (Chi et al. 2013; Dufu et al. 2010) and plays a central role in the coupling of successful mRNA maturation to the subsequent deposition of Aly/REF, an RNA-binding protein whose loading onto mRNA is the initiating step in bulk export and which acts as a key "mRNA export adaptor" for subsequent factors in the pathway (Rodrigues et al. 2001).

In metazoans, the most significant activator of mRNA export is the completion of splicing and, in particular, the deposition of the exon junction complex (EJC) on the maturing mRNA, which carries a checkpoint function for completed splicing (Le et al. 2001; Masuda et al. 2005). The core EJC components interact directly with several TREX components, including UAP56 and Aly/REF, and are required for the recruitment of these proteins and their subsequent loading onto the mRNA (Gerbracht and Gehring 2018; Gromadzka et al. 2016; Le et al. 2000; Viphakone et al. 2018). This loading mechanism distinguishes metazoans from *S. cerevisiae*, wherein the TREX complex becomes associated with mRNA co-transcriptionally *via* THO subcomplex interactions with the phosphorylated C-terminal domain of RNAPII and/or the CTD-loaded Prp19 splicing complex (Heath et al. 2016; Katahira 2012; Masuda et al. 2005; Zenklusen et al. 2002). This transition from transcription- to splicing-dependent loading likely evolved in response to the massive proliferation of splicing in metazoans; however, the observation that Aly/REF interacts with the RNAPII-CTD-binding Iws1:Spt6 complex suggests that at least some of the ancestral co-transcriptional loading machinery may have been retained in higher eukaryotes (Yoh et al. 2007). In addition to being loaded onto mRNA in a splicing-dependent fashion, Aly/REF and/or the TREX complex can be added via an interaction between Aly/REF and the CBP80 component of the cap-binding complex (CBC) once it is bound to a mature 5′ $N^7$-methylguanosine ($m^7G$) cap structure (Cheng et al. 2006). Lastly, an interaction between Aly/REF and CstF64, a component of the CstF complex required for mRNA 3′-end processing and polyadenylation, promotes loading of Aly/REF onto nascent mRNA (Shi et al. 2017), possibly in a mechanism similar to the Pcf11-dependent loading of Yra1 in *S. cerevisiae* (Johnson et al. 2009). Collectively, these mechanisms allow the loading of TREX and, as a result, Aly/REF onto the nascent mRNA (Shi et al. 2017; Viphakone et al. 2018) and, furthermore, ensure its coupling to several key steps in the maturation of nascent mRNAs including 5′ capping, splicing, and 3′-end processing/polyadenylation (Fig. 2.1a, b).

Following its deposition onto mRNA, Aly/REF and a component of the THO subcomplex, THOC5, act coordinately as adaptor/coadaptor pair to recruit the key mRNA export receptor, NXF1/TAP (Mex67 in *S. cerevisiae*), in a heterodimeric complex with its partner, NXT1/p15 (Mtr2) (Stutz et al. 2000). Unlike other nuclear export receptors of the importin-β subfamily, NXF1 does not require the GTP-binding protein Ran for NPC transit; instead, it exhibits a modular domain arrangement with an N-terminal Aly/REF/RNA-binding domain, a central NTF2L-like domain responsible for interactions with THOC5 and NXT1, and a C-terminal domain capable of directly interacting with FG regions within the NPC (Herold et al. 2000). Upon its recruitment to mRNA, NXF1 displaces UAP56 from Aly/REF, likely via steric effects resulting from their closely juxtaposed Aly/REF-binding sites (Hautbergue et al. 2008). Whether this displacement involves eviction of Aly/REF from TREX, or a more subtle rearrangement of TREX interactions, remains unknown (Fig. 2.1b).

In its free state, NXF1:NXT1 exhibits poor affinity for mRNA due to an intramolecular interaction between its N-terminal RNA-binding domain and central NTF2L domain which masks the RNA-binding surface. The binding of Aly/REF and THOC5 to NXF1:NXT1—to NXF1's RNA-binding domain and NTF2L domains, respectively—is able to disrupt this intramolecular interaction, opening up the NXF1 RNA-binding domain (Viphakone et al. 2012). As a consequence, an unusual rearrangement of the Aly/REF:NXF1:RNA complex occurs, in which the bound region of mRNA is released by Aly/REF and "handed off" to NXF1's now-available RNA-binding domain, while Aly/REF binds to a remote surface on NXF1's RNA-binding domain (Hautbergue et al. 2008; see Fig. 2.1c). This hand-off is facilitated by the methylation of key arginine residues in the Aly/REF RNA-binding surface by PRMT1, resulting in a decreased affinity for RNA relative to that for NXF1 (Hung et al. 2010).

Once deposited onto mRNA, NXF1:NXT1 must traverse the nucleoplasm with its mRNA cargo to reach the NPC. Detailed microscopic studies of single RNA molecules in living cells have suggested that mRNA transport through the nucleoplasm is a passive process involving diffusion through channels between chromatin domains (Mor and Shav-Tal 2010). Upon reaching the nuclear rim, NXF1:NXT1 interaction with the nuclear basket—an extended structure associated with the NPC protruding into the nucleoplasmic space and formed by TPR (Mlp1 in *S. cerevisiae*)—is believed to require the activity of chaperones (Wickramasinghe et al. 2010). The most prominent of these is TREX-2, a complex independent of TREX, which contains several subunits including the scaffolding GANP, ENY1, PCID2, DSS1, and several centrin proteins (Jani et al. 2012; Wickramasinghe et al. 2010). Like the TREX complex, TREX-2 is widely conserved but has evolved different functions from its *S. cerevisiae* ancestor, which is required for the tethering of transcriptionally active genes to the NPC (Cabal et al. 2006; Rodríguez-Navarro et al. 2004). In metazoans, TREX-2 loading onto NXF1:NXT1 is mediated by interaction of the NXF1 C-terminal domain with an N-terminal FG-Nup-like region on GANP (Jani et al. 2012; Umlauf et al. 2013; see Fig. 2.1c). It is believed that, following loading onto NXF1:NXT1, TREX-2 is then able to direct its mRNA cargo to the NPC via the interaction of GANP, ENY1, and/or PCID2—and, potentially, other proteins loaded onto the transported mRNP—with the nuclear basket component TPR (Fasken et al. 2008; Umlauf et al. 2013; Wickramasinghe et al. 2010; see Fig. 2.1d). However, some confusion remains regarding the exact timing and location of the loading of NXF1:NXT1 onto TREX-2, and the mechanism by which TREX-2 is recruited to nascent mRNAs (Jani et al. 2012; Umlauf et al. 2013). In addition to the TREX-2 complex, several other possible NXF1:NXT1 NPC chaperones have been reported, including the WD-repeat protein RAE1 in complex with the nucleoplasmic-mobile FG-Nup NUP98 (Blevins et al. 2003; Pritchard et al. 1999) and the inner nuclear membrane-embedded SUN1 protein (Li and Noegel 2015; Li et al. 2017b). It is important to note that these different chaperones may not be exclusive and may mediate different stages in the chaperoning of NXF1:NXT1-loaded mRNA to and through the NPC.

Interestingly, detailed microscopy studies have suggested that, upon reaching the nuclear basket, mRNPs are frequently returned to the nucleoplasm, with only a minority of mRNAs proceeding from basket binding to transit through the pore (Grünwald and Singer 2010; Ma et al. 2013; Siebrasse et al. 2012). Kinetic analyses of mRNA residency at the NPC in living cells have additionally shown a pause step at the nuclear basket, with mRNAs spending significant time resident on the basket prior to a relatively quick traversal of the nuclear pore (Grünwald and Singer 2010; Siebrasse et al. 2012), possibly due to a requirement for both quality control and remodeling of the mRNPs on the basket prior to mRNP transit through the nuclear pore (Grünwald and Singer 2010; Siebrasse et al. 2012). The precise mechanisms by which mRNAs transit the nuclear pore upon commitment to translocation remain a topic of considerable debate; however, it is generally agreed that successive interactions between the NXF1 C-terminal domain and the exposed tails of FG-Nup proteins throughout the channel mediate an mRNP's entry into and transit through the pore (Oeffinger and Zenklusen 2012; see Fig. 2.1e, f).

Upon reaching the cytoplasm, NXF1:NXT1-loaded mRNPs make contact with a complex of proteins loaded onto the cytoplasmic fibrils of the NPC that includes the helicase DDX19/Dbp5, Gle1, Nup42, and the activating signaling molecule inositol hexaphosphate ($IP_6$). This complex is responsible for remodeling the mRNP on the cytoplasmic face of the NPC, releasing the cargo mRNA from export factors including NXF1:NXT1, which are recycled back into the nucleus (Adams et al. 2017, 2018; Folkmann et al. 2011; see Fig. 2.1g). An important secondary function of this unloading is to act as a "ratchet" for directional translocation of the mRNA across the NPC; the interaction of NXF1 with FG-Nups has been found not to be inherently directional, allowing NXF1-mRNP complexes to move back and forth within the nuclear pore (Grünwald and Singer 2010). The eviction of NXF1:NXT1 from the mRNA prevents this backwards diffusion and ensures that mRNAs migrate through the NPC in a directional manner (Folkmann et al. 2011). The cytoplasmic mRNA cargo is then free to undergo any cytoplasmic-specific mRNA maturation steps prior to translation.

## 2.3 Orthology of Bulk mRNA Export Factors

Much of the early work on mRNA export was performed in *S. cerevisiae*, where the bulk export pathway is relatively canonical; the loading of Yra1 onto mRNA via TREX and subsequent recruitment of Mex67 are key steps in the export of most cellular mRNAs in this species, as evidenced by the lethal phenotypes of both *yra1Δ* and *mex67Δ* deletions and the severe mRNA export defects of hypomorphic mutations (Portman et al. 1997; Santos-Rosa et al. 1998; Segref et al. 1997; Zenklusen et al. 2002). While limited evidence has emerged of possible heterogeneity in mRNA export pathways in *S. cerevisiae*, including the description of a nonessential Yra1 homolog, Yra2, that can suppress Yra1 phenotypes when overexpressed, the linear and nonredundant bulk export pathway remains the major means of mRNA export in these cells (Hieronymus and Silver 2003; Okamura et al. 2015; Zenklusen et al. 2001).

In higher eukaryotes, however, several proteome-wide screens for mRNA export factors have demonstrated that while the bulk mRNA export pathway has been conserved from *S. cerevisiae*, metazoans have evolved numerous other mRNA export factors which are able to modify and/or complement this bulk pathway, greatly expanding the mechanistic and regulatory complexity of mRNA export (Delaleau and Borden 2015; Farny et al. 2008; Rehwinkel et al. 2004; Wickramasinghe and Laskey 2015). A major source of this variation has been the proliferation of orthologs of numerous key export factors (Delaleau and Borden 2015; Wickramasinghe and Laskey 2015). These "alternative factors" are able to functionally substitute for their orthologs in a constitutive and/or conditional fashion (Fig. 2.2a). Indeed, the identification of such factors is beginning to explain the long-time contradiction in the field that, while bulk export factors such as Yra1 are essential in *S. cerevisiae*, depletion of their metazoan homologs has only relatively mild effects on mRNA export efficiency in vivo (Gatfield and Izaurralde 2002; Katahira et al. 2009; Longman et al. 2003).

### *2.3.1 mRNA Export Adaptor Heterogeneity*

The relatively mild effects of Aly/REF depletion in metazoans led numerous groups to explore the possibility of alternative mRNA adaptors for the recruitment of NXF1: NXT1 to mRNA. A particularly successful approach to this was taken by Stuart Wilson's group, whose strategy of searching for conserved UAP56-binding motifs (UBMs) has identified several possible functional substitutes for Aly/REF in human cells (Chang et al. 2013; Hautbergue et al. 2009; Viphakone et al. 2015).

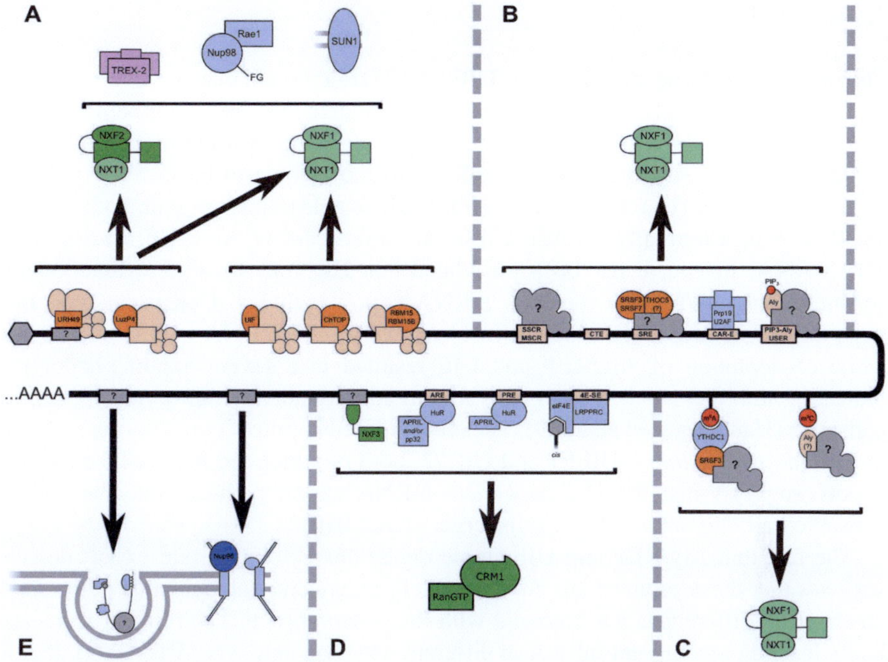


**Fig. 2.2** Schematic of characterized selective mRNA export paths in metazoans. Where *cis*- and/or *trans*-acting factors are unknown, they are shown in gray with a question mark. (**a**) Orthologs (dark orange/green) of bulk export pathway factors (light orange/green) may be loaded onto mRNAs and cooperate with bulk export pathway factors to promote selective mRNA export. Several different export receptor chaperones (purple/light blue) may act to direct the export receptor(s) to the nuclear pore. See Sect. 2.3. (**b**) *cis*-acting sequence/structural USER codes (light orange) can recruit *trans*-acting mRNA export adaptors/coadaptors from the bulk (light orange) or selective (dark orange) mRNA export pathways; alternatively, they may rely on repurposing of other cellular RBPs (light blue) or alteration of bulk export pathways by second messengers (red). See Sect. 2.4. (**c**) Posttranscriptionally added mRNA modifications (red) can act as sequence-nonspecific, *cis*-acting USER codes through recruitment of bulk (light orange) or selective (dark orange) export adaptors, with or without the assistance of mRNA modification reader proteins (light blue). See Sect. 2.5. (**d**) A number of *cis*-acting USER codes rely on binding to cellular RBPs (light blue) that lack canonical export adaptor/coadaptor activity, but can recruit an alternate mRNA export adaptor heterodimer, CRM1:RanGTP (dark green), via a leucine-rich nuclear export signal (NES). These include the atypical bulk mRNA export receptor ortholog, NXF3 (dark green). See Sects. 2.4 and 2.6. (**e**) Finally, mRNA export may also be modulated through the activity of specific nuclear pore components (dark blue) or may bypass the nuclear pore entirely through direct budding through the inner and outer nuclear membranes (gray lines). See Sect. 2.7. Given that one or more of these pathways may act in conjunction both with each other and with the bulk export pathway (see Fig. 2.1), they are depicted as being resident on the same mRNA (black line). See text for more details

#### 2.3.1.1 UIF

The first of these to be identified was UIF/FYTTD1, a protein that was evolutionarily unrelated to Aly/REF but shares many of its basic features, including being ubiquitously expressed throughout tissues and, upon tethering to a reporter mRNA, the ability to directly support export of mRNA to the cytoplasm in an NXF1:NXT1-dependent manner (Hautbergue et al. 2009). UIF is able to interact with both UAP56 and NXF1 in a mutually exclusive fashion reminiscent of Aly/REF loading onto TREX/RNA. Indeed, it has been postulated that UIF may be able to functionally substitute for Aly/REF in the bulk mRNA export pathway. Consistent with this hypothesis, UIF depletion resulted in a relatively mild mRNA export phenotype, while co-depletion of Aly/REF and UIF resulted in a severe export phenotype, indicating that these two proteins are able to act redundantly along mRNA export pathways (Hautbergue et al. 2009). The recent identification of two homologs of UIF in *Arabidopsis thaliana*, UIEF1 and UIEF2, both of which are required for mRNA export, suggests that the UIF-dependent mRNA export pathway may be widely conserved among eukaryotes (Ehrnsberger et al. 2019).

The fact that Aly/REF and UIF single depletions exhibited mild export defects suggests that these proteins are not completely redundant, and evidence of possible mechanistic differences has emerged with the observation that the two proteins are likely loaded onto nascent mRNA in different ways, namely, via SPT6/IWS1 and/or CstF64 for Aly/REF and via SSRP1 for UIF (Hautbergue et al. 2009; Shi et al. 2017; Yoh et al. 2007). Interestingly, overexpression of both Aly/REF and UIF also induced mRNA export defects, raising the possibility that the relative stoichiometry of these two proteins is important for mRNA export, though a possible indirect mechanism dependent on sequestration of other export factors such as TREX and/or NXF1:NXT1 cannot be excluded (Hautbergue et al. 2009).

#### 2.3.1.2 Luzp4

A similar UBM-searching strategy as for UIF was used to identify Luzp4/CT-8, a leucine zipper-containing protein (Viphakone et al. 2015). Like Aly/REF, Luzp4 localizes to the nuclear splicing speckles and is capable of interacting with the TREX complex, NXF1 and RNA (Türeci et al. 2002; Viphakone et al. 2015). Luzp4 overexpression was shown to completely rescue a modest mRNA export defect caused by Aly/REF depletion, confirming Luzp4 as a genuine mRNA export adaptor able to act redundantly with Aly/REF (Viphakone et al. 2015); however, it could only partially rescue the more severe defects of Aly/REF and UIF co-depletion, suggesting that it may not share complete functional redundancy with these ubiquitously expressed factors (Viphakone et al. 2015).

One major point of difference is that Luzp4 expression is tightly restricted to the testes during homeostatic growth, and its upregulation in numerous cancers marked it as a member of the cancer-testis antigen family of genes (Türeci et al. 2002;

Viphakone et al. 2015). In line with its testes-restricted expression pattern, Luzp4 was found to interact with NXF2, a testis-specific NXF1 ortholog discussed below, raising the possibility of an alternative, Luzp4-NXF2-based mRNA export pathway specific to the testes.

#### 2.3.1.3 Other Candidate Export Adaptors: SKAR and ZC11A

While UIF and Luzp4 remain the only definitively characterized Aly/REF substitutes in bulk mRNA export, several other export adapter candidates have been identified in other studies. Dufu et al. (2010) identified five previously unreported proteins in a proteomic screen for common interactors of UAP56, THOC5, and CIP29, suggesting that these factors may be novel components of the TREX complex; one of these, SKAR/POLDIP3, shows significant sequence and domain-arrangement homology with Aly/REF, while SKAR and another protein, ZC11A, associate with UAP56 in an ATP-dependent manner and induce an mRNA export defect upon depletion (Folco et al. 2012). While these proteins require further investigation prior to their confirmation as bona fide Aly/REF functional orthologs, their identification raises the possibility that the complete suite of mRNA export adaptors in human cells remains to be fully elucidated.

### *2.3.2 Export Coadaptor Heterogeneity*

While Aly/REF is essential for the recruitment and subsequent binding of NXF1: NXT1 to mRNA, it alone is not sufficient. The disruption of the RNA-binding domain-masking intramolecular interaction of NXF1 requires the combined activity of both an mRNA adaptor, such as Aly/REF, and a coadaptor, THOC5. While THOC5 is reported to interact with RNA less stringently than NXF1:NXT1, biochemical experiments showed its activity to be required for efficient handover of mRNA from Aly/REF to NXF1:NXT1 and consequent mRNA export, marking it as a core component of the bulk mRNA export pathway (Chang et al. 2013; Viphakone et al. 2012).

Despite this central role, however, THOC5 depletion, like that of Aly/REF, causes only modest defects in bulk mRNA export (Chang et al. 2013); indeed, experiments comparing the cytoplasmic versus nuclear distribution of RNAs by microarray found that only 0.7–2.9% of mRNAs showed a significant alteration in nuclear/cytoplasmic ratio following THOC5 knockdown (Guria et al. 2011; Rehwinkel et al. 2004). Furthermore, numerous reports have emerged identifying essential roles for THOC5 in the regulation of particular functional subsets of mRNA, including those involved in heat shock (Katahira et al. 2009), stem cell pluripotency (Ratnadiwakara et al. 2018; Wang et al. 2013), hematopoiesis (Mancini et al. 2010), and adipogenesis (Mancini et al. 2006). Collectively, these results suggest that, far from being a unique regulator of bulk mRNA export, THOC5, like Aly/REF, may simply be one of a set of mRNA export coadaptors with redundant and/or specialized roles in mRNA export.

#### 2.3.2.1 ChTOP

Like UIF and Luzp4, ChTOP was identified based on a canonical UBM required for interaction with UAP56 in the context of the TREX complex, suggesting a mutually exclusive loading onto TREX with Aly/REF (Chang et al. 2013). In support of an Aly/REF-esque role, ChTOP was shown to localize to splicing speckles, interact with RNA, and undergo a PRMT1 methylation-based hand-off interaction with NXF1:NXT1. However, examination of the interaction between ChTOP and other export components revealed that, unlike Aly/REF, UIF, and Luzp4, ChTOP interacted with the THOC5-binding site on NXF1's central NTF2L domain; furthermore, ChTOP bound NXF1:NXT1 cooperatively, rather than exclusively with Aly/REF, and, in so doing, additively enhanced NXF1's affinity for mRNA, identifying ChTOP as a novel alternative coadaptor for Aly/REF on NXF1 (Chang et al. 2013).

#### 2.3.2.2 RBM15 and RBM15B

RBM15 and its distantly related ortholog, RBM15B/OTT3, are RNA-binding proteins with diverse reported roles within cells, including the regulation of site-specific RNA methylation on $N^6$-adenosine ($m^6A$) and the transcriptional regulation of Notch signaling pathways (Ma et al. 2007; Patil et al. 2016). In the context of mRNA export, RBM15 has been reported to recruit the helicase DDX19 to NPC-transiting mRNAs for eviction of NXF1:NXT1 at the cytoplasmic face (Zolotukhin et al. 2009) and is a *trans*-acting factor responsible for the binding of a *cis*-acting mRNA export sequence element, the retroviral transport element (RTE), discussed below (Lindtner et al. 2006).

In addition to these myriad roles, a study by Uranishi et al. (2009) has reported the likely function of both RBM15 and RBM15B as alternative NXF1:NXT1 coadaptors. RBM15B, like RBM15 (Lindtner et al. 2006), can promote the export of tethered RNAs via a direct interaction with NXF1:NXT1. This interaction involved the C-terminal domains of RBM15/RBM15B binding to the central NTF2L domain of NXF1 in a fashion resembling that of THOC5 and ChTOP; the same domain of RBM15 also mediated an interaction with Aly/REF, suggesting that RBM15 is able to assemble a functional Aly/REF:RBM15/B:NXF1:NXT1 adaptor–coadaptor–receptor complex on bound mRNAs (Uranishi et al. 2009). While these observations remain incomplete with regard to a role for RBM15/RBM15B as true coadaptors in bulk mRNA export, and further experiments are essential to dissect the relative overlap and interplay of the multiple reported roles of RBM15/RBM15B, these observations suggest the possibility that RBM15 and/or RBM15B may act as additional Aly/REF-paired coadaptors, at least on a subset of mRNAs (Chang et al. 2013; Uranishi et al. 2009).

### 2.3.3 *Additional Heterogeneity Within the TREX Complex: UAP56 and URH49*

In *S. cerevisiae*, deletion of the TREX complex core helicase, Sub2, is lethal, likely due to a near-complete failure of bulk mRNA export (Sträßer et al. 2002). Similarly, depletion of the metazoan Sub2 homolog, UAP56, causes a significant export defect, supporting its identification as the major Sub2 homolog (Sträßer et al. 2002). However, in mammals, sequence analysis revealed that Sub2 in fact possesses two closely related orthologs, UAP56/DDX39B and URH49/DDX39, both of which are capable of partially rescuing a *sub2Δ* deletion phenotype in *S. cerevisiae*, raising the possibility that these two proteins may act in parallel (Kapadia et al. 2006; Pryor et al. 2004). In support of this observation, both UAP56 and URH49 are able to interact with Aly/REF, and their co-depletion results in a severe mRNA export defect greater than loss of either single protein.

However, a major point of difference between the UAP56/URH49 and the redundancy paradigm of Aly/REF and other export adaptors above is that depletions of UAP56 and URH49, while additive, do exhibit significant mRNA export defects by themselves, suggesting these proteins have at least partially independent functions in mRNA export. Indeed, a cytoplasmic RNA microarray analysis of the two proteins found that, while the depletion of each reduced the export of approximately 300 mRNAs, only approximately 60 of these were common between the two proteins (Yamazaki et al. 2010). Furthermore, phenotypic analysis of UAP56 and URH49 depletions revealed that, while they both induced mitotic dysfunction, they did so via different mechanisms, with UAP56 depletion causing premature sister chromatid separation during mitosis, while URH49 depletion prevented efficient chromosome arm resolution and cytokinesis (Yamazaki et al. 2010). It has also been suggested that these two helicases may have different affinities for non-Aly/REF TREX components, in particular the THO subcomplex and CIP29; however, subsequent work has disputed at least some of these findings (Dufu et al. 2010). Finally, URH49, but not UAP56, has been reported to exhibit highly regulated expression, both tissue-specific and growth-regulated, suggesting it may act as a stimulus-responsive complement to the constitutive activity of UAP56 (Pryor et al. 2004). Collectively, these observations suggest that, while UAP56 and URH49 likely play similar mechanistic roles in the regulation of mRNA export, their mRNA targets and the conditions under which they act may vary significantly.

### 2.3.4 The NXF1 Family of Export Receptors

NXF1 was the first mRNA export factor to be identified through its recruitment to *cis*-acting sequence elements in retroviral transcripts (see below; Grüter et al. 1998). Its clear homology to the essential yeast export receptor Mex67, and the severe export defect resulting from its depletion, has led to its characterization as the sole mRNA export receptor for constitutive bulk mRNA export (Björk and Wieslander 2017; Okamura et al. 2015). While this is likely to be true in the majority of cell types during constitutive growth, it is nonetheless notable that the *NXF1/Mex67* family has undergone significant diversification during eukaryotic evolution. While *S. cerevisiae* only possess one copy of the family (Mex67), there are two members in *C. elegans*, four in *D. melanogaster*, and five (including NXF1) in humans (Herold et al. 2000). In humans, these paralogs, termed NXF1-5, have generally retained the domain arrangement of Mex67/NXF1; however, NXF4 and NXF5 are likely not expressed due to the presence of multiple frameshifts and/or premature stop codons in their coding sequence, while NXF3 contains several truncations that prevent it from replicating the various interactions required for NXF1's export activity (Herold et al. 2000; Yang et al. 2001). Remarkably, NXF3 has in fact been found to be able to promote mRNA export via an entirely novel mechanism involving the export receptor CRM1, which will be discussed below (Yang et al. 2001).

NXF2 is one homolog that has retained a close similarity to NXF1 and the ability to interact with known mRNA export coadaptors, including ChTOP, as well as with mRNA itself (Chang et al. 2013; Herold et al. 2000). Immunofluorescence analyses showed that NXF2 localizes to the nucleoplasm and accumulates at the nuclear rim like NXF1 and is able to promote mRNA export in at least some assays, although the latter has been disputed by other groups working in heterologous systems (Herold et al. 2000; Yang et al. 2001). Interestingly, expression analysis determined that NXF2's expression, like that of LuzP4 and URH49, is tightly constrained to the testes, and reports that NXF2 cooperates with the RNA-binding protein (RBP) FMRP to cooperatively destabilize the *NXF1* mRNA and thus repress NXF1 expression suggest that, at least in the testes, NXF2 may take over as the predominant receptor in bulk mRNA export (Herold et al. 2000; Zhang et al. 2007). In addition to their observations on the NXF1 family, Herold et al. (2000) also noted the existence of a second NXT1 ortholog, termed p15-2. However, beyond identification of a broad expression profile with moderate upregulation in the testes by high-throughput tissue screening, the function of this ortholog has not yet been investigated (Herold et al. 2000).

### 2.3.5 *Chaperoning of NXF1:NXT1 to the NPC: Pervasive Mechanism or "Fast-Track" Selectivity?*

As mentioned above, the recruitment of mRNA-bound NXF1:NXT1 to the NPC is suggested to be mediated by a number of different chaperones. The best explored of these has been the TREX-2 complex, in which the scaffolding protein GANP binds the NXF1 C-terminal domain and directs NXF1 to the nuclear basket (Jani et al. 2012; Wickramasinghe et al. 2010). While this was assumed to be a universal process in bulk mRNA export, recent work by Wickramasinghe et al. (2014) profiled the mRNA export defect upon GANP depletion and found, surprisingly, that export of only approximately 50% of all NXF1 target mRNAs was negatively affected; furthermore, these 50% were predominantly abundant, short-lived mRNAs and were enriched for gene ontology annotations covering a variety of RNA processing functions. Strikingly, the export of those GANP-dependent RNAs was found to be significantly more rapid than those whose export was unaltered by GANP depletion, raising the possibility that GANP-mediated NXF1:NXT1 chaperoning may not represent a ubiquitous pathway but instead a "fast-track" export mechanism for a specific functional or regulatory subset of mRNAs (Okamoto et al. 2010; Wickramasinghe et al. 2014). It is not known whether GANP-independent mRNAs are dependent on other reported NXF1:NXT1-NPC chaperones, such as RAE1-NUP98 (Blevins et al. 2003; Pritchard et al. 1999) and/or SUN1 (Li and Noegel 2015; Li et al. 2017b), and, if so, what impact these chaperones have on the export rate of their mRNA cargoes; considerable further research is required to dissect if and how the different reported chaperone pathways interact. However, the finding that diffusion of mRNAs through the nucleoplasm and their loading onto the nuclear basket represent rate-limiting steps in mRNA export (Mor and Shav-Tal 2010; Shav-Tal et al. 2004) suggests that chaperoning of NXF1:NXT1 to the NPC may represent a good candidate target for the manipulation of mRNA export rate for subsets of mRNAs.

### 2.3.6 *Defining the Complete Suite of Bulk mRNA Export Orthologs and Parallel Pathways*

Despite the extensive work described above identifying orthologous bulk mRNA export factors acting on metazoan mRNA, our knowledge of the complete suite of bulk mRNA export orthologs likely remains incomplete, especially with regard to the existence of further tissue- and/or stimulus-specific factors. While biochemical testing of the mRNA export roles of these candidate orthologs remains important, additional alternative approaches are required to identify the complete suite of mRNA export factor orthologs in mammals.

The Wilson group recently used in silico homology screening of the key UBM interaction motif within Aly/REF to identify several novel mRNA export adaptors/

coadaptors including UIF, LuzP4, and ChTOP (Chang et al. 2013; Hautbergue et al. 2009; Viphakone et al. 2015) suggesting that in silico identification represents a viable strategy for the identification of future candidates. Alternatively, the ongoing development of high-content screening microscopy systems, which allow the automated imaging and quantitative analysis of thousands of immunofluorescence samples, enables the use of mRNA export readouts such as nuclear-cytoplasmic poly $(A)^+$ RNA ratios in the screening of large candidate protein classes to identify putative mRNA export factors (Mattiazzi Usaj et al. 2016). The recent use of these systems to generate a large immunofluorescence-based database colocalizing a library of approximately 300 RBPs with a range of different cellular structures emphasizes the viability of such experimental approaches and provides an important comparative database for the immunofluorescent localization of possible mRNA export factors identified by siRNA/overexpression screening as described above (Van Nostrand et al. 2018).

A second major question concerning orthologous bulk mRNA export adaptors/coadaptors is their degree of orthology. The relatively mild mRNA export defects resulting from the individual knockdowns of bulk mRNA export factors such as Aly/REF, UIF, and THOC5 coupled with the synthetic, severe export deficits upon co-depletion of pairs such as Aly/REF:UIF, Aly/REF:THOC5, and Aly/REF:ChTOP argue for near-complete overlap of these factors' export activities (Chang et al. 2013; Gatfield and Izaurralde 2002; Hautbergue et al. 2009; Katahira et al. 2009; Longman et al. 2003). However, several observations suggest that the system may be more complicated than these observations imply. While the pairwise testing of adaptor/coadaptor co-depletions is far from complete, several unexpected observations have emerged, such as that co-depletion of ChTOP:THOC5 causes no additive export defect, despite these two proteins supposedly acting in a redundant and compensatory mechanism as mRNA coadaptors (Chang et al. 2013). Furthermore, the co-depletion of both Aly/REF:THOC5 and Aly/REF:ChTOP resulted in severe export defects, despite the existence in each case of a hypothetically redundant adaptor-coadaptor pair (UIF:ChTOP and UIF:THOC5) (Chang et al. 2013). Other observations have shown that LuzP4 overexpression is able to completely rescue the export defect of Aly/REF depletion but can only partially rescue the defect of Aly/REF:UIF co-depletion, suggesting that UIF (or the combination of Aly/REF and UIF) performs functions that are not redundantly regulated by LuzP4 (Viphakone et al. 2015). Profiling of mRNA transcriptional upregulation upon depletion of particular mRNA export factors has revealed a complex but incomplete network of compensatory regulation in which, for example, depletion of either Aly/REF or UIF upregulates the expression of the other, while knockdown of Aly/REF does not induce expression of LuzP4; similarly, while ChTOP knockdown induces expression of UAP56, it does not alter expression of Aly/REF—despite knockdown of Aly/REF upregulating ChTOP (Chang et al. 2013; Hautbergue et al. 2008; Viphakone et al. 2015). Perhaps one of the most compelling observations speaking to this complexity, as discussed in the sections above, is that, while orthologous adaptors and coadaptors may act redundantly on most mRNAs, specific subsets of mRNAs seem to be regulated exclusively by a particular adaptor-coadaptor pairing. The prototypical example of this is the export of *HSP70* mRNA which remained unaltered by Aly/REF or THOC5 knockdown under constitutive growth conditions; however, upon induction of heat shock, Aly/REF and THOC5 became essential for its export (Guria et al. 2011; Katahira et al. 2009). Collectively, these observations suggest that, while many of the mRNA export adaptors and coadaptors share at least some redundancy in bulk mRNA export, many of these factors retain at least some partially or wholly unique functions. These observations also raise the possibility of an "adaptor-coadaptor code," in which particular combinations of adaptor(s) and coadaptor(s) govern the export of particular subsets of mRNA ("regulons"; Keene 2007). While this possibility is intriguing, future experiments will be required to test it, possibly in the form of complete pairwise testing of co-depletion or overexpression-rescue phenotypes to determine redundancy as well as the identification of the complete set of mRNAs bound not only to each adaptor and coadaptor (much of the data of which has already been generated from the large-scale ENCORE RBP iCLIP screening program; Van Nostrand et al. 2018) but specifically to each possible adaptor-coadaptor pairing.

## 2.4 Noncanonical Recruitment of NXF1 to mRNA: USER Codes and mRNA Regulons

While much work in the last 15 years has focused on the mechanisms by which mRNA adaptors and coadaptors can recruit NXF1:NXT1 to mRNA as part of the bulk export pathway, it is interesting to note that NXF1 was first identified by its ability to promote export of unspliced Mason-Pfizer monkey virus RNA via direct interaction, independently of adaptors or coadaptors, with a *cis-acting sequence element, termed the constitutive transport element* (CTE) (Braun et al. 1999; Bray et al. 1994; Grüter et al. 1998). Indeed, the unique requirement of retroviruses for parallel export of both spliced (coding) and unspliced (genomic) RNAs has led to the proliferation of numerous virally encoded sequence elements that coopt cellular mRNA export machinery in order to promote RNA export and/or bypass the mRNA quality control machinery (Cullen 1998; Hammarskjöld 1997). Some of these viral sequence elements, including the CTE itself, have subsequently been co-opted by host mRNAs to promote their specialized export (Li et al. 2006; see Fig. 2.2b).

As research on mRNA export pathways in metazoans progressed, it became clear that the use of *cis*-acting export specificity elements in RNA was not confined to viral RNAs, and that a range of different elements existed within metazoan transcriptomes that allowed coordinated export regulation of an mRNA, or set of mRNAs, through recruitment of mRNA export machinery independently of the bulk export pathway (Delaleau and Borden 2015; Wickramasinghe and Laskey 2015). This concept became formalized through the terminology of "USER" (*u*ntranslated *s*equence *e*lements for *r*egulation) codes, small sequence elements that direct the coordinated posttranscriptional regulation of a group of mRNAs, termed an “mRNA regulon” (Keene 2007). This section will focus on the discussion of USER codes and regulons that promote export of endogenous mRNAs via NXF1 (Fig. 2.2b) or the importin-β family member CRM1, while other mechanisms of mRNA export by CRM1 will be discussed in Sect. 2.6 (Fig. 2.2d).

### 2.4.1 USER Codes in the Export of Intronless mRNAs

One major question raised with the discovery of the link between splicing and mRNA export was how intronless mRNAs are exported (Le et al. 2001; Masuda et al. 2005). It was known that the export of such mRNAs was dependent on the bulk mRNA export machinery, including Aly/REF and NXF1 (Sträßer et al. 2002); however, the mechanism by which these factors were deposited on the mRNA was unclear. While subsequent observations on the recruitment of Aly/REF to bulk mRNA via interactions with the CBC (Cheng et al. 2006) and the 3′ processing factor CstF64 (Shi et al. 2017) have provided possible mechanisms for export, a parallel thread of research has revealed the widespread use of coding region-embedded USER codes to recruit the canonical mRNA export machinery to intronless mRNAs.

#### 2.4.1.1 The SRSF Family of Splicing Factors

The first intronless USER code in metazoans was identified through work on the replication-dependent histone mRNAs, a class of intronless mRNAs defined by a unique 3′ structure lacking a poly(A) tail but instead incorporating a conserved stem-loop structure (Dominski and Marzluff 2007). The *H2A* mRNA was found to contain a 22-nucleotide (nt) sequence element within its coding region that was necessary for export of this mRNA to the cytoplasm (Huang and Carmichael 1997; Huang and Steitz 2001). A UV-cross-linking approach was used to show that this sequence element was bound by two members of the SRSF family of splicing factors, SRSF3/SRp20 and SRSF7/9G8, and that these *trans*-acting factors were essential to the export of *H2A* mRNA in a fashion independent of their previously described roles in mRNA splicing (Huang and Steitz 2001). Upon binding to the *H2A* USER code, SRSF3 or SRSF7 is able to interact with NXF1 via its N-terminal domain and to mediate RNA hand-off to NXF1 in a fashion identical to Aly/REF, suggesting that SRSF3 and/or SRSF7 is able to functionally substitute to Aly/REF in NXF1:NXT1 loading of *H2A* mRNA onto NXF1:NXT1 (Hargous et al. 2006; Hautbergue et al. 2008; Huang et al. 2003). Given that both SRSF3 and SRSF7 bind via the Aly/REF-binding site, the identity of the coadaptor, if any, in the export of *H2A* remains unclear; however, the observations that SRSF3 is able to cooperate with coadaptors including THOC5 and YTHDC1 in other export pathways (Ratnadiwakara et al. 2018; Roundtree et al. 2017; Wang et al. 2013) and that the 3′-terminal stem-loop binding protein of replication-dependent histone mRNAs, SLBP, is also necessary for export (Sullivan et al. 2009) raise several possible candidates. Interestingly, this work also observed that a third SRSF protein, SRSF1/SF2, was able to interact with the NXF1 N-terminal domain redundantly with Aly/REF, raising the possibility that it too may regulate the export of as-yet-undetermined mRNA(s) (Huang et al. 2003; Lai and Tarn 2004; Tintaru et al. 2007).

While for many years *H2A* mRNA was considered to be the only mRNA export target of SRSF3/SRSF7, the recent emergence of high-resolution RNA-protein cross-linking methodologies has allowed the identification of putative RNA targets of export factors with unprecedented sensitivity and has redefined our understanding of SRSF protein-mediated mRNA export. In 2016, Müller-McNicoll et al. (2016) used the individual-nucleotide resolution cross-linking immunoprecipitation (iCLIP) methodology (König et al. 2010) to identify >1000 mRNAs whose export is dependent on SRSF proteins, including several hundred possible targets of SRSF3. These observations chime with earlier reports of SRSF protein being capable of binding and promoting the export of both intronless and spliced mRNAs, suggesting that the SRSF family may represent a large group of previously unknown mRNA export adaptors whose function requires further exploration (Masuyama et al. 2004). In addition, Müller-McNicoll et al. (2016) were able to demonstrate by iCLIP the closely juxtaposed binding of SRSF3/SRSF7 and NXF1 to mRNA targets, confirming an NXF1-dependent export pathway, and to identify sequence-specific binding of SRSF3 and SRSF7 to several degenerate motifs in last exons of mRNAs, suggesting the existence of as-yet-unidentified additional USER codes responsible for the recruitment of SRSF3 and/or SRSF7 to target mRNAs.

Consistent with the redefinition of SRSF3 as a potentially promiscuous USER code-directed mRNA export adaptor, several very recent studies have identified new pathways in which SRSF3 substitutes for Aly/REF in the export of specialized mRNA subsets. Roundtree et al. (2017) reported that SRSF3 cooperates with the $N^6$-methyladenosine ($m^6A$) reader protein YTHDC1 to promote the export of mRNAs marked by $m^6A$ modification (see below). In embryonic stem cells, SRSF3 cooperates with the known mRNA export adaptor THOC5 in a novel adaptor/coadaptor pairing to promote the export of key pluripotency maintenance factor mRNAs including *Nanog*, *Sox2*, *Klf4*, and *Esrrb* (Ratnadiwakara et al. 2018; Wang et al. 2013). THOC5 is highly expressed in embryonic stem cells and is downregulated over the course of differentiation; artificial expression of THOC5 is able to sustain stem cell pluripotency, while its depletion prevents reprogramming of somatic cells into induced pluripotent stem cells (iPSCs), suggesting that SRSF3/THOC5-mediated mRNA export regulation may represent an important and previously unknown axis of pluripotency control (Wang et al. 2013). While no USER code has yet been identified for these target mRNAs, it remains likely that such an element is responsible for the temporally and cell type-restricted regulation of these target mRNAs.

#### 2.4.1.2 The Prp19 Complex, U2AF, and the CAR-E Element

In 2011, a second, evolutionarily unrelated coding region-resident USER code, termed the cytoplasmic accumulation region (CAR), was identified in several independent intronless mRNAs, including *HSPB3*, *IFNα1*, and *IFNβ1*, as well as the previously reported *c-Jun* (Guang and Mertz 2005; Lei et al. 2011). The CAR element contains a variable number of copies of a short, 10nt USER code, termed the CAR element (CAR-E) which collectively promote the export of their resident mRNAs via the canonical TREX-NXF1:NXT1 export pathway (Lei et al. 2011, 2013). RNA immunoprecipitation-mass spectrometry (IP-MS) approaches revealed that CAR-E elements were bound by components of the Prp19 complex as well as by U2AF65/U2AF2 (Lei et al. 2013). The Prp19 complex and U2AF65 (as part of the heterodimeric U2AF complex with U2AF35) possess well-described, mRNA export-independent roles in the coordination of transcription with splicing (David et al. 2011); nevertheless, subsequent analyses suggested that the binding of the Prp19 complex and U2AF65 to CAR-E was necessary for the NXF1-mediated export of CAR-E-carrying intronless mRNAs, suggesting possible activity of the Prp19 complex and U2AF65 as export adaptors/coadaptors upstream of TREX (Lei et al. 2013). While a role for these splicing/transcription-regulatory factors in the coordination of specialized mRNA export seems surprising, it is noted that several previous papers have provided confirmatory evidence, reporting that the U2AF complex is able to recruit NXF1 to mRNA in both human cells and *Drosophila* (Blanchette et al. 2004; Zolotukhin et al. 2002), suggesting that this activity may represent a conserved mechanism both of intronless mRNA export and of coupling of export to the processes of splicing and transcription. While the extent of the CAR-E-mediated export pathway remains to be determined, it is notable that the model mRNAs tested here—*HSPB3*, *IFNα1*, *IFNβ1*, and *c-Jun*—derive from different functional classes and are not otherwise known to function as a regulon, supporting the possibility of CAR-E export being a widespread mechanism of intronless mRNA export.

### *2.4.2 The Signal Sequence Coding Region as a Dual-Functional Regulatory Element*

mRNAs that encode proteins destined for cellular export or embedding in the cellular membrane encode a short sequence, termed the signal sequence, immediately following the start codon of their ORFs; upon translation, this sequence targets the nascent polypeptide to the endoplasmic reticulum for subsequent membrane embedding or secretion (Aviram and Schuldiner 2017). Recently, it was revealed that, in addition to encoding the ER-targeting signal sequence, the signal sequence coding region (SSCR) is able to act in *cis* as a USER code governing export of mRNAs (Palazzo et al. 2007). The SSCR's nucleotide sequence is optimized not only to encode the requisite amino acid consensus for a function signal sequence but also to selectively exclude AMP nucleotides, enriching in particular for extended stretches of C/G bases; remarkably, silent coding mutations that introduce AMP residues to the SSCR prevent it from driving export of an unspliced reporter mRNA (Palazzo et al. 2007). Subsequent analysis of similar leader sequences on mRNAs encoding mitochondrially localized proteins, namely, mitochondrial signal coding regions (MSCRs), revealed this sequence is functionally identical to the SSCR in promoting mRNA export (Cenik et al. 2011). While much of the mechanism regulating SSCR/MSCR-dependent mRNA export remains to be characterized, it has been noted that the export activity of an SSCR/MSCR can be blocked by the presence of a 5′-proximal splice site, suggesting that proximity of the SSCR/MSCR to the cap and bound CBC may be important for the export (Cenik et al. 2011). Furthermore, SSCR/MSCR-containing intronless mRNAs have been observed transiting the splicing speckles, suggesting a role for this nuclear subdomain in the regulation and/or licensing of SSCR/MSCR-dependent mRNA export (Akef et al. 2013).

### 2.4.3 Posttranscriptional Control of USER Code Recognition by Aly/REF

While the examples outlined thus far in this section establish a pattern in which sequence-nonspecific bulk mRNA export factors are recruited by sequence-specific *trans*-acting factors, one recently described pathway has challenged this paradigm and suggested that the bulk mRNA export factors themselves may be modified in order to promote sequence-specific mRNA export.

IPMK is a multifunctional phosphatidylinositol kinase which synthesizes several inositol phosphate products including $IP_4$, $IP_5$, $IP_6$, and the second messenger phosphatidylinositol (3,4,5)-trisphosphate ($PIP_3$). IPMK localizes to splicing speckles in humans and has been identified as a regulatory hub for multiple aspects of nuclear RNA metabolism, including mRNA export (Kim et al. 2017; Salamon and Backer 2013). siRNA-mediated knockdown of IPMK resulted in the reduced export of approximately 13% of the mRNAs tested by cytoplasmic RNA microarray, including a significant enrichment for mRNAs that regulate processes including cell cycle progression and DNA double-strand break repair by homologous recombination but not nonhomologous end joining, suggesting function-specific regulons exist within the IPMK-regulated mRNA population (Wickramasinghe et al. 2013). The key regulatory step in this export pathway is the previously reported binding of the IPMK product $PIP_3$ to Aly/REF via an interaction surface on Aly/REF's RNA-binding domain (Okada et al. 2008). This interaction appears to alter the RNA-binding mode of Aly/REF, resulting in its interaction with a specific, 10nt USER code within the 3′-UTRs of target genes and thereby directing these mRNAs for NXF1:NXT1-dependent export. Supportive of this model, Aly/REF binding to the identified USER code was inhibited by IPMK but could be rescued by the exogenous addition of $PIP_3$ (Wickramasinghe et al. 2013).

Additional support for this model came from an independent study of two DNA damage-responsive splicing regulators, BCLAF1 and THRAP3 (Vohhodina et al. 2017). The semi-redundant BCLAF1 and THRAP3 were found to promote the export of a set of mRNAs in response to DNA damage response signaling that closely resembled the IPMK regulon. Crucially, in silico motif analysis of BCLAF1/THRAP3 mRNA export targets returned a highly enriched 3′-UTR motif almost identical to the USER code recognized by $PIP_3$-bound Aly/REF, suggesting that IPMK-regulated Aly/REF and BCLAF1/THRAP3 act via a common USER code in the 3′-UTRs of their target mRNAs (Vohhodina et al. 2017; Wickramasinghe et al. 2013). While the exact relationship between $PIP_3$-Aly/REF and BCLAF1/THRAP3 in the regulation of this pathway is not yet clear, the observation that both BCLAF1 and THRAP3 contain WxHD motifs reported to regulate interaction of Aly/REF with the EJC components eIF4A3 (Gromadzka et al. 2016; Vohhodina et al. 2017) and the identification of a BCLAF1-NXF1 interaction in a recent high-throughput proteomics study (Castello et al. 2012) make it tempting to speculate that $PIP_3$-Aly/REF in concert with BCLAF1 and/or THRAP3 may form an NXF1-regulatory adaptor/coadaptor pair on the 3′-UTR USER code of their target mRNAs. In addition to providing new insights into the mechanisms of USER code recognition in mRNA export, the IPMK-Aly/REF-BCLAF1/THRAP3 pathway also points toward a new layer of mRNA export regulation by second messenger-dependent recognition of cryptic USER codes by mRNA export adaptors.

### *2.4.4* Trans-*Acting Factors Without a USER Code: Candidate mRNA Regulon Export Factors*

While the above pathways represent examples of cases in which known *cis*-acting USER codes are able to direct the export of their carrier mRNAs, a number of RNA-binding proteins have been identified that are not assignable to known mRNA regulons or USER codes, but which represent candidate mRNA regulon export adaptors/factors that warrant further investigation.

In addition to its other roles in the regulation of mRNA export (discussed in Sects. 2.3.2.2 and 2.5.1), RBM15 and/or RBM15B has been reported to bind specific RNA sequences and mediate their RNA modification and export, including a role directing the $m^6A$ methyltransferase complex to specific sites on the *XIST* lncRNA (Patil et al. 2016), raising the possibility that RBM15/B may also recognize specific USER codes in endogenous metazoan mRNAs to mediate their metabolism and/or export. Moreover, hnRNP L, a multifunctional RNA-binding protein, was found to interact specifically with an RNA element in herpesvirus TK mRNA and promote the export of this mRNA (Liu and Mertz 1995) suggesting that such an action on endogenous mRNAs remains a possibility. Lastly, a more direct coupling of 3′-end processing of mRNA to mRNA export has been suggested by observation that CFIm68/CPSF6, a component of the CFI complex, is able to shuttle nucleocytoplasmically in a transcription-dependent fashion, bind to NXF1, and cause nuclear accumulation of polyadenylated mRNA upon its depletion (Ruepp et al. 2009). CFIm68 possesses mRNA-binding activity (Dettwiler et al. 2004) and is able to interact directly with THOC5 to regulate alternative polyadenylation (Katahira et al. 2013), raising the possibility that CFIm68 may act as an ortholog of Aly/REF on yet-to-be-identified mRNA targets.

### *2.4.5 The AU-Rich Element as a mRNA Export USER Code*

AU-rich elements (AREs) are among the best described of all *cis*-acting mRNA elements. They are bound by a range of different *trans*-acting RBPs, including HuR, TTP, and AUF1 among others, and these *trans*-acting factors control the fate of the mRNA throughout the cell, including at the levels of mRNA processing, stability, translation, subcellular localization, and mRNA export (Garcia-Mauriño et al. 2017; see Fig. 2.2d).

The ARE-binding protein HuR/ELAVL1 is known to stabilize bound mRNAs and promote their translation in the cytoplasm (Wu et al. 2018). In the nucleus, HuR bound to AREs interacts mutually exclusively with two accessory proteins, APRIL/TNFSF13 and pp32/ANP32A, both of which contain leucine-rich NESs and can shuttle nucleocytoplasmically in a manner dependent of the export receptor CRM1 (see Sect. 2.6). Importantly, inhibition of pp32 and/or APRIL shuttling using the CRM1 inhibitor leptomycin B (LMB) results in the accumulation of HuR:pp32, HuR:APRIL, and HuR:RNA complexes in the nucleus, suggesting that APRIL/pp32 are shuttled in the context of an assembled ARE-RNA:HuR:APRIL/pp32 complex (Brennan et al. 2000). While treatment with LMB induced no visible change in total nuclear polyadenylated RNA levels assayed by oligo(dT) FISH, testing of candidate ARE-containing mRNAs such as *c-fos*, *COX-2*, and *tp53* revealed that HuR-mediated shuttling by CRM1 was necessary for the efficient export of these mRNAs (Brennan et al. 2000; Dixon 2004; Jang et al. 2003; Nakamura et al. 2011). While the ubiquity of this export pathway on ARE-containing mRNAs is yet to be examined, the number of ARE-containing targets of HuR, and their myriad roles in the regulation of cellular growth, immune regulation, and cellular mobility, suggests that HuR-mediated mRNA export may represent an important regulator of cellular growth and oncogenesis (Garcia-Mauriño et al. 2017; Wu et al. 2018).

It should be noted that the ARE may not be the only USER code that can target mRNAs for export via HuR. The immune-regulatory *CD83* mRNA has been reported to undergo CRM1-mediated nuclear export in a manner dependent on the presence of HuR and APRIL (Fries et al. 2007; Prechtel et al. 2006). However, in this mRNA the element that recruits HuR was found not to be a 3′-UTR ARE but instead a novel sequence element, termed the posttranscriptional regulatory element (PRE), within the mRNA coding region (Prechtel et al. 2006). Similarly, while the export of the ARE-containing *IFNα1* mRNA was found to be dependent on CRM1, the ARE was found to be dispensable for this export, suggesting an alternative USER code (Kimura et al. 2004). It is noted that this latter paper did not test whether *IFNα1* mRNA export was dependent on HuR, and that it is not clear how, if at all, this CRM1-dependent export pathway overlaps or interacts with the Prp19:U2AF65-NXF1-dependent export pathway for *IFNα1* discussed above (Sect. 2.4.1.2).

## 2.5 Posttranslational Formation of USER Codes by mRNA Modification

The modification of mRNAs in order to alter their biochemical and functional properties has been known about since the early days of research on mRNAs. Indeed, several mRNA modifications—the $m^7G$ cap, the poly(A) tail, and the splicing out of introns—are so well established and ubiquitous that they are intricately woven into the cell's definition of functional, translatable mRNAs, and loss of any one of these modifications has drastic effects on the functionality of almost all mRNAs (Chan et al. 2011; Ramanathan et al. 2016; Will and Lührmann 2011). However, these canonical modifications are far from being the only ones that may occur on mRNA; indeed, over 100 nucleotide and/or base modifications have now been identified from cellular mRNA, exhibiting a wide range of frequencies, specificities, and biochemical and/or functional consequences (Boccaletto et al. 2018). Furthermore, some of these modifications are both regulatable and reversible, leading to the concept of "epitranscriptomics" in which a code of mRNA modifications is able to direct the posttranscriptional fate of mRNAs in a manner analogous to epigenetic marks controlling regulation of a gene (Eckmann et al. 2011; Mauer et al. 2016; Trotman and Schoenberg 2019).

While early research on mRNA modifications was limited by the low-resolution and/or piecemeal experimental techniques available, a recent explosion of next-generation sequencing-based techniques has allowed the mapping of particular RNA modifications with unprecedented detail and sensitivity (Li et al. 2016). In so doing, these techniques have revealed a plethora of insights into how mRNA modifications influence numerous posttranscriptional pathways of mRNA regulation, including mRNA export. Specifically, several mRNA modifications have been found to be specifically bound by "reader" proteins that subsequently recruit the mRNA export machinery to direct export of these mRNAs (Roundtree et al. 2017; Yang et al. 2017), suggesting a new regulatory paradigm of "inducible USER codes"—the use of mRNA modifications to posttranscriptionally mark a subset of mRNAs with a *cis*-acting sequence/structural element and thereby allow mRNA export (or other posttranscriptional metabolic processes; see Fig. 2.2c).

It should be noted that mRNA adenosine-inosine deamination editing has been extensively reported to regulate the retention of mRNAs carrying extended double-stranded regions in the subnuclear paraspeckle domains (Chen and Carmichael

2009; Chen et al. 2008; Fox and Lamond 2010; Fox et al. 2018; Prasanth et al. 2005). However, an extended discussion of the mechanics of mRNA nuclear retention is beyond the scope of this chapter, and interested readers are encouraged to consult several excellent recent reviews on the topic (Fox et al. 2018; Palazzo and Lee 2018; Wegener and Müller-McNicoll 2018).

### 2.5.1 N$^6$-Methyladenosine in the Regulation of mRNA Export

The addition of a methyl group to the 6-nitrogen ($N^6$) of adenosine bases in RNA to generate $N^6$-methyladenose (m$^6$A) is the most common mRNA modification in human cells outside of the canonical m$^7$G cap, poly(A) tail, and splicing. The development of sequencing methodologies capable of specifically mapping this methylation event revealed its presence on approximately 30% of the human transcriptome, with a specific deposition on [A/G][A/G]A∗C[A/C/U] consensus motifs predominantly, though not exclusively, found in RNA 3′-UTRs and within long internal exons (Dominissini et al. 2012; Harper et al. 1990; Meyer et al. 2012). This modification is deposited by the m$^6$A methyltransferase/"writer" complex containing a catalytic heterodimer (METTL3/METTL14), several scaffolding/accessory proteins (ZC3H13, WTAP, and KIAA1429), and, at least in some cases, the RNA-binding proteins RBM15/RBM15B (Duan et al. 2019; Lesbirel and Wilson 2019) and can be removed by one of the two m$^6$A demethylases/"erasers," FTO (which recognizes cap-proximal m$^6$A) and ALKBH5 (which acts on the gene body; Hess et al. 2013; Jia et al. 2011; Ke et al. 2017; Mauer et al. 2016; Zheng et al. 2013). The final components of the m$^6$A regulatory system are the "readers," a set of nuclear (hnRNP A2B1, YTHDC1, YTHDC2, and, indirectly, hnRNP C) and cytoplasmic (YTHDF1,2,3 and eIF3) m$^6$A-binding proteins that promote different downstream metabolic processes such as splicing, mRNA destabilization, and mRNA translation, as well as the mRNA export discussed herein (Meyer and Jaffrey 2017).

Several early pieces of evidence pointed to a possible role for the regulation of mRNA export by m$^6$A, including the findings that inhibition of total mRNA methylation reduced mRNA export (Camper et al. 1984). Furthermore, knockdown of the m$^6$A methyltransferase factor METTL3 slowed the circadian clock due to reduced export of circadian clock-related mRNAs (Fustin et al. 2013), while knockdown of the m$^6$A "eraser" ALKBH5 caused increased translocation of poly(A)$^+$ RNA to the cytoplasm, suggesting an enhancement of mRNA export activity (Zheng et al. 2013). However, the molecular mechanism by which m$^6$A regulates mRNA export was not delineated until 2017, when detailed work by Roundtree et al. (2017) demonstrated that, in the nucleus, m$^6$A sites in mRNA were bound by the nuclear "reader" protein YTHDC1, and that m$^6$A-bound YTHDC1 is able to recruit the known USER code-interacting mRNA export adaptor SRSF3. SRSF3 loaded onto YTHDC1 subsequently recruits NXF1 to promote export of m$^6$A-modified mRNAs, though YTHDC1 and NXF1 do not themselves interact; the loss of any one of these elements was sufficient to inhibit mRNA export of candidate mRNA targets. A combination of sequencing-based RNA-protein interaction mapping techniques was used to identify greater than 700 endogenous mRNAs whose export may be regulated by the YTHDC1-SRSF3-NXF1 pathway, suggesting the existence of a large mRNA regulon targeted by these proteins (Roundtree et al. 2017). Independent work has additionally reported direct interactions between components of the TREX complex and the catalytic METTL3/METTL14 subunits of the $m^6A$ methyltransferase complex (Lesbirel et al. 2018), suggesting that the closely juxtaposed association of TREX and SRSF3:YTHDC1 with mRNA may enhance $m^6A$-dependent mRNA export; however, the relative contributions and mechanism of cooperation, if any, between these two pathways have yet to be tested.

Interestingly, while the above observations deal with the deposition of $m^6A$ within the mRNA body, a functionally discrete pathway has evolved to regulate mRNAs that initiate transcription with an AMP residue, placing AMP at the N1 position immediately 3′ to the $m^7$GpppN cap ($m^7$GpppA). These N1-AMP residues undergo opposed $N^6$-methylation and demethylation by their own dedicated factors, namely the methyltransferase CAPAM/PCIF1 and the demethylase FTO (Akichika et al. 2019; Boulias et al. 2018; Cowling 2019; Hess et al. 2013; Jia et al. 2011; Mauer et al. 2016; Sendinc et al. 2018; Sun et al. 2019). Since the first two nucleotides 3′ to the cap in almost all mammalian mRNA undergo ribose 2′-OH methylation by CMTR1 and CMTR2, respectively, these modifications result in the establishment of $N^6$, 2′-*O*-dimethylated AMP ($m^6A_m$) at the N1 position of a large number of cellular mRNAs (Linder et al. 2015). The predominant biological consequence of $m^6A_m$ formation at the N1 nucleotide is the inhibition of decapping enzyme Dcp2 binding to the $m^7$G cap, resulting in stabilization of these mRNAs (Mauer et al. 2016). However, early biochemical studies of the canonical CBC found that, while $m^7$GpppA caps bound CBC approximately 50% less well than $m^7$GpppG, the addition of an $N^6$-methyl group on the N1 nucleotide ($m^7$Gpppm$^6$A) restored CBC binding to a level comparable to that of $m^7$GpppG (Worch et al. 2005). While a role for $m^6A_m$-dependent modulation of CBC-binding efficiency in mRNA export remains purely theoretical at this point, the central role that CBC binding to the $m^7$GpppN cap plays in the initiation and control of mRNA export (Cheng et al. 2006; Ohno et al. 2002; Shi et al. 2017), as well as the considerable variation in $m^6A_m$ deposition between tissues and transcripts (Kruse et al. 2011), makes this an intriguing hypothesis worthy of further testing.

### 2.5.2 $N^5$-*Methylcytosine as a Possible Regulator of Aly/REF-Dependent Recruitment to mRNA*

Like $m^6A$, $N^5$-methylcytosine ($m^5C$) is a highly abundant RNA modification that has recently been implicated in mRNA export regulation. While $m^5C$ was initially identified only in highly stable ncRNAs such as tRNAs and rRNAs (García-Vílchez et al. 2019), the development of a range of $m^5C$-focused sequencing strategies including bisulfite sequencing have allowed the identification of $m^5C$ in mRNA (Schaefer et al. 2009). Aza-IP-Seq (Khoddami and Cairns 2013) and miCLIP (methylation iCLIP) (Hussain et al. 2013) have revealed a distribution of $m^5C$ throughout mRNAs, with a potential enrichment in C/G-rich stretches in the vicinity of the ATG start codon (Amort et al. 2017; Yang et al. 2017). These analyses have also uncovered extensive variation of $m^5C$ site choice and saturation across tissues and developmental stages (Yang et al. 2017). The deposition of $m^5C$ in mRNA is predominantly mediated by the methyltransferase NSUN2, although other NSUN family members are also likely to act on mRNA (García-Vílchez et al. 2019; Yang et al. 2017). However, while $m^5C$ in the context of mRNA has been posited to regulate mRNA stability and/or translation, little hard evidence of a functional role has been forthcoming (García-Vílchez et al. 2019).

Recently, Yang et al. (2017) have reported a potential role for posttranscriptionally deposited $m^5C$ in the regulation of mRNA export. They used a comparative IP-MS protocol with a short RNA probe with or without CMP $N^5$-methylation to unexpectedly identify Aly/REF as a reader protein for $m^5C$ both in vitro and in vivo. Supporting a role for this interaction in a putative $m^5C$:Aly-dependent mRNA export pathway, the authors reported that depletion of the $m^5C$ methyltransferase NSUN2 resulted in decreased Aly/REF binding to candidate $m^5C$ target mRNAs, an accumulation of Aly/REF in splicing speckles, and reduced trafficking to the cytoplasm and an accumulation of polyadenylated RNA in the nucleus (Yang et al. 2017). It should be noted that this study omitted several important supporting experiments, including testing for a requirement for NXF1: NXT1 for $m^5C$-driven mRNA export; also, other groups have disputed the finding that Aly/REF is able to interact directly with $m^5C$ (Lesbirel and Wilson 2019). However, while the proposal that Aly/REF is able to bind directly to $m^5C$ in mRNA is markedly different from the paradigm of reader proteins recruiting downstream mRNA export factors established for $m^6A$ by Roundtree et al. (2017), an earlier observation that Aly/REF is able to modify its RNA-binding specificity in response to binding of intracellular signaling molecules (Wickramasinghe et al. 2013) suggests that the implementation of a specific binding mode of Aly/REF on $m^5C$ is not beyond the realms of possibility.

## 2.6 Regulation of mRNA Export by the Protein Export Receptor CRM1

CRM1 is a member of the importin-β superfamily of nuclear transport receptors which, like other members of this family but unlike the canonical mRNA export receptor NXF1, relies on successive cycles of GTP hydrolysis by its binding partner Ran to drive its nucleocytoplasmic shuttling through the nuclear pore. CRM1 binds specifically to leucine-rich nuclear export signals (NESs) in proteins to promote their export from the nucleus; indeed, CRM1:RanGTP is the dominant receptor for leucine-rich NES export from the nucleus (Fornerod et al. 1997; Hutten and Kehlenbach 2007).

In addition to its role in the export of proteins from the nucleus, CRM1:RanGTP is required for the export of a number of highly structured endogenous ncRNAs from the nucleus, including U snRNAs and maturing ribosomal subunits. In order to achieve export of these RNAs, CRM1LRanGTP relies on a mechanism reminiscent of that of NXF1:NXT1 in which RBP adaptor(s) bind directly to the target RNA, then recruit CRM1:RanGTP via one or more canonical leucine-rich NESs to assemble an export-competent RNA:adaptor:receptor complex (Nerurkar et al. 2015; Okamura et al. 2015). Also like NXF1:NXT1, CRM1:RanGTP has been repurposed by viral proteins and/or *cis*-acting sequences to promote export of unspliced mRNAs; the most well-characterized example of this is the export of HIV-1 RNAs in response to the NES-containing viral protein Rev binding to the *cis*-acting Rev response element (RRE) in the viral RNA (Suhasini and Reddy 2009).

Early studies in *S. cerevisiae* and *Xenopus* oocytes concluded that CRM1 was not required for the export of mRNA in eukaryotes (Fischer et al. 1994; Jarmolowski et al. 1994; Neville and Rosbash 1999). However, a range of subsequent studies have refined this view, finding that, while CRM1:RanGTP is indeed dispensable for the constitutive export of bulk mRNA, it is capable of mediating the export of several functional mRNA regulons through the action of discrete mRNA USER codes and *trans*-acting export adaptors (see Fig. 2.2d). Export of ARE-containing mRNAs via HuR:CRM1:RanGTP has been discussed in Sect. 2.4.5.

### 2.6.1 A Nuclear Role for the Cytoplasmic Cap-Binding Protein eIF4E

eIF4E is an $m^7G$ cap-binding protein that, in complex with the helicase eIF4A and the scaffolding protein eIF4G, assembles the heterotrimeric complex eIF4F. eIF4F is a key activator of translation in the cytoplasm, coordinating the loading of the 40S ribosomal subunit onto mRNAs (Pelletier et al. 2015). Given the importance of this role, it is not surprising that cytoplasmic eIF4E is a major regulatory node in the coordination of mRNA translation and/or metabolism in the cytoplasm and is an important oncogene in numerous cancers (Ho and Lee 2016; Pelletier et al. 2015). While eIF4A and eIF4G are predominantly cytoplasmic proteins, a significant portion of cellular eIF4E localizes to the nucleus, where it is additionally able to regulate nuclear mRNA dynamics, and in particular mRNA export, by a mechanism independent of its cytoplasmic translation-regulation activity (Osborne and Borden 2015; Rousseau et al. 1996). eIF4E's roles in mRNA export regulation contribute to its function as an oncogene, emphasizing the pathological importance of the eIF4E-mediated mRNA export pathway (Osborne and Borden 2015; Pelletier et al. 2015).

mRNA export via eIF4E was found to require two key *cis*-acting elements: the RNA $m^7G$ cap and a structurally but not sequence-conserved USER code within the

3′-UTR of target mRNAs, termed the 4E-sensitivity element (4E-SE) (Culjkovic et al. 2005, 2006). IP-MS analyses have revealed that, while eIF4E predominantly binds the cap structure on its target mRNAs, the 4E-SE is bound by an accessory protein, LRPPRC, which binds mRNA cooperatively with eIF4E and promotes stability of the ternary RNA:eIF4E:LRPPRC complex (Topisirovic et al. 2009). Unlike in the NXF1:NXT1 export pathway, eIF4E:LRPPRC-bound mRNAs are not recruited to the splicing speckles; instead, they accumulate in discrete foci throughout the nuclei enriched in eIF4E, termed "eIF4E granules," which exhibit partial overlap with PML bodies (discussed below; Cohen et al. 2001; Culjkovic et al. 2005; Topisirovic et al. 2002). The export receptor complex CRM1:RanGTP is recruited to the ternary RNA:eIF4E:LRPPRC complex via prototypical leucine-rich NES elements within LRPPRC and mediates translocation of the cargo mRNA through the nuclear pore via an as-yet poorly characterized mechanism. This mechanism culminates in the binding of CRM1:RanGTP to the cytoplasmic fibril protein Nup358/RanBP2 which, with RanGAP, promotes release of cargo mRNA through RanGTP→RanGDP hydrolysis (Culjkovic et al. 2006; Hutten and Kehlenbach 2007; Volpon et al. 2017).

The first identified mRNA target of the eIF4E-dependent export pathways was the key cell cycle-regulatory mRNA *CYCD1* (Rousseau et al. 1996), and it was this RNA on which many of the mechanistic dissections of the eIF4E-dependent export pathway were conducted. However, a subsequent RNA immunoprecipitation (RIP)-differential display study was able to identify hundreds of mRNAs that interacted with eIF4E within the nucleus and therefore represent candidate mRNA export cargoes of eIF4E:LRPPRC (Culjkovic et al. 2006). These mRNAs were markedly enriched for a range of pro-growth factors including cell cycle components and regulators of the pro-growth AKT signaling pathway, among others; the strongly proliferative nature of the mRNAs within the eiF4E regulon provides significant insight in to the mechanisms by which eIF4E-dependent mRNA export can promote oncogenesis (Culjkovic et al. 2006, 2008; Osborne and Borden 2015).

Given the highly proliferative nature of the eIF4E export regulon, it is not surprising that mechanisms have evolved to constrain and regulate its export activity in vivo. It was observed in early studies that a subset of nuclear eIF4E granules overlapped with PML bodies and that eIF4E binding to mRNAs in the nucleus specifically targeted them to eIF4E granules that lacked significant PML staining (Cohen et al. 2001; Culjkovic et al. 2005; Topisirovic et al. 2002). Consistent with this, PML was found to be a negative regulator of eIF4E-driven mRNA export, through a direct interaction with eIF4E that reduced its affinity for the $m^7G$ cap in the nucleus (Cohen et al. 2001; Culjkovic et al. 2006). The PML-eIF4E regulatory axis allows the direct regulation of pro-growth mRNA export by environmental stimuli, as evidenced by the finding that cadmium treatment and IFNγ treatment have positive and negative effects, respectively, on the PML-eIF4E interaction and consequently on the extent of PML-dependent inhibition of eIF4E mRNA export (Topisirovic et al. 2002). Similarly, the myeloid cell line-specific homeobox transcription factor, PRH, is able to bind eIF4E and negatively influence its ability to export mRNAs (Topisirovic et al. 2003). In fact, eIF4E-binding motifs were found to be conserved across almost 25% of the human cell's 803 homeobox proteins (Topisirovic et al. 2003), suggesting that homeobox proteins may represent a large class of eIF4E modulators; indeed several other homeobox proteins have been found to modulate eIF4E's activity, either at the level of mRNA export (HOXA9; Topisirovic et al. 2005) or translation initiation (Emx2, Otx2, and En-2; Brunet et al. 2005; Nédélec et al. 2004).

A number of questions remain to be answered regarding the mechanism of eIF4E-mediated mRNA export, not least the question of how and when the bulk mRNA cap-binding complex, CBC, is evicted to allow access of eIF4E, and how the large suite of possible eIF4E-regulatory homeobox proteins may contribute to the efficacy of this export path across different tissues and developmental stages (Osborne and Borden 2015).

### 2.6.2 *Teaching an Old Dog New Tricks: CRM1-Dependent Export Via NXF3*

As discussed in Sect. 2.3.4, the NXF1 family of bulk mRNA export adaptors has diversified from its single *S. cerevisiae* ancestor, Mex67, to include multiple paralogs in metazoans, including five (NXF1-5) in humans and mice (Herold et al. 2000). Of these five members, NXF4 and NXF5 possess corrupted coding sequences and are unlikely to be expressed, while NXF2 bears a close resemblance to NXF1 and appears to duplicate NXF1 function in a tissue-specific fashion, though its functionality in mRNA export has been questioned by other groups (Herold et al. 2000; Yang et al. 2001).

NXF3 represents an anomaly among the NXF1 family in that, while it is an intact ORF and is expressed in humans, albeit in a tissue-restricted fashion (GTEx Consortium 2015), it contains several truncations, including of 32 amino acids in its RNA-binding domain and the loss of its C-terminal FG-Nup-binding domain, that are expected to prevent it acting as a NXF1-like mRNA export receptor (Herold et al. 2000). Despite these observations, tethering of NXF3 to a reporter RNA in human cells (Yang et al. 2001) but not in quail cells (Herold et al. 2000) unexpectedly resulted in the robust export of the reporter mRNA; however, this export, instead of proceeding via a prototypical NXF1-like export pathway, was reliant on the binding of CRM1 to a conserved leucine-rich NES in the NXF3 sequence, suggesting that NXF3 has evolved alternative mechanisms to promote mRNA export of target mRNAs in the absence of the FG-Nup-binding capability typical of its protein family (Yang et al. 2001).

While NXF3 was reported to interact with poly(A)$^+$ mRNA in vivo, the mechanism of this binding, given that NXF3 lacks a significant portion of its RNA-binding domain, remains unclear (Yang et al. 2001). Similarly, unresolved are the identity of NXF3's mRNA export cargo(es), which await a detailed analysis with high-sensitivity cross-linking methods such as iCLIP for identification. Interestingly, both NXF2 and NXF3 show strong tissue-specific expression in the testes (Herold et al. 2000; Yang et al. 2001) in which NXF1 expression is repressed (Zhang et al. 2007), suggesting that these two NXF1 orthologs may act together or independently to reproduce NXF1's mRNA export activities in this tissue. The recent identification of an NXF3:CRM1-mediated export pathway for snoRNAs in response to stress conditions also suggests that NXF3 may regulate multiple classes of endogenous RNAs in humans (Li et al. 2017a).

### *2.6.3 CRM1 Versus NXF1 in the Export of mRNAs in Metazoans*

Much like the cases of bulk mRNA export orthology and NXF1-dependent USER codes discussed above, the current suite of known CRM1-mediated export pathways is likely to only scratch the surface of the total CRM1-dependent export program in metazoan cells; characterization of this complete program is likely to face many of the same challenges and require many of the same approaches, as have been described in Sect. 2.3.6. Other unresolved points raised by the pathways above are specific to the regulation of CRM1, such as the mechanism by which nuclear eIF4E mediates eviction of the canonical CBC to promote CRM1-mediated export at the likely expense of NXF1-depdendent pathways.

## 2.7 Roles for the NPC in mRNA Export Heterogeneity

The nuclear pore complex (NPC) is one of the largest protein assemblies in the human cell, containing 8–64 copies of more than 30 different "nucleoporins" in a precisely arranged eightfold-symmetric structure weighing in excess of 100 MDa (Beck and Hurt 2017). The pore contains several regions, including a partially ordered nuclear "basket," a central pore, and an array of cytoplasmic nucleoporins, each of which plays an essential role in the transport of proteins and RNAs between the nucleus and the cytoplasm (Beck and Hurt 2017; Oeffinger and Zenklusen 2012; Okamura et al. 2015). The NPC is the site of a complicated series of mRNP metabolic steps, including RNA quality control and remodeling on the nuclear basket, NXF1:NXT1/CRM1:RanGTP-chaperoned transit through the central pore, and release of the mRNA cargo from its export adaptors on the cytoplasmic side of the NPC (Beck and Hurt 2017; Oeffinger and Zenklusen 2012).

Given its status as the sole conduit for the nuclear export of the vast majority of cellular mRNAs, and the many and varied protein-RNA interactions that are required for an mRNA to transit, the NPC represents an ideal candidate for mRNA export heterogeneity regulation. However, the considerable technical difficulties of investigating a structure of the size and complexity of the NPC, as well as the highly intertwined interaction networks and rapid, reversible kinetics of mRNA transit through the NPC, have limited our ability to dissect the role of the NPC in specialized and/or regulated mRNA export (Grünwald and Singer 2010; Ma et al. 2013; Saroufim et al. 2015; Shav-Tal et al. 2004; Siebrasse et al. 2012). Nevertheless, one example of Nup-dependent mRNA export heterogeneity has been reported, providing a possible precedent for NPC regulation of mRNA export heterogeneity (Fig. 2.2e).

### 2.7.1 *NUP96 in Cell Cycle and Interferon Regulation*

NUP96 is a conserved nucleoporin that is a component of the Y-complex (also termed the Nup107-160 complex) subdomain of the NPC. The Y-complex contains seven different Nups in stoichiometric ratios and is an essential factor in the assembly of functional NPCs. In mature NPCs, the Y-complex forms a double ring at both the nuclear and cytoplasmic surfaces of the central NPC channel, ideally positioning NUP96 as one of the first Nups able to come into contact with mRNPs that have left the basket to transit the central channel (Beck and Hurt 2017). Upon disassembly of the NPC during mitosis, NUP96 is also reported to play an independent role in the regulation of mitotic kinetochore dynamics (Belgareh et al. 2001; Mishra et al. 2010).

Several papers have found *NUP96* to be an essential gene in both *S. cerevisiae* and in mice, consistent with a central role in NPC assembly and transport (Dockendorff et al. 1997; Faria et al. 2006). However, the depletion rather than knockout of NUP96 in heterozygous *Nup96*$^{+/-}$ mouse models was found to have a variable effect of the export of different mRNAs, with a subset of interferon-inducible mRNAs including *MHC-I* and *-II*, *ICAM-1*, and *CD86* showing particularly robust export inhibition (Faria et al. 2006). A separate paper reported that NUP96, unlike other members of the Y-complex, undergoes rapid proteasomal degradation during mitosis, resulting in depletion of NUP96 during the following $G_1$ phase; this depletion was required for the timely progression of the cell cycle, with $G_1$-specific overexpression of NUP96 causing a severe cell cycle progression arrest (Chakraborty et al. 2008). This arrest resulted from a NUP96 overexpression-specific inhibition of the export of several cell cycle-regulatory mRNAs including *CYCD1*, *CDK6*, and *IκBα*, suggesting that, contrary to the regulation reported for IFN-regulated transcripts above, NUP96 actually represses the export of these transcripts during $G_1$ (Chakraborty et al. 2008; Faria et al. 2006).

These observations raise a number of significant questions regarding the mechanism by which NUP96 specifically regulates the export of these target mRNAs, including: how and why is the polarity of NUP96 action on the export of IFN-induced versus cell cycle-regulatory mRNAs reversed? If NUP96 is a core structural component of the NPC, what happens to NPC structures when NUP96 is proteasomally degraded during mitosis? Does NUP96 act within the context of the NPC, or is it a nucleoplasmically mobile Nup like the putative NXF1:NXT1 chaperone NUP98? And lastly, how does NUP96 identify the mRNAs which undergo selective mRNA export modulation?

### *2.7.2 Bypassing the NPC: Nuclear Membrane Budding for the Export of Large mRNP Complexes*

A long- and firmly-held belief in the field of mRNA export was that the nuclear pore represents the exclusive port of nuclear egress of exported mRNAs, and that all mRNA export pathways, however divergent, must ultimately end with translocation of their mRNA cargoes through the nuclear pore. However, this creed has recently been challenged by a remarkable pathway recently reported in *Drosophila* muscle cells (Fig. 2.2e).

The Wnt signaling pathways are a class of related signaling cascades that regulate a diverse range of cellular processes including embryonic pattern formation, cellular differentiation, and the formation of synaptic junctions, among other functions (Wiese et al. 2018). Wnt pathway activation at *Drosophila* neuromuscular junctions involves the binding of the protein ligand Wg to the DFz2 receptor; this receptor is then internalized, and the cytoplasmic C-terminal tail (DFz2C) is released by proteolysis to govern downstream signaling pathways throughout the cytoplasm and nucleus (He et al. 2018).

Studies of the DFz2C moiety revealed that, upon Wnt activation, this signaling factor is specifically localized to the nucleus, where it assembles large granules containing numerous mRNAs and a range of mRNA-associated proteins; notably, the size of these granules—on average, 192 nm in diameter, compared to a total diameter of the NPC of ~120 nm—suggests that they are likely to be unable to transport to the cytoplasm via the NPC (Speese et al. 2012). Remarkably, super-resolution microscopy approaches combined with live-cell imaging and rapid fixation/co-staining were able to reveal an export pathway that did not require the NPC, but in which the large DFz2C granules were able to bud directly from the inner nuclear membrane into the perinuclear space. This process required phosphorylation-dependent disruption of the nuclear lamina and was followed by fusion of the nascent vesicle with the outer nuclear membrane to allow release of the contents into the cytoplasm (Speese et al. 2012). mRNAs targeted to the DFz2C granules in the nucleus were found to form a specific mRNA regulon enriched for regulators of synaptic structure and synaptogenesis, consistent with a key role for DFz2C signaling in the formation of functional neuromuscular synapses (He et al. 2018; Speese et al. 2012).

While it appears to be a unique and highly unusual pathway upon first inspection, it should be noted that large RNA-containing granules have been serendipitously observed in the perinuclear space in other organisms, ranging from plants to humans, suggesting that this may in fact represent a conserved pathway for the export of mRNPs above the size exclusion limit of the NPC (Speese et al. 2012). In addition, this export mechanism bears remarkable resemblance to that used by human pathogenic herpesviruses and, potentially, cytomegaloviruses to export their large DNA-containing capsids from the nucleus to the cytoplasm by cell-internal capsid envelopment/de-envelopment across the nuclear membranes (Buser et al. 2007; Lee and Chen 2010). When first identified, these pathways were thought to be exclusive to the viruses and to be mediated exclusively by viral proteins; however, the observations here suggest that these pathways may instead represent the repurposing of a latent cellular export pathway for large nucleic acid-containing complexes (Buser et al. 2007; Lee and Chen 2010; Speese et al. 2012).

## 2.8 Outstanding Questions in mRNA Export Heterogeneity

Above we described the overlapping and complementary mechanisms of mRNA export from the nucleus to the cytoplasm. While these studies have provided considerable insights into the guiding principles of mRNA export so far, a number of important questions regarding their function still remain unanswered and will be discussed here.

### *2.8.1 Stoichiometry of Export Adaptors/Coadaptors on mRNA*

One fundamental question is whether mRNAs might contain competing signals that drive export or retention toward a specific export pathway and whether the deposition of a single or multiple export complexes stimulates this process. This can be best illustrated by the question of how many Aly/REF molecules are actually deposited on an mRNA, and whether this number influences export. Early investigations of Aly/REF deposition on mRNA reported noncontradictory interactions between Aly/REF (in the context of TREX) and the CBC-bound 5′ $m^7G$ cap and/or EJCs, leading to the hypothesis that Aly/REF binding (and subsequent NXF1:NXT1 loading) at these site(s) was sufficient for export (Cheng et al. 2006; Masuda et al. 2005). However, subsequent characterization of the DDX19/Gle1/Nup42 complex revealed a mechanism of nuclear pore transit in which DDX19 helicase bound to the NPC's cytoplasmic fibrils evicts the FG-Nup-binding NXF1:NXT1 complex from cargo mRNAs upon transit of the central NPC pore, acting as a "ratchet" to instill directionality to the transit of mRNPs through the NPC (Adams et al. 2017, 2018; Folkmann et al. 2011). A key requirement of this model is that NXF1:NXT1, and therefore Aly/REF, must be deposited throughout the length of the mRNA in order to allow regular interactions of the RNA:NXF1:NXT1 complex with DDX19 during NPC transit; indeed, several cross-linking-immunoprecipitation studies of Aly/REF found binding along the length of mRNA, albeit with enrichment at the 5′- and

3′-ends (Shi et al. 2017; Viphakone et al. 2018). A possible explanation for these apparently contradictory observations emerged from the observation that the core helicase of the TREX complex, UAP56, must bind ATP in order to assemble the TREX complex, and that Aly/REF binding to UAP56 promotes its ATPase activity (Dufu et al. 2010). It was further found that UAP56's ATPase activity was required for deposition of Aly/REF onto mRNA, leading to a compromise model in which TREX is recruited to 5′ caps and/or EJCs and then undergoes successive rounds of ATP hydrolysis to deposit Aly/REF along the target mRNA (Taniguchi and Ohno 2008). While this model satisfies extant observations regarding Aly/REF-binding mechanisms on mRNA, it was complicated by the finding that Aly/REF and THOC5 were coordinately required for the activation of NXF1:NXT1 RNA binding during bulk mRNA export (Viphakone et al. 2012), since under this model THOC5 was expected to comigrate with UAP56 rather than remaining in close coordination with Aly/REF on the mRNA. The model was further complicated by the discovery of numerous orthologous export adaptors/coadaptors that are able to act with and/or substitute for Aly/REF in the bulk export pathway (Chang et al. 2013; Hautbergue et al. 2009; Uranishi et al. 2009; Viphakone et al. 2015). The observation that these proteins typically bind TREX and NXF1:NXT1 in an identical manner to the Aly/REF:THOC5 suggests that they may also be deposited by UAP56, suggesting a possible heterogeneous series of adaptors/coadaptors along the length of mRNAs destined for bulk mRNA export.

Given these contradictory observations and their self-evident relevance to our understanding of the mechanisms and regulation of bulk mRNA export, it is imperative to establish the relative position, order, and stoichiometry of export adaptors and coadaptors on mRNAs in vivo. However, currently available protein-RNA-focused sequencing techniques such as iCLIP are not capable of answering this question due to their generation of population-wide averages of mRNA site binding (van Dijk et al. 2018), and alternative approaches are required to address these questions. One alternative avenue might be the use of sequential IPs of serried adaptor/coadaptor pairs followed by iCLIP analysis to demonstrate the binding patterns of a given factor when co-deposited with another. Other possible approaches could make use of third-generation sequencing technologies which allow the sequencing of entire mRNA molecules without fragmentation (van Dijk et al. 2018). Pairing such an approach with a protein proximity-dependent RNA tagging strategy such as APEX-Seq or, alternatively, fusions of specific proteins with an mRNA-modifying enzyme such as NSUN2 in order to promote the specific modification of mRNAs with diagnostic sequence modifications in the immediate vicinity of an mRNA-bound protein of interest would be particularly powerful (Padròn et al. 2018). Finally, imaging-based approaches that allow the determination of stoichiometry of proteins within complexes using stepwise photobleaching of fluorescently labeled components upon purification and immobilization onto microscopy coverslips might be useful in complementing sequencing-based methods (Jain et al. 2011). Such future investigations will be required to define the binding patterns of export factors on candidate mRNAs and thereby provide new insides into how the stoichiometry of adaptors and coadaptors may coordinate mRNA export.

### 2.8.2 Posttranslational Regulation of mRNA Export Pathways in Response to Cellular Stimuli

In addition to the already formidable complexity of mRNA export engendered by the variety of pathways thus far reported, evidence is emerging that mRNA export pathways are highly sensitive to environmental, tissue-specific, and developmental cues and that their activity can be altered by posttranslational signaling. Early reports regarding PML-mediated suppression of eIF4E:CRM1:RanGTP-mediated export found that this regulatory axis could be manipulated by exposure of cells either to the stressor cadmium or to the signaling molecule IFNγ (Topisirovic et al. 2002), while NXF1:NXT1-dependent export of the *HSP70* mRNA appears to gain an absolute requirement for Aly/REF and THOC5 upon transition from constitutive to heat shock conditions (Guria et al. 2011; Katahira et al. 2009). In addition to its role as a bulk mRNA export coadaptor, THOC5 has been shown to play specific roles in hematopoiesis and adipogenesis in mice (Mancini et al. 2006, 2010). Most strikingly, numerous export factors, including LuzP4, URH49, CIP29, NXF2, and NXF3, all exhibit highly restricted expression in the testes, suggesting the possible existence of an entirely distinct and parallel mRNA export pathway in this tissue (Kapadia et al. 2006; Pryor et al. 2004; Viphakone et al. 2015; Yang et al. 2001; Zhang et al. 2007). The regulation of mRNA export by posttranslational factors is exemplified by the IPMK:Aly/REF export pathway in which the secondary messenger $PIP_3$ binds to Aly/REF and modulates its mRNA-binding motif (Okada et al. 2008; Wickramasinghe et al. 2013). Interestingly, the binding of $PIP_3$ to Aly/REF can itself be modulated by AKT-mediated phosphorylation of Aly/REF, suggesting this mechanism may represent a regulatory hub for several signaling pathways (Wickramasinghe et al. 2013). Prior to its identification as an mRNA export factor, THOC5 (a.k.a. FMIP) was found to be phosphorylated by the GM-CSF and PKC signaling pathways, resulting in its partitioning to the cytoplasm (Mancini et al. 2004; Tamura et al. 1999); similarly, the HuR cofactor APRIL can be phosphorylated at a site adjacent to its nuclear localization signal, resulting in accumulation in the cytoplasm (Fries et al. 2007). Lastly, it has been found both SRSF1 and SRSF3 must be in a hypophosphorylated state to participate in their mRNA export activities, possibly as a means of isolating these activities from their other roles in splicing regulation (Lai and Tarn 2004; Roundtree et al. 2017). Collectively, these results underline the highly regulatable activities of the mRNA export machinery. Detailed investigation of the cellular- and tissue-specific expression patterns and PTM-dependent regulatory pathways of mRNA export will require a detailed and interdisciplinary approach; however, it is noted that recent large-scale collaborative projects cataloging proteome-wide cell- and tissue-specific expression profiles represent important first ports of call for investigation of novel or established export factors (GTEx Consortium 2015; Uhlén et al. 2015) while new highly sensitive mass spectrometry-based methods for the profiling of posttranslational modifications are emerging on a regular basis, providing a powerful experimental tool to dissect regulatory mechanisms contacting the mRNA export machinery (Ke et al. 2016).

### 2.8.3 *Cooperation or Competition of Export Pathways on Common mRNAs*

Given the remarkable diversity and ubiquity of specific mRNA export pathways described herein, it is almost inevitable that these pathways may intersect on particular mRNAs. Understanding how these different pathways may coordinately or competitively regulate export of a common mRNA target will provide key insights into our understanding of the complex network of mRNA export pathways described above. While a thorough description of the extent and significance of export pathway overlap must await exhaustive characterization of the mRNA targets of all known mRNA export pathways, several instances of pathway interface have already been reported serendipitously. A key feature of the eIF4E:CRM1-mediated export pathway is the binding of eIF4E to the 5′ $m^7G$ cap in concert with LRPPRC; however, this interaction necessitates the eviction of the canonical CBC from the 5′-end of mRNA and its replacement by eIF4E via an as-yet undescribed mechanism (Culjkovic et al. 2006). Given the significant though nonessential role that CBC plays in the promotion of bulk export, it seems likely that partitioning of mRNA into the eIF4E export pathway is likely to abolish or attenuate bulk mRNA export (Cheng et al. 2006). Two different mRNA export pathways have also been found to target the *IFNα1* mRNA which can undergo both Prp19 complex/U2AF65-mediated export via a coding-region USER code, the CAR-E, and can also be exported via an incompletely described pathway dependent on NUP96 (Faria et al. 2006; Lei et al. 2011, 2013). More generally, several distinct mRNA regulons have been reported to contain highly similar functional groups of mRNAs, including the observation that cell cycle-regulatory factors are enriched in the regulons of IPMK:Aly/REF, eIF4E: CRM1, and NUP96-dependent mRNA export pathways (Chakraborty et al. 2008; Culjkovic et al. 2006; Wickramasinghe et al. 2013). Finally, the observation that eIF4E is able to promote the export and expression of key components of the AKT signaling pathway, a key regulator of Aly/REF binding to $PIP_3$ in the IPMK:Aly/REF export pathway, suggests that distinct export pathways may utilize cellular signaling pathways to balance their activities in vivo (Culjkovic et al. 2008; Okada et al. 2008; Wickramasinghe et al. 2013).

The first and most fundamental step in characterizing cross-talk between mRNA export pathways will be to define the complete set of mRNA targets of all known export pathways; while candidate-specific XL-Seq methodologies such as iCLIP will likely prove useful for this, a major resource is likely to be the database of iCLIP profiles for >300 RBPs generated by the ENCODE consortium (Van Nostrand et al. 2018). Once a common mRNA target of two export pathways is identified, a suite of exploratory techniques will likely be required to define the protein interactome of these mRNAs (Castello et al. 2012; Ramanathan et al. 2018), their behavior within the nucleoplasm and at the NPC, and the relative contributions of the two or more export pathways to the mRNA's export kinetics in living cells (Grünwald and Singer 2010; Heinrich et al. 2017; Ma et al. 2013; Siebrasse et al. 2012). Given the likely number and complexity of interactions between export pathways in metazoan cells, it is expected that new, network-level experimental techniques will be required to dissect the complex interplay of these pathways.

## 2.9 Concluding Remarks

Historically, research into the mechanisms of protein expression heterogeneity have focused on transcriptional and translational regulation, control of splicing, and the means of mRNA degradation in the cytoplasm. However, mRNA export is far from being a passive player in this process and is itself a significant source of heterogeneity in the gene expression program and an important regulatory hub. As further details of mRNA export pathways emerge, it is expected that they will provide new insights into the means by which cells regulate their expression program and identify new possible therapeutic angles by which mRNA export may contribute to the pathogenesis and/or treatment of human disease.

**Acknowledgments** The authors thank Daniel Zenklusen for critical reading of the manuscript, as well as Eric Lécuyer and François Robert for comments on the manuscript during preparation. This work has been supported by the Natural Sciences and Engineering Research Council of Canada (NSERC) and Fonds de recherche du Québec—Santé (FRQ-S Chercheur-boursier Junior 2) for MO.

## References

Adams RL, Mason AC, Glass L, Aditi, Wente SR (2017) Nup42 and $IP_6$ coordinate Gle1 stimulation of Dbp5/DDX19B for mRNA export in yeast and human cells. Traffic (Copenhagen, Denmark) 18:776–790

Adams RL, Mason AC, Glass L, Aditi, Wente SR (2018) Correction to: Nup42 and $IP_6$ coordinate Gle1 stimulation of Dbp5/DDX19B for mRNA export in yeast and human cells. Traffic (Copenhagen, Denmark) 19:650

Akef A, Zhang H, Masuda S, Palazzo AF (2013) Trafficking of mRNAs containing ALREX-promoting elements through nuclear speckles. Nucleus 4:326–340

Akichika S, Hirano S, Shichino Y, Suzuki T, Nishimasu H, Ishitani R, Sugita A, Hirose Y, Iwasaki S, Nureki O et al (2019) Cap-specific terminal $N^6$-methylation of RNA by an RNA polymerase II–associated methyltransferase. Science 363:eaav0080

Amort T, Rieder D, Wille A, Khokhlova-Cubberley D, Riml C, Trixl L, Jia XY, Micura R, Lusser A (2017) Distinct 5-methylcytosine profiles in poly(A) RNA from mouse embryonic stem cells and brain. Genome Biol 18:1

Aviram N, Schuldiner M (2017) Targeting and translocation of proteins to the endoplasmic reticulum at a glance. J Cell Sci 130:4079–4085

Beck M, Hurt E (2017) The nuclear pore complex: understanding its function through structural insight. Nat Rev Mol Cell Biol 18:73–89

Belgareh N, Rabut G, Baï SW, van Overbeek M, Beaudouin J, Daigle N, Zatsepina OV, Pasteau F, Labas V, Fromont-Racine M et al (2001) An evolutionarily conserved NPC subcomplex, which redistributes in part to kinetochores in mammalian cells. J Cell Biol 154:1147–1160

Björk P, Wieslander L (2017) Integration of mRNP formation and export. Cell Mol Life Sci 74:2875–2897

Blanchette M, Labourier E, Green RE, Brenner SE, Rio DC (2004) Genome-wide analysis reveals an unexpected function for the *Drosophila* splicing factor U2AF$^{50}$ in the nuclear export of intronless mRNAs. Mol Cell 14:775–786

Blevins MB, Smith AM, Phillips EM, Powers MA (2003) Complex formation among the RNA export proteins Nup98, Rae1/Gle2, and TAP. J Biol Chem 278:20979–20988

Boccaletto P, Machnicka MA, Purta E, Piątkowski P, Bagiński B, Wirecki TK, de Crécy-Lagard V, Ross R, Limbach PA, Kotter A et al (2018) MODOMICS: a database of RNA modification pathways. 2017 update. Nucleic Acids Res 46:D303–D307

Boulias K, Toczydlowska-Socha D, Hawley BR, Liberman-Isakov N, Takashima K, Zaccara S, Guez T, Vasseur J-J, Debart F, Aravind L et al (2018) Identification of the m6Am methyltransferase PCIF1 reveals the location and functions of m6Am in the transcriptome. bioRxiv 485862

Braun IC, Rohrbach E, Schmitt C, Izaurralde E (1999) TAP binds to the constitutive transport element (CTE) through a novel RNA-binding motif that is sufficient to promote CTE-dependent RNA export from the nucleus. EMBO J 18:1953–1965

Bray M, Prasad S, Dubay JW, Hunter E, Jeang KT, Rekosh D, Hammarskjöld ML (1994) A small element from the Mason-Pfizer monkey virus genome makes human immunodeficiency virus type 1 expression and replication Rev-independent. PNAS 91:1256–1260

Brennan CM, Gallouzi IE, Steitz JA (2000) Protein ligands to HuR modulate its interaction with target mRNAs *in vivo*. J Cell Biol 151:1–14

Brunet I, Weinl C, Piper M, Trembleau A, Volovitch M, Harris W, Prochiantz A, Holt C (2005) The transcription factor Engrailed-2 guides retinal axons. Nature 438:94–98

Buser C, Walther P, Mertens T, Michel D (2007) Cytomegalovirus primary envelopment occurs at large infoldings of the inner nuclear membrane. J Virol 81:3042–3048

Cabal GG, Genovesio A, Rodríguez-Navarro S, Zimmer C, Gadal O, Lesne A, Buc H, Feuerbach-Fournier F, Olivo-Marin JC, Hurt EC et al (2006) SAGA interacting factors confine sub-diffusion of transcribed genes to the nuclear envelope. Nature 441:770–773

Camper SA, Albers RJ, Coward JK, Rottman FM (1984) Effect of undermethylation on mRNA cytoplasmic appearance and half-life. Mol Cell Biol 4:538–543

Castello A, Fischer B, Eichelbaum K, Horos R, Beckmann BM, Strein C, Davey NE, Humphreys DT, Preiss T, Steinmetz LM et al (2012) Insights into RNA biology from an atlas of mammalian mRNA-binding proteins. Cell 149:1393–1406

Cenik C, Chua HN, Zhang H, Tarnawsky SP, Akef A, Derti A, Tasan M, Moore MJ, Palazzo AF, Roth FP (2011) Genome analysis reveals interplay between 5′UTR introns and nuclear mRNA export for secretory and mitochondrial genes. PLoS Genet 7:e1001366

Chakraborty P, Wang Y, Wei J-H, van Deursen J, Yu H, Malureanu L, Dasso M, Forbes DJ, Levy DE, Seemann J et al (2008) Nucleoporin levels regulate cell cycle progression and phase-specific gene expression. Dev Cell 15:657–667

Chan S, Choi EA, Shi Y (2011) Pre-mRNA 3′-end processing complex assembly and function. Wiley Interdiscip Rev RNA 2:321–335

Chang C-T, Hautbergue GM, Walsh MJ, Viphakone N, van Dijk TB, Philipsen S, Wilson SA (2013) Chtop is a component of the dynamic TREX mRNA export complex. EMBO J 32:473–486

Chen LL, Carmichael GG (2009) Altered nuclear retention of mRNAs containing inverted repeats in human embryonic stem cells: functional role of a nuclear noncoding RNA. Mol Cell 35:467–478

Chen LL, DeCerbo JN, Carmichael GG (2008) Alu element-mediated gene silencing. EMBO J 27:1694–1705

Cheng H, Dufu K, Lee CS, Hsu JL, Dias A, Reed R (2006) Human mRNA export machinery recruited to the 5′ end of mRNA. Cell 127:1389–1400

Chi B, Wang Q, Wu G, Tan M, Wang L, Shi M, Chang X, Cheng H (2013) Aly and THO are required for assembly of the human TREX complex and association of TREX components with the spliced mRNA. Nucleic Acids Res 41:1294–1306
Cohen N, Sharma M, Kentsis A, Perez JM, Strudwick S, Borden KL (2001) PML RING suppresses oncogenic transformation by reducing the affinity of eIF4E for mRNA. EMBO J 20:4547–4559
Cowling VH (2019) CAPAM: the mRNA cap adenosine N6-methyltransferase. Trends Biochem Sci 44:183–185
Culjkovic B, Topisirovic I, Skrabanek L, Ruiz-Gutierrez M, Borden KLB (2005) eIF4E promotes nuclear export of *cyclin D1* mRNAs via an element in the 3′UTR. J Cell Biol 169:245–256
Culjkovic B, Topisirovic I, Skrabanek L, Ruiz-Gutierrez M, Borden KLB (2006) eIF4E is a central node of an RNA regulon that governs cellular proliferation. J Cell Biol 175:415–426
Culjkovic B, Tan K, Orolicki S, Amri A, Meloche S, Borden KLB (2008) The eIF4E RNA regulon promotes the Akt signaling pathway. J Cell Biol 181:51–63
Cullen BR (1998) Retroviruses as model systems for the study of nuclear RNA export pathways. Virology 249:203–210
David CJ, Boyne AR, Millhouse SR, Manley JL (2011) The RNA polymerase II C-terminal domain promotes splicing activation through recruitment of a U2AF65-Prp19 complex. Genes Dev 25:972–983
Delaleau M, Borden KL (2015) Multiple export mechanisms for mRNAs. Cell 4:452–473
Dettwiler S, Aringhieri C, Cardinale S, Keller W, Barabino SM (2004) Distinct sequence motifs within the 68-kDa subunit of cleavage factor $I_m$ mediate RNA binding, protein-protein interactions, and subcellular localization. J Biol Chem 279:35788–35797
Dixon DA (2004) Dysregulated post-transcriptional control of *COX-2* gene expression in cancer. Curr Pharm Des 10:635–646
Dockendorff TC, Heath CV, Goldstein AL, Snay CA, Cole CN (1997) C-terminal truncations of the yeast nucleoporin Nup145p produce a rapid temperature-conditional mRNA export defect and alterations to nuclear structure. Mol Cell Biol 17:906–920
Dominissini D, Moshitch-Moshkovitz S, Schwartz S, Salmon-Divon M, Ungar L, Osenberg S, Cesarkas K, Jacob-Hirsch J, Amariglio N, Kupiec M et al (2012) Topology of the human and mouse $m^6A$ RNA methylomes revealed by $m^6A$-seq. Nature 485:201–206
Dominski Z, Marzluff WF (2007) Formation of the 3′ end of histone mRNA: getting closer to the end. Gene 396:373–390
Duan HC, Wang Y, Jia G (2019) Dynamic and reversible RNA $N^6$-methyladenosine methylation. Wiley Interdiscip Rev RNA 10:e1507
Dufu K, Livingstone MJ, Seebacher J, Gygi SP, Wilson SA, Reed R (2010) ATP is required for interactions between UAP56 and two conserved mRNA export proteins, Aly and CIP29, to assemble the TREX complex. Genes Dev 24:2043–2053
Eckmann CR, Rammelt C, Wahle E (2011) Control of poly(A) tail length. Wiley Interdiscip Rev RNA 2:348–361
Ehrnsberger HF, Pfaff C, Hachani I, Flores-Tornero M, Soerensen BB, Langst G, Sprunck S, Grasser M, Grasser KD (2019) The UAP56-interacting export factors UIEF1 and UIEF2 function in mRNA export. Plant Physiol 179(4):1525–1536
Faria AMC, Levay A, Wang Y, Kamphorst AO, Rosa MLP, Nussenzveig DR, Balkan W, Chook YM, Levy DE, Fontoura BMA (2006) The nucleoporin Nup96 is required for proper expression of interferon-regulated proteins and functions. Immunity 24:295–304
Farny NG, Hurt JA, Silver PA (2008) Definition of global and transcript-specific mRNA export pathways in metazoans. Genes Dev 22:66–78
Fasken MB, Stewart M, Corbett AH (2008) Functional significance of the interaction between the mRNA-binding protein, Nab2, and the nuclear pore-associated protein, Mlp1, in mRNA export. J Biol Chem 283:27130–27143
Fischer U, Meyer S, Teufel M, Heckel C, Lührmann R, Rautmann G (1994) Evidence that HIV-1 Rev directly promotes the nuclear export of unspliced RNA. EMBO J 13:4105–4112

Folco EG, Lee CS, Dufu K, Yamazaki T, Reed R (2012) The proteins PDIP3 and ZC11A associate with the human TREX complex in an ATP-dependent manner and function in mRNA export. PLoS One 7:e43804

Folkmann AW, Noble KN, Cole CN, Wente SR (2011) Dbp5, Gle1-IP6 and Nup159: a working model for mRNP export. Nucleus 2:540–548

Fornerod M, Ohno M, Yoshida M, Mattaj IW (1997) CRM1 is an export receptor for leucine-rich nuclear export signals. Cell 90:1051–1060

Fox AH, Lamond AI (2010) Paraspeckles. Cold Spring Harb Perspect Biol 2:a000687

Fox AH, Nakagawa S, Hirose T, Bond CS (2018) Paraspeckles: where long noncoding RNA meets phase separation. Trends Biochem Sci 43:124–135

Fries B, Heukeshoven J, Hauber I, Grüttner C, Stocking C, Kehlenbach RH, Hauber J, Chemnitz J (2007) Analysis of nucleocytoplasmic trafficking of the HuR ligand APRIL and its influence on CD83 expression. J Biol Chem 282:4504–4515

Fustin JM, Doi M, Yamaguchi Y, Hida H, Nishimura S, Yoshida M, Isagawa T, Morioka MS, Kakeya H, Manabe I et al (2013) RNA-methylation-dependent RNA processing controls the speed of the circadian clock. Cell 155:793–806

Garcia-Mauriño SM, Rivero-Rodríguez F, Velázquez-Cruz A, Hernández-Vellisca M, Díaz-Quintana A, De la Rosa MA, Díaz-Moreno I (2017) RNA binding protein regulation and cross-talk in the control of AU-rich mRNA fate. Front Mol Biosci 4:71

García-Vílchez R, Sevilla A, Blanco S (2019) Post-transcriptional regulation by cytosine-5 methylation of RNA. Biochim Biophys Acta 1862:240–252

Gatfield D, Izaurralde E (2002) REF1/Aly and the additional exon junction complex proteins are dispensable for nuclear mRNA export. J Cell Biol 159:579–588

Gerbracht JV, Gehring NH (2018) The exon junction complex: structural insights into a faithful companion of mammalian mRNPs. Biochem Soc Trans 46:153–161

Gromadzka AM, Steckelberg A-L, Singh KK, Hofmann K, Gehring NH (2016) A short conserved motif in ALYREF directs cap- and EJC-dependent assembly of export complexes on spliced mRNAs. Nucleic Acids Res 44:2348–2361

Grünwald D, Singer RH (2010) *In vivo* imaging of labelled endogenous *β-actin* mRNA during nucleocytoplasmic transport. Nature 467:604–607

Grüter P, Tabernero C, von Kobbe C, Schmitt C, Saavedra C, Bachi A, Wilm M, Felber BK, Izaurralde E (1998) TAP, the human homolog of Mex67p, mediates CTE-dependent RNA export from the nucleus. Mol Cell 1:649–659

GTEx Consortium, T (2015) The Genotype-Tissue Expression (GTEx) pilot analysis: multitissue gene regulation in humans. Science 348:648–660

Guang S, Mertz JE (2005) Pre-mRNA processing enhancer (PPE) elements from intronless genes play additional roles in mRNA biogenesis than do ones from intron-containing genes. Nucleic Acids Res 33:2215–2226

Guria A, Tran DD, Ramachandran S, Koch A, El Bounkari O, Dutta P, Hauser H, Tamura T (2011) Identification of mRNAs that are spliced but not exported to the cytoplasm in the absence of THOC5 in mouse embryo fibroblasts. RNA 17:1048–1056

Hammarskjöld ML (1997) Regulation of retroviral RNA export. Semin Cell Dev Biol 8:83–90

Hargous Y, Hautbergue GM, Tintaru AM, Skrisovska L, Golovanov AP, Stevenin J, Lian L-Y, Wilson SA, Allain FHT (2006) Molecular basis of RNA recognition and TAP binding by the SR proteins SRp20 and 9G8. EMBO J 25:5126–5137

Harper JE, Miceli SM, Roberts RJ, Manley JL (1990) Sequence specificity of the human mRNA N6-adenosine methylase *in vitro*. Nucleic Acids Res 18:5735–5741

Hautbergue GM, Hung M-L, Golovanov AP, Lian L-Y, Wilson SA (2008) Mutually exclusive interactions drive handover of mRNA from export adaptors to TAP. Proc Natl Acad Sci USA 105:5154–5159

Hautbergue GM, Hung M-L, Walsh MJ, Snijders APL, Chang C-T, Jones R, Ponting CP, Dickman MJ, Wilson SA (2009) UIF, a new mRNA export adaptor that works together with REF/ALY, requires FACT for recruitment to mRNA. Curr Biol 19:1918–1924

He CW, Liao CP, Pan CL (2018) Wnt signalling in the development of axon, dendrites and synapses. Open Biol 8:180116

Heath CG, Viphakone N, Wilson SA (2016) The role of TREX in gene expression and disease. Biochem J 473:2911–2935

Heinrich S, Derrer CP, Lari A, Weis K, Montpetit B (2017) Temporal and spatial regulation of mRNA export: single particle RNA-imaging provides new tools and insights. Bioessays 39:1600124

Herold A, Suyama M, Rodrigues JP, Braun IC, Kutay U, Carmo-Fonseca M, Bork P, Izaurralde E (2000) TAP (NXF1) belongs to a multigene family of putative RNA export factors with a conserved modular architecture. Mol Cell Biol 20:8996–9008

Hess ME, Hess S, Meyer KD, Verhagen LAW, Koch L, Brönneke HS, Dietrich MO, Jordan SD, Saletore Y, Elemento O et al (2013) The fat mass and obesity associated gene (*Fto*) regulates activity of the dopaminergic midbrain circuitry. Nat Neurosci 16:1042–1048

Hieronymus H, Silver PA (2003) Genome-wide analysis of RNA-protein interactions illustrates specificity of the mRNA export machinery. Nat Genet 33:155–161

Ho JJD, Lee S (2016) A cap for every occasion: alternative eIF4F complexes. Trends Biochem Sci 41:821–823

Huang Y, Carmichael GG (1997) The mouse histone H2a gene contains a small element that facilitates cytoplasmic accumulation of intronless gene transcripts and of unspliced HIV-1-related mRNAs. Proc Natl Acad Sci USA 94:10104–10109

Huang Y, Steitz JA (2001) Splicing factors SRp20 and 9G8 promote the nucleocytoplasmic export of mRNA. Mol Cell 7:899–905

Huang Y, Gattoni R, Stévenin J, Steitz JA (2003) SR splicing factors serve as adapter proteins for TAP-dependent mRNA export. Mol Cell 11:837–843

Hung M-L, Hautbergue GM, Snijders APL, Dickman MJ, Wilson SA (2010) Arginine methylation of REF/ALY promotes efficient handover of mRNA to TAP/NXF1. Nucleic Acids Res 38:3351–3361

Hussain S, Sajini AA, Blanco S, Dietmann S, Lombard P, Sugimoto Y, Paramor M, Gleeson JG, Odom DT, Ule J et al (2013) NSun2-mediated cytosine-5 methylation of vault noncoding RNA determines its processing into regulatory small RNAs. Cell Rep 4:255–261

Hutten S, Kehlenbach RH (2007) CRM1-mediated nuclear export: to the pore and beyond. Trends Cell Biol 17:193–201

Jain A, Liu R, Ramani B, Arauz E, Ishitsuka Y, Ragunathan K, Park J, Chen J, Xiang YK, Ha T (2011) Probing cellular protein complexes using single-molecule pull-down. Nature 473:484–488

Jang BC, Muñoz-Najar U, Paik JH, Claffey K, Yoshida M, Hla T (2003) Leptomycin B, an inhibitor of the nuclear export receptor CRM1, inhibits *COX-2* expression. J Biol Chem 278:2773–2776

Jani D, Lutz S, Hurt E, Laskey RA, Stewart M, Wickramasinghe VO (2012) Functional and structural characterization of the mammalian TREX-2 complex that links transcription with nuclear messenger RNA export. Nucleic Acids Res 40:4562–4573

Jarmolowski A, Boelens W, Izaurralde E, Mattaj I (1994) Nuclear export of different classes of RNA is mediated by specific factors. J Cell Biol 124:627–635

Jia G, Fu Y, Zhao X, Dai Q, Zheng G, Yang Y, Yi C, Lindahl T, Pan T, Yang Y-G et al (2011) N6-methyladenosine in nuclear RNA is a major substrate of the obesity-associated FTO. Nat Chem Biol 7:885–887

Johnson SA, Cubberley G, Bentley DL (2009) Cotranscriptional recruitment of the mRNA export factor Yra1 by direct interaction with the 3′ end processing factor Pcf11. Mol Cell 33:215–226

Kapadia F, Pryor A, Chang TH, Johnson LF (2006) Nuclear localization of poly(A)+ mRNA following siRNA reduction of expression of the mammalian RNA helicases UAP56 and URH49. Gene 384:37–44

Katahira J (2012) mRNA export and the TREX complex. Biochim Biophys Acta 1819:507–513

Katahira J, Inoue H, Hurt E, Yoneda Y (2009) Adaptor Aly and co-adaptor Thoc5 function in the Tap-p15-mediated nuclear export of *HSP70* mRNA. EMBO J 28:556–567

Katahira J, Okuzaki D, Inoue H, Yoneda Y, Maehara K, Ohkawa Y (2013) Human TREX component Thoc5 affects alternative polyadenylation site choice by recruiting mammalian cleavage factor I. Nucleic Acids Res 41:7060–7072

Ke M, Shen H, Wang L, Luo S, Lin L, Yang J, Tian R (2016) Identification, quantification, and site localization of protein posttranslational modifications via mass spectrometry-based proteomics. In: Mirzaei H, Carrasco MA (eds) Modern proteomics – sample preparation, analysis and practical applications. Springer, Berlin, pp 345–382

Ke S, Pandya-Jones A, Saito Y, Fak JJ, Vågbø CB, Geula S, Hanna JH, Black DL, Darnell JE, Darnell RB (2017) m6A mRNA modifications are deposited in nascent pre-mRNA and are not required for splicing but do specify cytoplasmic turnover. Genes Dev 31:990–1006

Keene JD (2007) RNA regulons: coordination of post-transcriptional events. Nat Rev Genet 8:533–543

Khoddami V, Cairns BR (2013) Identification of direct targets and modified bases of RNA cytosine methyltransferases. Nat Biotechnol 31:458–464

Kim E, Ahn H, Kim MG, Lee H, Kim S (2017) The expanding significance of inositol polyphosphate multikinase as a signaling hub. Mol Cells 40:315–321

Kimura T, Hashimoto I, Nagase T, Fujisawa J-I (2004) CRM1-dependent, but not ARE-mediated, nuclear export of *IFN-α1* mRNA. J Cell Sci 117:2259–2270

König J, Zarnack K, Rot G, Curk T, Kayikci M, Zupan B, Turner DJ, Luscombe NM, Ule J (2010) iCLIP reveals the function of hnRNP particles in splicing at individual nucleotide resolution. Nat Struct Mol Biol 17:909–915

Kruse S, Zhong S, Bodi Z, Button J, Alcocer MJC, Hayes CJ, Fray R (2011) A novel synthesis and detection method for cap-associated adenosine modifications in mouse mRNA. Sci Rep 1:126

Lai MC, Tarn WY (2004) Hypophosphorylated ASF/SF2 binds TAP and is present in messenger ribonucleoproteins. J Biol Chem 279:31745–31749

Le Hir H, Izaurralde E, Maquat LE, Moore MJ (2000) The spliceosome deposits multiple proteins 20-24 nucleotides upstream of mRNA exon-exon junctions. EMBO J 19:6860–6869

Le Hir H, Gatfield D, Izaurralde E, Moore MJ (2001) The exon-exon junction complex provides a binding platform for factors involved in mRNA export and nonsense-mediated mRNA decay. EMBO J 20:4987–4997

Lee C-P, Chen M-R (2010) Escape of herpesviruses from the nucleus. Rev Med Virol 20:214–230

Legrain P, Rosbash M (1989) Some *cis*- and *trans*-acting mutants for splicing target pre-mRNA to the cytoplasm. Cell 57:573–583

Lei H, Dias AP, Reed R (2011) Export and stability of naturally intronless mRNAs require specific coding region sequences and the TREX mRNA export complex. Proc Natl Acad Sci USA 108:17985–17990

Lei H, Zhai B, Yin S, Gygi S, Reed R (2013) Evidence that a consensus element found in naturally intronless mRNAs promotes mRNA export. Nucleic Acids Res 41:2517–2525

Lesbirel S, Wilson SA (2019) The $m^6A$-methylase complex and mRNA export. Biochim Biophys Acta 1862:319–328

Lesbirel S, Viphakone N, Parker M, Parker J, Heath C, Sudbery I, Wilson SA (2018) The $m^6A$-methylase complex recruits TREX and regulates mRNA export. Sci Rep 8:13827

Li P, Noegel AA (2015) Inner nuclear envelope protein SUN1 plays a prominent role in mammalian mRNA export. Nucleic Acids Res 43:9874–8988

Li Y, Bor YC, Misawa Y, Xue Y, Rekosh D, Hammarskjold ML (2006) An intron with a constitutive transport element is retained in a *Tap* messenger RNA. Nature 443:234–237

Li X, Xiong X, Yi C (2016) Epitranscriptome sequencing technologies: decoding RNA modifications. Nat Methods 14:23–31

Li MW, Sletten AC, Lee J, Pyles KD, Matkovich SJ, Ory DS, Schaffer JE (2017a) Nuclear export factor 3 regulates localization of small nucleolar RNAs. J Biol Chem 292:20228–20239

Li P, Stumpf M, Müller R, Eichinger L, Glöckner G, Noegel AA (2017b) The function of the inner nuclear envelope protein SUN1 in mRNA export is regulated by phosphorylation. Sci Rep 7:9157

Linder B, Grozhik AV, Olarerin-George AO, Meydan C, Mason CE, Jaffrey SR (2015) Single-nucleotide-resolution mapping of $m^6A$ and $m^6A_m$ throughout the transcriptome. Nat Methods 12:767–772

Lindtner S, Zolotukhin AS, Uranishi H, Bear J, Kulkarni V, Smulevitch S, Samiotaki M, Panayotou G, Felber BK, Pavlakis GN (2006) RNA-binding motif protein 15 binds to the RNA transport element RTE and provides a direct link to the NXF1 export pathway. J Biol Chem 281:36915–36928

Liu X, Mertz JE (1995) HnRNP L binds a *cis*-acting RNA sequence element that enables intron-dependent gene expression. Genes Dev 9:1766–1780

Longman D, Johnstone IL, Cáceres JF (2003) The Ref/Aly proteins are dispensable for mRNA export and development in *Caenorhabditis elegans*. RNA 9:881–891

Ma X, Renda MJ, Wang L, Cheng EC, Niu C, Morris SW, Chi AS, Krause DS (2007) Rbm15 modulates Notch-induced transcriptional activation and affects myeloid differentiation. Mol Cell Biol 27:3056–3064

Ma J, Liu Z, Michelotti N, Pitchiaya S, Veerapaneni R, Androsavich JR, Walter NG, Yang W (2013) High-resolution three-dimensional mapping of mRNA export through the nuclear pore. Nat Commun 4:2414

Mancini A, Koch A, Whetton AD, Tamura T (2004) The M-CSF receptor substrate and interacting protein FMIP is governed in its subcellular localization by protein kinase C-mediated phosphorylation, and thereby potentiates M-CSF-mediated differentiation. Oncogene 23:6581–6589

Mancini A, El Bounkari O, Norrenbrock AF, Scherr M, Schaefer D, Eder M, Banham AH, Pulford K, Lyne L, Whetton AD et al (2006) FMIP controls the adipocyte lineage commitment of C2C12 cells by downmodulation of C/EBPalpha. Oncogene 26:1020

Mancini A, Niemann-Seyde SC, Pankow R, El Bounkari O, Klebba-Färber S, Koch A, Jaworska E, Spooncer E, Gruber AD, Whetton AD et al (2010) THOC5/FMIP, an mRNA export TREX complex protein, is essential for hematopoietic primitive cell survival in vivo. BMC Biol 8:1

Masuda S, Das R, Cheng H, Hurt E, Dorman N, Reed R (2005) Recruitment of the human TREX complex to mRNA during splicing. Genes Dev 19:1512–1517

Masuyama K, Taniguchi I, Kataoka N, Ohno M (2004) SR proteins preferentially associate with mRNAs in the nucleus and facilitate their export to the cytoplasm. Genes Cells 9:959–965

Mattiazzi Usaj M, Styles EB, Verster AJ, Friesen H, Boone C, Andrews BJ (2016) High-content screening for quantitative cell biology. Trends Cell Biol 26:598–611

Mauer J, Luo X, Blanjoie A, Jiao X, Grozhik AV, Patil DP, Linder B, Pickering BF, Vasseur J-J, Chen Q et al (2016) Reversible methylation of m6Am in the 5′ cap controls mRNA stability. Nature 541:371

Meyer KD, Jaffrey SR (2017) Rethinking $m^6A$ readers, writers, and erasers. Annu Rev Cell Dev Biol 33:319–342

Meyer KD, Saletore Y, Zumbo P, Elemento O, Mason CE, Jaffrey SR (2012) Comprehensive analysis of mRNA methylation reveals enrichment in 3′ UTRs and near stop codons. Cell 149:1635–1646

Mishra RK, Chakraborty P, Arnaoutov A, Fontoura BMA, Dasso M (2010) The Nup107-160 complex and $\gamma$-TuRC regulate microtubule polymerization at kinetochores. Nat Cell Biol 12:164

Mor A, Shav-Tal Y (2010) Dynamics and kinetics of nucleo-cytoplasmic mRNA export. Wiley Interdiscip Rev RNA 1:388–401

Müller-McNicoll M, Botti V, de Jesus Domingues AM, Brandl H, Schwich OD, Steiner MC, Curk T, Poser I, Zarnack K, Neugebauer KM (2016) SR proteins are NXF1 adaptors that link alternative RNA processing to mRNA export. Genes Dev 30:553–566

Nakamura H, Kawagishi H, Watanabe A, Sugimoto K, Maruyama M, Sugimoto M (2011) Cooperative role of the RNA-binding proteins Hzf and HuR in p53 activation. Mol Cell Biol 31:1997–2009

Nédélec S, Foucher I, Brunet I, Bouillot C, Prochiantz A, Trembleau A (2004) Emx2 homeodomain transcription factor interacts with eukaryotic translation initiation factor 4E (eIF4E) in the axons of olfactory sensory neurons. PNAS 101:10815–10820

Nerurkar P, Altvater M, Gerhardy S, Schütz S, Fischer U, Weirich C, Panse VG (2015) Eukaryotic ribosome assembly and nuclear export. Int Rev Cell Mol Biol 319:107–140

Neville M, Rosbash M (1999) The NES-Crm1p export pathway is not a major mRNA export route in *Saccharomyces cerevisiae*. EMBO J 18:3746–3756

Oeffinger M, Zenklusen D (2012) To the pore and through the pore: a story of mRNA export kinetics. Biochim Biophys Acta 1819:494–506

Ohno M, Segref A, Kuersten S, Mattaj IW (2002) Identity elements used in export of mRNAs. Mol Cell 9:659–671

Okada M, Jang S-W, Ye K (2008) Akt phosphorylation and nuclear phosphoinositide association mediate mRNA export and cell proliferation activities by ALY. Proc Natl Acad Sci USA 105:8649–8654

Okamoto N, Kuwahara K, Ohta K, Kitabatake M, Takagi K, Mizuta H, Kondo E, Sakaguchi N (2010) Germinal center-associated nuclear protein (GANP) is involved in mRNA export of *Shugoshin-1* required for centromere cohesion and in sister-chromatid exchange. Genes Cells 15:471–484

Okamura M, Inose H, Masuda S (2015) RNA export through the NPC in eukaryotes. Genes 6:124–149

Osborne MJ, Borden KLB (2015) The eukaryotic translation initiation factor eIF4E in the nucleus: taking the road less traveled. Immunol Rev 263:210–223

Padròn A, Iwasaki S, Ingolia N (2018) Proximity RNA labeling by APEX-Seq reveals the organization of translation initiation complexes and repressive RNA granules. bioRxiv 454066

Palazzo AF, Lee ES (2018) Sequence determinants for nuclear retention and cytoplasmic export of mRNAs and lncRNAs. Front Genet 9:440

Palazzo AF, Springer M, Shibata Y, Lee C-S, Dias AP, Rapoport TA (2007) The signal sequence coding region promotes nuclear export of mRNA. PLoS Biol 5:e322

Patil DP, Chen CK, Pickering BF, Chow A, Jackson C, Guttman M, Jaffrey SR (2016) $m^6A$ RNA methylation promotes *XIST*-mediated transcriptional repression. Nature 537:369–373

Pelletier J, Graff J, Ruggero D, Sonenberg N (2015) Targeting the eIF4F translation initiation complex: a critical nexus for cancer development. Cancer Res 75:250–263

Portman DS, O'Connor JP, Dreyfuss G (1997) *YRA1*, an essential *Saccharomyces cerevisiae* gene, encodes a novel nuclear protein with RNA annealing activity. RNA 3:527–537

Prasanth KV, Prasanth SG, Xuan Z, Hearn S, Freier SM, Bennett CF, Zhang MQ, Spector DL (2005) Regulating gene expression through RNA nuclear retention. Cell 123:249–263

Prechtel AT, Chemnitz J, Schirmer S, Ehlers C, Langbein-Detsch I, Stülke J, Dabauvalle M-C, Kehlenbach RH, Hauber J (2006) Expression of CD83 is regulated by HuR via a novel *cis*-active coding region RNA element. J Biol Chem 281:10912–10925

Pritchard CE, Fornerod M, Kasper LH, van Deursen JM (1999) RAE1 is a shuttling mRNA export factor that binds to a GLEBS-like NUP98 motif at the nuclear pore complex through multiple domains. J Cell Biol 145:237–254

Pryor A, Tung L, Yang Z, Kapadia F, Chang TH, Johnson LF (2004) Growth-regulated expression and $G_0$-specific turnover of the mRNA that encodes URH49, a mammalian DExH/D box protein that is highly related to the mRNA export protein UAP56. Nucleic Acids Res 32:1857–1865

Ramanathan A, Robb GB, Chan SH (2016) mRNA capping: biological functions and applications. Nucleic Acids Res 44:7511–7526

Ramanathan M, Majzoub K, Rao DS, Neela PH, Zarnegar BJ, Mondal S, Roth JG, Gai H, Kovalski JR, Siprashvili Z et al (2018) RNA-protein interaction detection in living cells. Nat Methods 15:207–212

Ratnadiwakara M, Archer SK, Dent CI, Ruiz De Los Mozos I, Beilharz TH, Knaupp AS, Nefzger CM, Polo JM, Anko M-L (2018) SRSF3 promotes pluripotency through *Nanog* mRNA export and coordination of the pluripotency gene expression program. Elife 7:e37419

Rehwinkel J, Herold A, Gari K, Köcher T, Rode M, Ciccarelli FL, Wilm M, Izaurralde E (2004) Genome-wide analysis of mRNAs regulated by the THO complex in *Drosophila melanogaster*. Nat Struct Mol Biol 11:558–566

Rodrigues JP, Rode M, Gatfield D, Blencowe BJ, Carmo-Fonseca M, Izaurralde E (2001) REF proteins mediate the export of spliced and unspliced mRNAs from the nucleus. Proc Natl Acad Sci USA 98:1030–1035

Rodríguez-Navarro S, Fischer T, Luo MJ, Antúnez O, Brettschneider S, Lechner J, Pérez-Ortin JE, Reed R, Hurt E (2004) Sus1, a functional component of the SAGA histone acetylase complex and the nuclear pore-associated mRNA export machinery. Cell 116:75–86

Roundtree IA, Luo G-Z, Zhang Z, Wang X, Zhou T, Cui Y, Sha J, Huang X, Guerrero L, Xie P et al (2017) YTHDC1 mediates nuclear export of $N^6$-methyladenosine methylated mRNAs. Elife 6: e31311

Rousseau D, Kaspar R, Rosenwald I, Gehrke L, Sonenberg N (1996) Translation initiation of ornithine decarboxylase and nucleocytoplasmic transport of *cyclin D1* mRNA are increased in cells overexpressing eukaryotic initiation factor 4E. Proc Natl Acad Sci USA 93:1065–1070

Ruepp MD, Aringhieri C, Vivarelli S, Cardinale S, Paro S, Schumperli D, Barabino SM (2009) Mammalian pre-mRNA 3′ end processing factor CF $I_m$68 functions in mRNA export. Mol Biol Cell 20:5211–5223

Salamon RS, Backer JM (2013) Phosphatidylinositol-3,4,5-trisphosphate: tool of choice for class I PI 3-kinases. Bioessays 35:602–611

Santos-Rosa H, Moreno H, Simos G, Segref A, Fahrenkrog B, Panté N, Hurt E (1998) Nuclear mRNA export requires complex formation between Mex67p and Mtr2p at the nuclear pores. Mol Cell Biol 18:6826–6838

Saroufim MA, Bensidoun P, Raymond P, Rahman S, Krause MR, Oeffinger M, Zenklusen D (2015) The nuclear basket mediates perinuclear mRNA scanning in budding yeast. J Cell Biol 211:1131–1140

Schaefer M, Pollex T, Hanna K, Lyko F (2009) RNA cytosine methylation analysis by bisulfite sequencing. Nucleic Acids Res 37:e12

Segref A, Sharma K, Doye V, Hellwig A, Huber J, Lührmann R, Hurt E (1997) Mex67p, a novel factor for nuclear mRNA export, binds to both poly(A)$^+$ RNA and nuclear pores. EMBO J 16:3256–3271

Sendinc E, Valle-Garcia D, Dhall A, Chen H, Henriques T, Sheng W, Adelman K, Shi Y (2018) PCIF1 catalyzes m6Am mRNA methylation to regulate gene expression. bioRxiv 484931

Shav-Tal Y, Darzacq X, Shenoy SM, Fusco D, Janicki SM, Spector DL, Singer RH (2004) Dynamics of single mRNPs in nuclei of living cells. Science 304:1797–1800

Shi M, Zhang H, Wu X, He Z, Wang L, Yin S, Tian B, Li G, Cheng H (2017) ALYREF mainly binds to the 5′ and the 3′ regions of the mRNA *in vivo*. Nucleic Acids Res 45:9640–9653

Siebrasse JP, Kaminski T, Kubitscheck U (2012) Nuclear export of single native mRNA molecules observed by light sheet fluorescence microscopy. PNAS 109:9426–9431

Speese SD, Ashley J, Jokhi V, Nunnari J, Barria R, Li Y, Ataman B, Koon A, Chang Y-T, Li Q et al (2012) Nuclear envelope budding enables large ribonucleoprotein particle export during synaptic Wnt signaling. Cell 149:832–846

Sträßer K, Masuda S, Mason P, Pfannstiel J, Oppizzi M, Rodriguez-Navarro S, Rondón AG, Aguilera A, Struhl K, Reed R et al (2002) TREX is a conserved complex coupling transcription with messenger RNA export. Nature 417:304–308

Stutz F, Bachi A, Doerks T, Braun IC, Seraphin B, Wilm M, Bork P, Izaurralde E (2000) REF, an evolutionary conserved family of hnRNP-like proteins, interacts with TAP/Mex67p and participates in mRNA nuclear export. RNA 6:638–650

Suhasini M, Reddy TR (2009) Cellular proteins and HIV-1 Rev function. Curr HIV Res 7:91–100

Sullivan KD, Mullen TE, Marzluff WF, Wagner EJ (2009) Knockdown of SLBP results in nuclear retention of histone mRNA. RNA 15:459–472

Sun H, Zhang M, Li K, Bai D, Yi C (2019) Cap-specific, terminal $N^6$-methylation by a mammalian m6Am methyltransferase. Cell Res 29:80–82

Tamura T, Mancini A, Joos H, Koch A, Hakim C, Dumanski J, Weidner KM, Niemann H (1999) FMIP, a novel Fms-interacting protein, affects granulocyte/macrophage differentiation. Oncogene 18:6488–6495

Taniguchi I, Ohno M (2008) ATP-dependent recruitment of export factor Aly/REF onto intronless mRNAs by RNA helicase UAP56. Mol Cell Biol 28:601–608

Tintaru AM, Hautbergue GM, Hounslow AM, Hung M-L, Lian L-Y, Craven CJ, Wilson SA (2007) Structural and functional analysis of RNA and TAP binding to SF2/ASF. EMBO Rep 8:756–762

Topisirovic I, Capili AD, Borden KLB (2002) Gamma interferon and cadmium treatments modulate eukaryotic initiation factor 4E-dependent mRNA transport of cyclin D1 in a PML-dependent manner. Mol Cell Biol 22:6183–6198

Topisirovic I, Culjkovic B, Cohen N, Perez JM, Skrabanek L, Borden KLB (2003) The proline-rich homeodomain protein, PRH, is a tissue-specific inhibitor of eIF4E-dependent *cyclin D1* mRNA transport and growth. EMBO J 22:689–703

Topisirovic I, Kentsis A, Perez JM, Guzman ML, Jordan CT, Borden KL (2005) Eukaryotic translation initiation factor 4E activity is modulated by HOXA9 at multiple levels. Mol Cell Biol 25:1100–1112

Topisirovic I, Siddiqui N, Lapointe VL, Trost M, Thibault P, Bangeranye C, Piñol-Roma S, Borden KLB (2009) Molecular dissection of the eukaryotic initiation factor 4E (eIF4E) export-competent RNP. EMBO J 28:1087–1098

Trotman JB, Schoenberg DR (2019) A recap of RNA recapping. Wiley Interdiscip Rev RNA 10: e1504

Türeci O, Sahin U, Koslowski M, Buss B, Bell C, Ballweber P, Zwick C, Eberle T, Zuber M, Villena-Heinsen C et al (2002) A novel tumour associated leucine zipper protein targeting to sites of gene transcription and splicing. Oncogene 21:3879–3888

Uhlén M, Fagerberg L, Hallström BM, Lindskog C, Oksvold P, Mardinoglu A, Sivertsson Å, Kampf C, Sjöstedt E, Asplund A et al (2015) Tissue-based map of the human proteome. Science 347:1260419

Umlauf D, Bonnet J, Waharte F, Fournier M, Stierle M, Fischer B, Brino L, Devys D, Tora L (2013) The human TREX-2 complex is stably associated with the nuclear pore basket. J Cell Sci 126:2656–2667

Uranishi H, Zolotukhin AS, Lindtner S, Warming S, Zhang GM, Bear J, Copeland NG, Jenkins NA, Pavlakis GN, Felber BK (2009) The RNA-binding motif protein 15B (RBM15B/OTT3) acts as cofactor of the nuclear export receptor NXF1. J Biol Chem 284:26106–26116

van Dijk EL, Jaszczyszyn Y, Naquin D, Thermes C (2018) The third revolution in sequencing technology. Trends Genet 34:666–681

Van Nostrand EL, Freese P, Pratt GA, Wang X, Wei X, Blue SM, Dominguez D, Cody NAL, Olson S, Sundararaman B et al. (2018) A large-scale binding and functional map of human RNA binding proteins. bioRxiv 179648

Viphakone N, Hautbergue GM, Walsh M, Chang C-T, Holland A, Folco EG, Reed R, Wilson SA (2012) TREX exposes the RNA-binding domain of Nxf1 to enable mRNA export. Nat Commun 3:1006

Viphakone N, Cumberbatch MG, Livingstone MJ, Heath PR, Dickman MJ, Catto JW, Wilson SA (2015) Luzp4 defines a new mRNA export pathway in cancer cells. Nucleic Acids Res 43:2353–2366

Viphakone N, Sudbery IM, Heath C, Sims D, Wilson SA (2018) Co-transcriptional loading of RNA export factors shapes the human transcriptome. bioRxiv 318709

Vohhodina J, Barros EM, Savage AL, Liberante FG, Manti L, Bankhead P, Cosgrove N, Madden AF, Harkin DP, Savage KI (2017) The RNA processing factors THRAP3 and BCLAF1 promote the DNA damage response through selective mRNA splicing and nuclear export. Nucleic Acids Res 45:12816–12833

Volpon L, Culjkovic-Kraljacic B, Sohn HS, Blanchet-Cohen A, Osborne MJ, Borden KLB (2017) A biochemical framework for eIF4E-dependent mRNA export and nuclear recycling of the export machinery. RNA 23:927–937

Wang L, Miao Y-L, Zheng X, Lackford B, Zhou B, Han L, Yao C, Ward JM, Burkholder A, Lipchina I et al (2013) The THO complex regulates pluripotency gene mRNA export and controls embryonic stem cell self-renewal and somatic cell reprogramming. Cell Stem Cell 13:676–690

Wegener M, Müller-McNicoll M (2018) Nuclear retention of mRNAs – quality control, gene regulation and human disease. Semin Cell Dev Biol 79:131–142

Wickramasinghe VO, Laskey RA (2015) Control of mammalian gene expression by selective mRNA export. Nat Rev Mol Cell Biol 16:431

Wickramasinghe VO, McMurtrie PI, Mills AD, Takei Y, Penrhyn-Lowe S, Amagase Y, Main S, Marr J, Stewart M, Laskey RA (2010) mRNA export from mammalian cell nuclei is dependent on GANP. Curr Biol 20:25–31

Wickramasinghe VO, Savill JM, Chavali S, Jonsdottir AB, Rajendra E, Grüner T, Laskey RA, Babu MM, Venkitaraman AR (2013) Human inositol polyphosphate multikinase regulates transcript-selective nuclear mRNA export to preserve genome integrity. Mol Cell 51:737–750

Wickramasinghe VO, Andrews R, Ellis P, Langford C, Gurdon JB, Stewart M, Venkitaraman AR, Laskey RA (2014) Selective nuclear export of specific classes of mRNA from mammalian nuclei is promoted by GANP. Nucleic Acids Res 42:5059–5071

Wiese KE, Nusse R, van Amerongen R (2018) Wnt signalling: conquering complexity. Development 145:dev165902

Will CL, Lührmann R (2011) Spliceosome structure and function. Cold Spring Harb Perspect Biol 3:a003707

Worch R, Niedzwiecka A, Stepinski J, Mazza C, Jankowska-Anyszka M, Darzynkiewicz E, Cusack S, Stolarski R (2005) Specificity of recognition of mRNA 5′ cap by human nuclear cap-binding complex. RNA 11:1355–1363

Wu M, Tong CWS, Yan W, To KKW, Cho WCS (2018) The RNA binding protein HuR: a promising drug target for anticancer therapy. Curr Cancer Drug Targets [Epub ahead of print]

Yamazaki T, Fujiwara N, Yukinaga H, Ebisuya M, Shiki T, Kurihara T, Kioka N, Kambe T, Nagao M, Nishida E et al (2010) The closely related RNA helicases, UAP56 and URH49, preferentially form distinct mRNA export machineries and coordinately regulate mitotic progression. Mol Biol Cell 21:2953–2965

Yang J, Bogerd HP, Wang PJ, Page DC, Cullen BR (2001) Two closely related human nuclear export factors utilize entirely distinct export pathways. Mol Cell 8:397–406

Yang X, Yang Y, Sun B-F, Chen Y-S, Xu J-W, Lai W-Y, Li A, Wang X, Bhattarai DP, Xiao W et al (2017) 5-methylcytosine promotes mRNA export – NSUN2 as the methyltransferase and ALYREF as an $m^5C$ reader. Cell Res 27:606–625

Yoh SM, Cho H, Pickle L, Evans RM, Jones KA (2007) The Spt6 SH2 domain binds Ser2-P RNAPII to direct Iws1-dependent mRNA splicing and export. Genes Dev 21:160–174

Zenklusen D, Vinciguerra P, Strahm Y, Stutz F (2001) The yeast hnRNP-Like proteins Yra1p and Yra2p participate in mRNA export through interaction with Mex67p. Mol Cell Biol 21:4219–4232

Zenklusen D, Vinciguerra P, Wyss JC, Stutz F (2002) Stable mRNP formation and export require cotranscriptional recruitment of the mRNA export factors Yra1p and Sub2p by Hpr1p. Mol Cell Biol 22:8241–8253

Zhang M, Wang Q, Huang Y (2007) Fragile X mental retardation protein FMRP and the RNA export factor NXF2 associate with and destabilize *Nxf1* mRNA in neuronal cells. Proc Natl Acad Sci USA 104:10057–10062

Zheng G, Dahl JA, Niu Y, Fedorcsak P, Huang CM, Li CJ, Vågbø CB, Shi Y, Wang WL, Song SH et al (2013) ALKBH5 is a mammalian RNA demethylase that impacts RNA metabolism and mouse fertility. Mol Cell 49:18–29

Zolotukhin AS, Tan W, Bear J, Smulevitch S, Felber BK (2002) U2AF participates in the binding of TAP (NXF1) to mRNA. J Biol Chem 277:3935–3942

Zolotukhin AS, Uranishi H, Lindtner S, Bear J, Pavlakis GN, Felber BK (2009) Nuclear export factor RBM15 facilitates the access of DBP5 to mRNA. Nucleic Acids Res 37:7151–7162

# Chapter 3
# View from an mRNP: The Roles of SR Proteins in Assembly, Maturation and Turnover

**Marius Wegener and Michaela Müller-McNicoll**

**Abstract** Serine- and arginine-rich proteins (SR proteins) are a family of multitasking RNA-binding proteins (RBPs) that are key determinants of messenger ribonucleoprotein (mRNP) formation, identity and fate. Apart from their essential functions in pre-mRNA splicing, SR proteins display additional pre- and post-splicing activities and connect nuclear and cytoplasmic gene expression machineries. Through changes in their post-translational modifications (PTMs) and their subcellular localization, they provide functional specificity and adjustability to mRNPs. Transcriptome-wide UV crosslinking and immunoprecipitation (CLIP-Seq) studies revealed that individual SR proteins are present in distinct mRNPs and act in specific pairs to regulate different gene expression programmes. Adopting an mRNP-centric viewpoint, we discuss the roles of SR proteins in the assembly, maturation, quality control and turnover of mRNPs and describe the mechanisms by which they integrate external signals, coordinate their multiple tasks and couple subsequent mRNA processing steps.

**Keywords** SR proteins · mRNPs · iCLIP · PTMs · Gene expression · RNA-binding proteins · Splicing

## 3.1 General Introduction

Alternative pre-mRNA processing generates an astonishing variety of mRNAs from a limited pool of genes. Transcripts can differ at their 5′ and 3′ ends as well as throughout the transcript body, allowing for the expression of several distinct protein products per gene. Around 95% of human genes with multiple exons undergo alternative pre-mRNA splicing (AS), and the majority of transcript isoforms are

M. Wegener · M. Müller-McNicoll (✉)
RNA Regulation Group, Cluster of Excellence 'Macromolecular Complexes', Institute of Cell Biology and Neuroscience, Goethe University Frankfurt, Frankfurt/Main, Germany
e-mail: mueller-mcnicoll@bio.uni-frankfurt.de

 
M. Oeffinger, D. Zenklusen (eds.), *The Biology of mRNA: Structure and Function*, Advances in Experimental Medicine and Biology 1203,
https://doi.org/10.1007/978-3-030-31434-7_3

present at different levels in various cells and tissues (Pan et al. 2008). Moreover, at least 70% of genes produce mRNAs with different 3′UTRs by alternative polyadenylation [APA (Tian and Manley 2017)]. Alternative mRNA processing is achieved through the binding and action of numerous RNA-binding proteins (RBPs) that form large messenger ribonucleoprotein (mRNP) particles. From transcription to decay, mRNPs undergo a constant remodelling, whereby RBPs are gained, lost and/or post-translationally modified. The mRNP composition is critical for the fate of bound transcripts and influences gene expression at several steps including transcription, 5′ capping, splicing, polyadenylation, mRNA export, mRNA stability, mRNA localization and translation (Müller-McNicoll and Neugebauer 2013). Moreover, continuous binding of particular RBPs can assist in the coupling of subsequent steps in gene expression and provide checkpoints to ensure mRNP quality. In the past decade, we have made enormous progress and characterized the composition of endogenous mRNPs in different cell types, tissues and animals. A multitude of transcriptome-wide RNA-RBP interaction maps and RNA-bound proteomes have been reported, and recent reviews suggest that more than 1000 proteins expressed in our cells bind to RNA (mouse: 1914; human: 1393) (Hentze et al. 2018; Wheeler et al. 2017). However, we still do not understand how nascent mRNPs assemble in cells, how they mature through different cellular compartments and how they are turned over. This is particularly difficult, as mRNPs involve not only RNA-protein interactions (ranging from low to high affinity and specificity) but also protein-protein interactions (Jankowsky and Harris 2015). The in vivo composition of individual mRNPs and their spatial organization have only recently begun to be unravelled (Adivarahan et al. 2018; Khong and Parker 2018; Metkar et al. 2018), and the functions of individual mRNP components and the influence of PTMs in the expression of bound mRNAs remain enigmatic.

Arginine-serine-rich proteins (SR proteins) are a family of multifunctional RBPs that bind to mRNAs throughout their journey from transcription in the nucleus to translation in the cytoplasm. SR proteins are well known for their essential roles in constitutive splicing and as regulators of alternative splicing. Through additional activities beyond splicing, SR proteins have emerged as key determinants of mRNP formation, fate and identity (Änkö 2014). Individual SR proteins are present in distinct mRNPs and regulate different gene expression programmes in response to external stimuli through changes in their post-translational modifications (PTMs) (Änkö et al. 2010; Bjork et al. 2009). Several transcriptome-wide UV crosslinking and immunoprecipitation (CLIP-Seq) studies that analysed SR protein-RNA interactions in different cell types provided a wealth of information about in vivo consensus binding motifs, binding site distributions and endogenous mRNA targets of individual SR proteins (Änkö et al. 2012; Botti et al. 2017; Bradley et al. 2015; Brugiolo et al. 2017; Krchnakova et al. 2018; Müller-McNicoll et al. 2016; Pandit et al. 2013; Ratnadiwakara et al. 2018; Sanford et al. 2009; Van Nostrand et al. 2016). Based on these findings, we will illuminate the roles of SR proteins in the assembly, maturation, quality control and turnover of mRNPs and discuss the mechanisms by which they may coordinate their different tasks and couple nuclear and cytoplasmic gene expression machineries from an mRNP-centric viewpoint.

## 3.2 The Family of Classical SR Proteins

The SR protein family comprises 12 canonical members, which are evolutionarily conserved and structurally related (Busch and Hertel 2012). They are encoded by separate genes and are named SRSF1 to SRSF12 in the order of their discovery (Manley and Krainer 2010). The first mammalian SR protein was discovered by two different groups in 1990 and was named ASF/SF2 (now SRSF1) (Ge and Manley 1990; Krainer et al. 1990). In 1991, it was reported that a monoclonal antibody (mAb104) recognizes a conserved phospho-epitope on a group of proteins with molecular masses of 20, 30, 40, 55 and 75 kDa (Roth et al. 1991). One year later, these proteins were classified as members of a larger family of splicing factors that included SRSF1 and SC35 (now SRSF2), which was characterized in the same year (Fu and Maniatis 1992). The term SR proteins was coined as all members contained a C-terminal domain (CTD) with repeated Ser (S) and Arg (R) dipeptides [SRp20 (SRSF3), SRp75 (SRSF4), SRp40 (SRSF5) and SRp55 (SRSF6) (Zahler et al. 1992)]. Two years later, the seventh canonical SR protein, which contained an additional CCCH zinc knuckle, was characterized and named 9G8 (SRSF7) (Cavaloc et al. 1994). In 1995, SRp30c (SRSF9) was discovered and described as the family member with the shortest RS domain (Screaton et al. 1995). SRp46 (SRSF8) was discovered in 1998 and is encoded by a functional SRSF2 pseudogene that is not present in the mouse genome and is functionally different from SRSF2 (Soret et al. 1998). SRp38 (SRSF10) and SRrp35 (SRSF12) were discovered in 2001 (Cowper et al. 2001) and found to be very different from typical SR proteins, as they act as general inhibitors of splicing, particularly under changing cellular conditions such as mitosis, heat stress and neural differentiation (Liu and Harland 2005; Shin et al. 2004; Shin and Manley 2002). SRp54 (SRSF11) had already been discovered in 1991 (Chaudhary et al. 1991), but it was functionally characterized only in 1996, and still only little is known about its functions in vivo (Zhang and Wu 1996).

All classical SR proteins contain one or two N-terminal RNA recognition motifs (RRMs), a glycine–arginine-rich spacer region and a C-terminal RS domain of at least 50 amino acids with more than 40% RS dipeptide content (Manley and Krainer 2010) (Fig. 3.1). The RS domain is the defining feature of the SR protein family, mediates mostly protein-protein interactions and serves as nuclear localization signal (NLS) (Caceres et al. 1997). Differences between family members include the length and amino acid composition of their spacer region, which is the binding site for the nuclear mRNA export receptor 1 (NXF1) (Botti et al. 2017; Hargous et al. 2006; Tintaru et al. 2007), the length and composition of their RS domains, the presence of additional functional domains embedded within the RS domain (Ko and Gunderson 2002), the additional zinc knuckle in SRSF7 and a second RRM domain in SRSF1, SRSF4, SRSF5, SRSF6 and SRSF9. This second RRM is also called RRM homolog (RRMH) or pseudo-RRM (ΨRRM), as it reveals an atypical RRM-fold and binds to RNA in a unique manner involving a conserved SWQDLKD heptapeptide (Tintaru et al. 2007). Although RS domains are present in more than 50 splicing factors, most of them are not considered classical SR proteins as they either lack RRMs, contain

| SR protein | Alias | aa | RS | RS content |
|---|---|---|---|---|
| SRSF1 | SF2, ASF,SRp30a | 249 | 198-248 | 75% |
| SRSF2 | SC35, PR264 | 222 | 116-208 | 79% |
| SRSF3 | SRp20 | 164 | 86-164 | 69% |
| SRSF4 | SRp75 | 494 | 179-494 | 59% |
| SRSF5 | SRp40;HRS | 272 | 180-267 | 90% |
| SRSF6 | SRp55, B52 | 345 | 188-344 | 66% |
| SRSF7 | 9G8 | 238 | 121-238 | 76% |
| SRSF8 | SRp46 | 282 | 98-274 | 65% |
| SRSF9 | SRp30c | 221 | 188-200 | 66% |
| SRSF10 | SRp38 | 262 | 106-260 | 53% |
| SRSF11 | SRp54, p54 | 484 | 245-373 | 66% |
| SRSF12 | SRrp35 | 261 | 105-254 | 52% |

**Fig. 3.1** Domain structure of 12 canonical SR proteins. SR proteins contain one or two N-terminal RNA recognition motifs (RRM and ΨRRM) separated by a glycine/arginine-rich spacer (S) region as well as a C-terminal arginine–serine-rich (RS) domain of at least 50 amino acids with more than 40% RS content. Here, these domains were drawn to scale for all 12 canonical SR protein family members. For historical reasons, various names have been used in the literature (*Alias* column), but they have recently been renamed SRSF1 to SRSF12. Additionally shown here are the length in amino acids of each SR protein (*aa*), the position (*RS*) and percentage of arginines and serines (*RS content*) of their RS domains. See text for more details and references

additional domains, contain RRM and RS domains in the reverse order or are inactive in splicing (Manley and Krainer 2010).

While SR proteins act redundantly in constitutive splicing, they exhibit functional specificity in their regulation of alternative exons and in post-splicing steps of nuclear and cytoplasmic gene expression. SRSF1 and SRSF2, the two best-studied members of the SR protein family, have been implicated in the regulation of a multitude of cellular processes including transcription (Lin et al. 2008; Paz et al. 2014), mRNA stability (Lemaire et al. 2002), microRNA processing (Ratnadiwakara et al. 2017), nuclear speckle architecture (Tripathi et al. 2012), selective nuclear mRNA export and retention (Hautbergue et al. 2017; Müller-McNicoll et al. 2016; Zhou et al. 2017), nonsense-mediated decay (NMD) (Sato et al. 2008; Zhang and Krainer 2004), mRNA translation (Michlewski et al. 2008; Sanford et al. 2004), protein degradation (Pelisch et al. 2010) and maintenance of genomic stability (Li and Manley 2005), essentially influencing the fate of specific mRNAs from transcription in the nucleus to translation in the cytoplasm. Perhaps owing to such non-redundant functions, all knockout (KO) mouse models generated to date died early during embryonic development [reviewed in Änkö (2014)].

## 3.3 Regulation of SR Protein Activity

The activities of SR proteins are regulated via post-translational modifications (PTMs), providing the possibility to rapidly integrate signals from cellular pathways to coordinate gene expression (Table 3.1). For example, SR proteins display distinct phosphorylation states, which is caused by the reversible phosphorylation and dephosphorylation of Ser residues within their RS domains through the interplay of various protein kinases and phosphatases (Zhou and Fu 2013). Best studied are the CDC2-like kinases (CLKs) such as CLK1/4 and SRPK1/2. Both types of kinases differ in their subcellular localization, substrate specificity and mechanism of phosphorylation. For SRSF1, it was shown that SRPKs phosphorylate Ser-Arg dipeptides within the N-terminal part of its RS domain in a processive manner. In contrast, CLKs phosphorylate Ser-Lys, Ser-Pro and Ser-Arg dipeptides in the C-terminal part of the RS domain in a distributive fashion (Ghosh and Adams 2011; Zhou and Fu 2013). The phosphorylation state of SR proteins influences their interaction with components of the pre-spliceosome (Zhou and Fu 2013), their RNA-binding affinities and specificities (Tacke et al. 1997), their subnuclear distribution (Aubol et al. 2016; Misteli and Spector 1996), their mRNA export (Botti et al. 2017; Cazalla et al. 2002), as well as their nuclear re-import (Lai et al. 2001) (Fig. 3.2). SRPK1/2 are retained in the cytoplasm by molecular chaperones, but under certain conditions they translocate to the nucleus and regulate SR protein activities, for example, during osmotic stress, epidermal growth factor (EGF) signalling, mammalian target of rapamycin complex 1 (mTORC1) signalling or during G2/M phase of the cell cycle (Gui et al. 1994; Lee et al. 2017; Zhou et al. 2012). Nuclear CLK1/4 kinases are also activated by osmotic stress and heat shock and are responsible for the re-phosphorylation of SR proteins during stress recovery (Ninomiya et al. 2011). SR protein kinases and phosphatases act together to integrate external signals and ensure appropriate phosphorylation levels of SR proteins (Aubol et al. 2016).

Other kinases that regulate SR protein activities and alternative splicing include the Ser/Thr kinase AKT2, which was shown to phosphorylate SRSF5 on $Ser^{86}$ in response to insulin as well as SRSF1 and SRSF7 in response to growth factor signalling (Blaustein et al. 2005; Patel et al. 2005). The cAMP-dependent protein kinase A (PKA) phosphorylates SRSF1 and SRSF7 in response to cAMP and modulates inclusion of exon 10 of the *TAU* mRNA (Shi et al. 2011). DNA topoisomerase I phosphorylates SRSF1 within its RS domain and ensures the coordination between transcription and splicing (Malanga et al. 2008; Soret et al. 2003). Phosphorylation of SRSF1 and SRSF7 by the dual-specificity tyrosine phosphorylation-regulated kinase 1A (DYRK1A) induces their cytoplasmic translocation, whereas phosphorylation of SRSF2 and SRSF6 causes their release from nuclear speckles (Naro and Sette 2013).

SR protein activities are also regulated by other PTMs, including phosphorylation of Pro residues within the RS domain of SRSF1 by CLK1, which enhances its splicing-promoting activity (Keshwani et al. 2015) (Table 3.1). Di-methylation of Arg residues within the glycine-rich spacer region improves interaction between SRSF5 and NXF1 and influences the subcellular localization of SRSF1 and SRSF9

**Table 3.1** Post-translational modifications of SR proteins, enzymes and functions

| Protein | Modified residues | Modification | Enzymes localization | Function | References |
|---|---|---|---|---|---|
| SRSF1–SRSF7 | RS dipeptides N-terminal | Phosphorylation | SRPK1/2—cytoplasm | Nuclear re-import by transportin 2, localization to nuclear speckles | Lai et al. 2001 |
| SRSF1–SRSF7 | RS dipeptides C-terminal | Phosphorylation | CLKs—nuclear | Release from nuclear speckles, recruitment to pre-mRNA, splicing | Ghosh and Adams 2011 |
| SRSF1 | SP dipeptides $S^{227}$, $S^{237}$, $S^{238}$ | Phosphorylation | CLKs—nuclear | Enhances splicing activity | Keshwani et al. 2015 |
| SRSF1 | RS dipeptides | Phosphorylation | Topoisomerase I | Changes in AS | Soret et al. 2003; Malanga et al. 2008 |
| SRSF1 | $Arg^{93}$, $Arg^{97}$, $Arg^{109}$ | Di-methylation | PRMT1 | Influences subcellular localization, enhances NMD, changes in AS | Sinha et al. 2010; Bressan et al. 2009 |
| SRSF1 and SRSF7 | RS dipeptides | Phosphorylation | AKT2 | Changes in AS in response to growth factor signalling | Blaustein et al. 2005 |
| SRSF1 | RS dipeptides | Phosphorylation | PKA | Changes in AS in response to cAMP, e.g. Tau exon10 splicing | Shi et al. 2011 |
| SRSF1 and SRSF7 | RS dipeptides | Phosphorylation | DYRK1A | Cytoplasmic translocation | Naro and Sette 2013 |
| SRSF2 and SRSF6 | RS dipeptides | Phosphorylation | DYRK1A | Release from nuclear speckles. Changes in AS | Naro and Sette 2013 |
| SRSF2 | $Arg^{31}$, $Arg^{33}$, $Arg^{47}$, $Arg^{55}$, $Arg^{66}$, $Arg^{94}$, $Arg^{109}$, $Arg^{117}$ | Mono-methylation | PRMT5, CARM1 | Enhances RNA-binding capacity, changes localization to NS | Larsen et al. 2016 |
| SRSF2 | $Lys^{52}$ | Acetylation | Tip60 | Decreases protein levels in response to cisplatin. Changes in AS | Yin et al. 2018 |

(continued)

**Table 3.1** (continued)

| Protein | Modified residues | Modification | Enzymes localization | Function | References |
|---|---|---|---|---|---|
| SRSF2 | $Pro^{6}$, $Pro^{7}$ | Hydroxylation | Prolyl hydroxylases | Decreases protein stability in hypoxia. Changes in AS | Stoehr et al. 2016 |
| SRSF3 | $Lys^{85}$ | Neddylation | NEDD8-conjugating system | Response to arsenite stress, assembly of SGs | Jayabalan et al. 2016 |
| SRSF5 | $Ser^{86}$ | Phosphorylation | Akt | Alternative splicing in response to insulin | Patel et al. 2005 |
| SRSF5 | $Arg^{88}$, $Arg^{92}$, $Arg^{93}$ | Di-methylation | PRMT5 | Improves NXF1 interaction and nucleo-cytoplasmic shuttling | Botti et al. 2017; Larsen et al. 2016 |
| SRSF5 | $Lys^{125}$ | Acetylation | Tip60 | Changes in AS, promotes tumour growth | Chen et al. 2018 |
| SRSF5 | $Lys^{125}$ | Ubiquitination | Smurf1 | Protein degradation upon glucose starvation | Chen et al. 2018 |
| SRSF7 | RS dipeptides | Phosphorylation | PKA | Changes Tau exon10 splicing upon cAMP | Shi et al. 2011 |
| SRSF9 | ND | Di-methylation | PRMT1 | Influences subcellular localization | Bressan et al. 2009 |

(Botti et al. 2017; Bressan et al. 2009; Sinha et al. 2010). Mono-methylation of Arg residues within the RRM of SRSF2 enhances its RNA-binding capacity (Larsen et al. 2016). Acetylation of $Lys^{52}$ decreases SRSF2 protein levels, whereas deacetylation, either in response to genotoxic stress or through the activity of the deacetylase SIRT1, stabilizes SRSF2, whose higher levels cause a switch in *CASP8* pre-mRNA splicing and programmed cell death and promote inclusion of exon 10 in the *TAU* mRNA in frontotemporal dementia (FTD) (Edmond et al. 2011; Yin et al. 2018). Finally, PTMs with generally poorly characterized functions were also shown to affect SR protein activity, such as hydroxylation of two Pro residues within the RRM of SRSF2 [which also decreases its stability (Stoehr et al. 2016)]. SRSF5 is acetylated on $Lys^{125}$, which antagonizes ubiquitination at the same residue and this way inhibits the degradation of SRSF5 and promotes tumour growth (Chen et al. 2018). SRSF3 is neddylated on $Lys^{85}$ in response to arsenite stress, which appears to be crucial for the assembly of cytoplasmic stress granules (SGs) (Jayabalan et al. 2016).

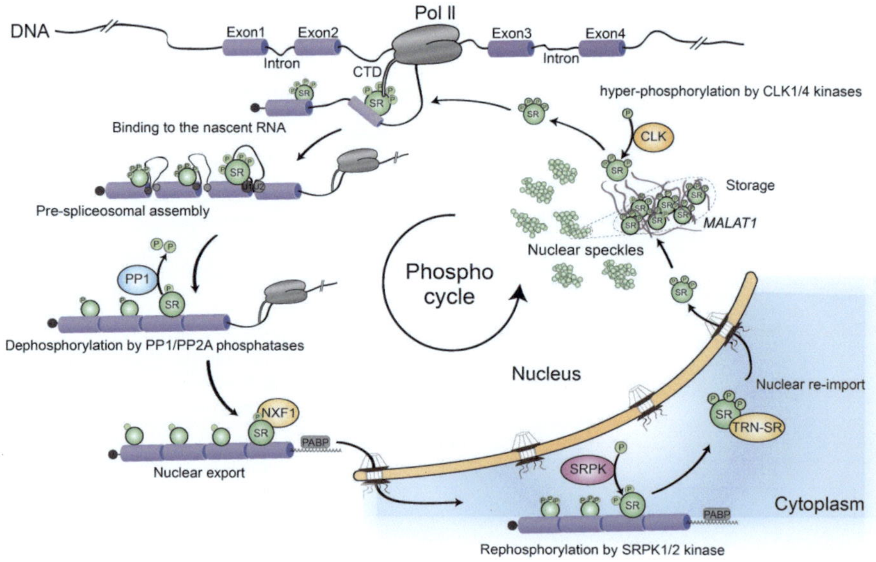


**Fig. 3.2** Phosphorylation and splicing cycle of SR proteins. Upon hyper-phosphorylation by CLK1/4, SR proteins are released from nuclear speckles and co-transcriptionally bind exons in pre-mRNAs, where they recruit spliceosomal components U1 and U2 snRNPs to 5′ and 3′ splice sites. During the splicing reaction, SR proteins are partially dephosphorylated by PP1 and PP2A, which enables them to act as export adaptor through recruitment of export factor NXF1. Most SR proteins escort their bound mRNAs to the cytoplasm, where they can regulate mRNA localization, stability and translation before being released, partly re-phosphorylated by SRPK1/2 and re-imported by TRN-SR into the nucleus, where they are stored again in nuclear speckles, awaiting the next round of splicing. See text for more details and references

## 3.4 SR Protein Recruitment to Pre-mRNAs

In the nucleus, phosphorylated SR proteins accumulate in nuclear speckles, highly dynamic membrane-less compartments where pre-mRNA splicing factors are stored, assembled and modified (Galganski et al. 2017). Upon activation of transcription, SR proteins are released from speckles and recruited to polymerase II (Pol II) transcription sites, which requires hyper-phosphorylation of the RS domain on Ser-Arg (SR) and Pro-Arg (PR) dipeptides by CLK1/4 kinases (Colwill et al. 1996; Keshwani et al. 2015; Misteli et al. 1998; Zhou and Fu 2013) (Fig. 3.2). RS domain hyper-phosphorylation may break low-affinity interactions between SR proteins and other RS domain-containing proteins in nuclear speckles, allowing SR proteins to diffuse to the edges, so-called perichromatin fibrils, where transcription and co-transcriptional splicing are thought to take place (Sanchez-Hernandez et al. 2017). Thus, recruitment of SR proteins to nascent pre-mRNAs is likely favoured by their high concentration in speckles and the short distance to transcription sites.

Each SR protein binds to single-stranded RNA in a sequence-specific manner via its RRMs. A typical RRM consists of four anti-parallel β-sheets connected to two α-helices. The amino acids present in the β-sheets are thought to be non-selective (Maris et al. 2005). Sequence-specific binding is achieved because other domains—e.g. the ΨRRMs, the zinc knuckle, the glycine-rich linker and the phosphorylated RS domain—all contribute to sequence specificity and regulate binding affinity (Botti et al. 2017; Cavaloc et al. 1999; Cho et al. 2011a; Phelan et al. 2012; Tacke et al. 1997) (see Fig. 3.1). For example, SRSF2 has only one RRM, which does not bind efficiently to RNA, but it contains an extended glycine-rich linker that is absent in all other SR proteins and folds into an additional L3 loop (Phelan et al. 2012). This loop is crucial for selective binding to AG, while the C-terminus of the RRM contacts U residues (Phelan et al. 2012). SRSF7 contains a zinc knuckle that contributes to the specific recognition of GAYGAY motifs (Cavaloc et al. 1999; Müller-McNicoll et al. 2016). SRSF5 has two RRMs, and both contribute to the recognition of specific enhancer sequences in vitro and in vivo, but only when its RS domain is phosphorylated (Botti et al. 2017; Tacke et al. 1997). For SRSF1, it was shown that both RRMs and the intervening glycine-rich linker together mediate the cooperative binding to exonic splicing enhancers (ESEs). The linker brings both RRMs in close proximity to optimally accommodate target RNAs (Cho et al. 2011a).

High-affinity in vitro binding motifs have been identified for all 12 classical SR proteins by systematic evolution of ligands by exponential enrichment [SELEX; reviewed in Änkö (2014)]. Structural studies using recombinant SR proteins and short synthetic RNAs with high-affinity binding motifs have provided valuable mechanistic insights into specific target recognition (Clery et al. 2013; Daubner et al. 2012; Hargous et al. 2006; Phelan et al. 2012; Tintaru et al. 2007). However, high-affinity binding is not preferable for dynamic processes such as pre-mRNA splicing, where binding and release of SR proteins must be flexible. Moreover, in vivo binding sites depend on the local RNA structure (i.e. might be inaccessible in double-stranded RNA), on the cooperation and competition with other RBPs and on PTMs that may influence binding specificity of SR proteins (Edmond et al. 2011; Larsen et al. 2016; Tacke et al. 1997; Yin et al. 2018). Thus, the relevance of SELEX for the prediction of in vivo binding sites is limited. More recently, several transcriptome-wide CLIP-Seq studies have identified in vivo binding sites of most SR proteins in different cell types (Änkö et al. 2012; Botti et al. 2017; Bradley et al. 2015; Brugiolo et al. 2017; Müller-McNicoll et al. 2016; Pandit et al. 2013; Sanford et al. 2009) (Fig. 3.3). These data partially confirmed the SELEX motifs, but also revealed interesting differences. For example, most SR proteins bind to a broad spectrum of sequences with loose consensus, but their binding motifs differ depending on the bound transcript class, transcript region and between constitutive and alternative exons.

The in vivo binding motifs of SRSF1, SRSF4 and SRSF6 are purine rich, suggesting that the RRM, which binds preferentially pyrimidine-rich sequences, does not contribute to the binding specificity in these proteins (Änkö et al. 2012; Müller-McNicoll et al. 2016; Pandit et al. 2013; Sanford et al. 2009). Indeed, the ΨRRM of SRSF1 binds efficiently to GGA and is sufficient to regulate splicing of many target transcripts (Clery et al. 2013). This finding contrasts with an earlier report showing that both RRMs and the linker of SRSF1 are all essential for ESE

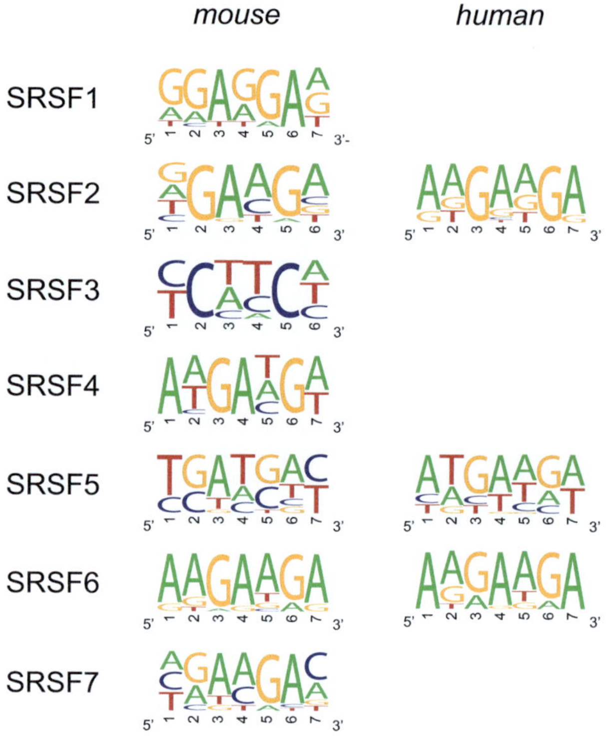


**Fig. 3.3** In vivo binding motif of seven SR proteins (mouse and human) derived from individual-nucleotide resolution crosslinking and immunoprecipitation studies. See text for more details and references

binding (Cho et al. 2011a). It is possible that the RRM and the linker only contribute to binding when high specificity is needed. Indeed, access of its RRM is regulated through phosphorylation and intra-protein interactions with the RS domain of SRSF1 (Aubol et al. 2017; Cho et al. 2011b) (Fig. 3.3).

SRSF5 displays a more complex binding motif and has fewer targets than other family members, suggesting that both its RRMs contribute to RNA binding (Botti et al. 2017; Müller-McNicoll et al. 2016). SRSF2 has the most divergent binding motif of all SR proteins (SSNG), recognizing purines and pyrimidines almost equally well (Daubner et al. 2012; Pandit et al. 2013). Such promiscuous binding might be advantageous for constitutive splicing, but seems counterproductive for the regulation of AS. Interestingly, SRSF2 binds very often in close proximity to other SR proteins and co-regulates bound exons or competes for binding—for example with SRSF5 (Botti et al. 2017), SRSF7 (Preussner et al. 2017), SRSF1 (Pandit et al. 2013) and SRSF6 (Chandradas et al. 2010). Given that its RRM accommodates a vast amount of different sequences (Daubner et al. 2012; Pandit et al. 2013), SRSF2 binding might be dictated through cooperative interactions with other SR proteins (Fu and Ares 2014). It is also conceivable that SRSF2 binds constitutively and early during transcription to all pre-mRNAs to distinguish them from other Pol II-derived RNPs or regulate their downstream fate. In line with such a function, the activity of SRSF2 is very adjustable and is regulated under various stresses and changing cellular conditions through PTMs, including Ser and Pro phosphorylation, Lys acetylation, Pro hydroxylation and Arg methylation (Botti et al. 2017; Edmond et al. 2011; Larsen et al. 2016; Preussner et al. 2017; Stoehr et al. 2016). Through changes in PTMs, SRSF2 was shown to influence transcription efficiency, AS and nucleo-cytoplasmic shuttling of mRNPs.

In addition to PTMs on proteins, modifications of the pre-mRNA itself may also modulate SR protein binding. For example, the nuclear reader protein YTHDC1 recognizes the modification of $N^6$-methyladenosine ($m^6A$) in RNAs and recruits SRSF3 to these sites, while blocking the binding of SRSF10. In this way, YTHDC1 and $m^6A$ modulate the access of SR proteins to their cognate binding sites in target mRNAs (Xiao et al. 2016). Moreover, iron ions reduce the RNA-binding capacity of SRSF7, likely by replacing zinc in the Zn knuckle (Tejedor et al. 2015). Finally, SR protein binding to pre-mRNAs is also regulated through the long non-coding RNA *MALAT1*, which resides in nuclear speckles and sequesters phosphorylated SR proteins through non-specific interactions in these compartments (Tripathi et al. 2010). Hyper-phosphorylation of SR proteins may break these low-affinity interactions and enable SR proteins to escape from nuclear speckles.

## 3.5 Co-transcriptional Splicing

SR proteins bind to pre-mRNAs as soon as the first splice sites emerge from Pol II and engage in co-transcriptional splicing (Bjork et al. 2009; Mabon and Misteli 2005; Sapra et al. 2009). This rapid recruitment to nascent RNA suggests that SR proteins piggyback on transcribing Pol II, although it is not clear whether their recruitment occurs before or after initiation of transcription. There is evidence for both scenarios. In agreement with the former, SR proteins co-localize with Pol II in nuclear speckles, and this interaction is mediated by the C-terminal domain (CTD) of Pol II (Kim et al. 1997; Yuryev et al. 1996). Truncation of the CTD or selective mutations prevent their targeting to transcription sites and inhibit pre-mRNA splicing (de la Mata and Kornblihtt 2006; Du and Warren 1997; Misteli and Spector 1999), suggesting that the CTD pre-assembles with SR proteins in nuclear speckles in the absence of pre-mRNA. In contrast, chromatin immunoprecipitation (ChIP) experiments demonstrated that the interaction between most SR proteins and Pol II was dependent on RNA and ongoing transcription (Sapra et al. 2009).

Recruitment of SR proteins to Pol II affects its elongation rate as depletion of SRSF1 and SRSF2 leads to a dramatic decrease in the production of nascent RNA. Particularly, SRSF2 was shown to promote Pol II elongation in a subset of genes that contain SRSF2-binding sites close to the transcription start site (Ji et al. 2013; Lin et al. 2008; Mo et al. 2013). SRSF2 binds to the non-coding RNA 7SK, which is part of stalled Pol II complexes that pause near the promoter of genes. When SRSF2-binding sites emerge, SRSF2 is transferred from 7SK onto the nascent pre-mRNA, which triggers a coordinated release of the transcriptional regulator positive transcription elongation factor, subsequent phosphorylation of Pol II and its release from promoter-proximal pausing (Ji et al. 2013). This example highlights a close coupling of transcription and splicing through SR proteins, which may allow selective transcription and splicing of specific pre-mRNAs. Binding of SRSF2 labels the transcribed Pol II product as a specific pre-mRNA and feeds the information back to the transcription machinery. This may enhance the elongation rate of transcribing Pol II and at the same time alter the recognition of alternative exons in this pre-mRNA (de la Mata et al. 2003). The physical coupling also enables a rapid recruitment of SRSF2 to the first splice site for efficient co-transcriptional spliceosome assembly. At the same time, enzymes associated with chromatin might modify the SR proteins, either to change their binding/splicing activities in downstream exons or to integrate external signals (Hnilicova et al. 2011).

Removal of non-coding introns and the ligation of coding exons are catalysed by the spliceosome, which consists of five small nuclear ribonucleoproteins (snRNPs) called U1, U2, U4, U5 and U6, and of approximately 200 associated proteins (Scheres and Nagai 2017). The spliceosome assembles de novo onto each intron. SR proteins contribute to the recognition of splice sites (ss) during both constitutive and alternative splicing through the binding of ESEs or intronic splicing enhancer (ISE) sequences. ESEs and ISEs are short 4-8-nt-long degenerate sequences that are located in close proximity to splice sites. Several CLIP studies revealed that SR proteins bind similarly to $5'$ and $3'$ss in a distance of ~60 nt (Änkö et al. 2012; Bradley et al. 2015; Müller-McNicoll et al. 2016; Pandit et al. 2013). Binding to ESEs activates neighbouring splice sites and facilitates the recognition of exon-intron boundaries. Interactions between different SR proteins via their RS domains bring $5'$ and $3'$ss in close proximity while intronic or exonic sequences loop out. In this way, SR proteins participate in intron and exon definition (De Conti et al. 2013). During spliceosome assembly, SR proteins stabilize the binding of the U1 snRNP component U170K at the $5'$ and of U2AF35 at the $3'$ splice site through RS domain interactions. The RS domain also contacts RNA directly at the splice sites and enhances binding of improperly paired U2 snRNA ($3'$ss) and U6 snRNA ($5'$ss) (Shen and Green 2006). Indeed, minor binding peaks of SR proteins were observed in these regions in several CLIP studies, but it is unclear whether RS domains are responsible (Müller-McNicoll et al. 2016; Pandit et al. 2013).

In nuclear speckles, SRSF1 is prevented from binding to RNAs through intra-molecular contacts between its N-terminally phosphorylated RS domain and its RRM (Cho et al. 2011b; Serrano et al. 2016) (Fig. 3.4). Hyper-phosphorylation of the C-terminal RS domain by CLK1 was shown to cause a switch from a floppy disordered state to a more ordered and rigid arc-like structure (Xiang et al. 2013). This likely induces a conformational change in the SR protein, breaking the interactions between RRMs and RS domain and making both domains become available for interactions (Cho et al. 2011b; Serrano et al. 2016). The RRMs bind to ESEs or interact with other RRM-containing proteins, while the RS domain contacts U1 and U2 snRNP components (Zhou and Fu 2013). Assembly of early and active spliceosomes only occurs when SR proteins are hyper-phosphorylated (Keshwani et al. 2015).

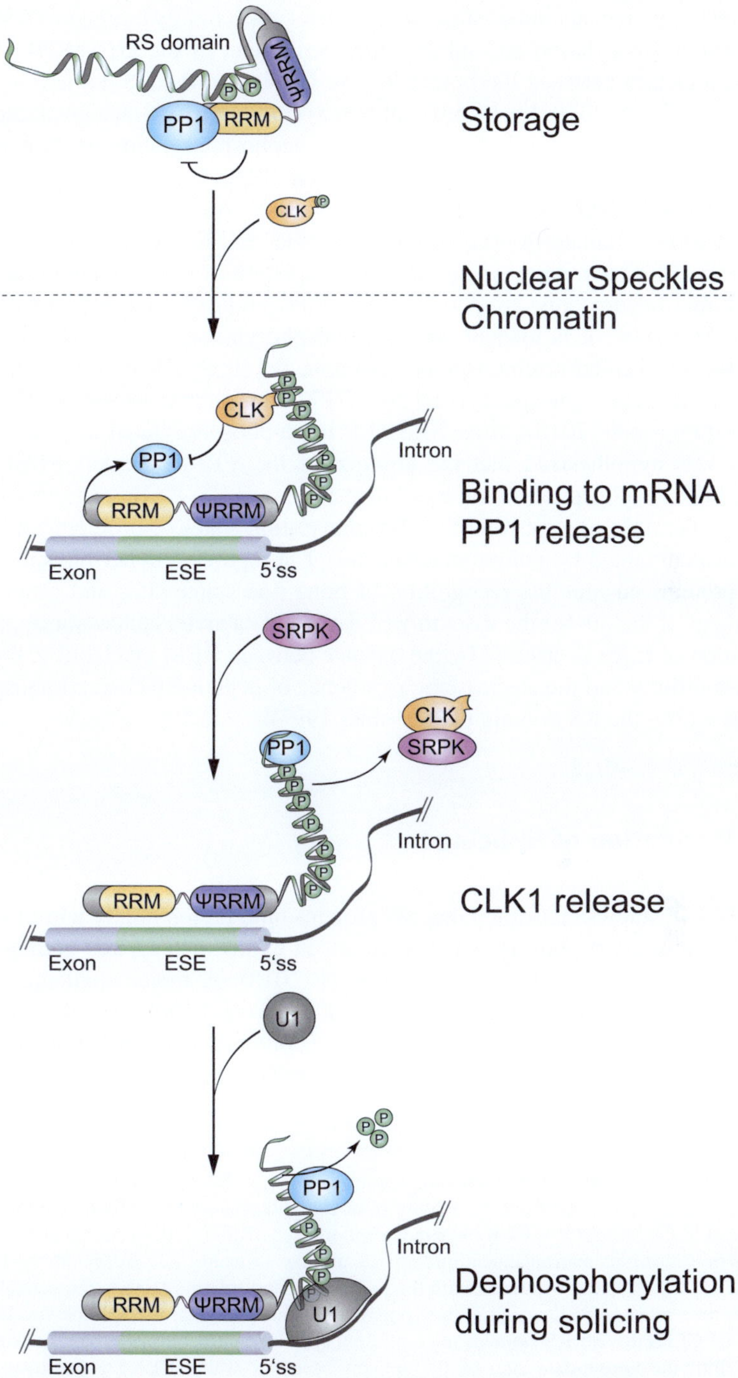


**Fig. 3.4** Storage of SRSF1 in nuclear speckles and its recruitment to pre-mRNAs are regulated through an interplay between kinases and phosphatases. SRSF1 is stored in nuclear speckles in a partly phosphorylated state where its floppy, intrinsically disordered RS domain is free to interact

Interestingly, recent studies suggest that CLK1 stays tightly bound to SRSF1 after its hyper-phosphorylation and inhibits the recruitment of the U1 snRNP due to a strong interaction between the N-terminus of CLK1 and the RS domain of SRSF1 (Aubol et al. 2014, 2016; Keshwani et al. 2015) (Fig. 3.4). This sustained interaction might protect the RS domain from premature dephosphorylation by protein phosphatase 1 (PP1) prior to splicing. Only when SRPK1 enters the complex and removes CLK1, SRSF1 is able to recruit U1 snRNP (Aubol et al. 2016, 2017) (see section below). It remains to be determined whether SRPK1, which is predominantly cytoplasmic at steady state, is required for completion of most splicing reactions or whether this occurs only at specific splice sites and/or under particular circumstances. Moreover, it is unclear whether phosphorylation of other SR proteins is controlled by a similar mechanism. For example, the single RRM-containing SRSF3 appears to be rather phosphorylated by SRPK2 throughout its entire RS domain in vitro (Long et al. 2018). Since SRSF3 is hypo-phosphorylated in cells at steady state, it was hypothesized that the absence of the ΨRRM renders SRSF3 more susceptible to dephosphorylation by phosphatases.

Altogether, this suggests that the phosphorylation states of different SR proteins might be maintained by unique mechanisms. The regulated hyper-phosphorylation of SR proteins ensures the recognition of bona fide splice sites and promotes the recruitment of factors for the assembly of early and catalytic spliceosomes. Specific recognition of ESEs is ensured by the balance between RRM availability, their high sequence affinity and the electrostatic repellence of negatively charged phospho-Ser residues within the RS domain (Tacke et al. 1997).

## 3.6 Regulation of Splicing

Several CLIP studies confirmed that SR proteins bind preferentially within exons of protein-coding genes, but are notably absent in exons of long non-coding RNAs, which are poorly spliced (Krchnakova et al. 2018). A major challenge remains to understand how different combinations of SR proteins influence the splicing of specific exons. Whereas constitutive exons are bound by many SR proteins, unique binding is prevalent in alternative exons, untranslated regions (UTRs) and introns

---

**Fig. 3.4** (continued) with its RRM domain, thereby masking its RNA-binding site. In this state, the RS domain is also protected from the activity of bound phosphatase PP1. Upon co-transcriptional recruitment to the pre-mRNA, CLK1 hyper-phosphorylates SRSF1 in the C-terminal portion of its RS domain, causing its rearrangement into a rigid, arc-like structure. This frees both the RRM and RS domains, which can now interact with the pre-mRNA and splicing factors, respectively, and at the same time releases PP1, which is now poised to dephosphorylate the RS domain. Prolonged binding of CLK1 to the RS domain after phosphorylation might further protect it from PP1's activity until the appropriate step of the splicing reaction, where CLK1 is removed through recruitment and binding of additional factors such as SRPK1. See text for more details and references

(Änkö et al. 2012; Bradley et al. 2015; Müller-McNicoll et al. 2016; Pandit et al. 2013). Most exons are bound by at least one SR protein, but a significant level of binding site co-occupancy by particular SR proteins is indicative of competition or cooperation. For example, SRSF5 and SRSF2 often co-bind within the same exons and can be co-immunoprecipitated within the same mRNPs, indicating that they likely coordinate their functions (Botti et al. 2017). SRSF1 and SRSF2 appear to have a very complex relationship, either cooperating or competing for binding sites depending on the context (Pandit et al. 2013). In contrast, SRSF3 and SRSF4 share only few common targets (Änkö et al. 2012).

In general, SR proteins act as splicing activators. Binding to alternative exons with weak splice sites often leads to their inclusion in the mature mRNA, and there is an inverse correlation between the extent of SR protein binding and the strength of the splice site and the extent of exon skipping after depletion (Bradley et al. 2015). However, several transcriptome-wide studies have reported that depletion of individual SR proteins leads to similar proportions of exon inclusion and skipping events (Bradley et al. 2015; Müller-McNicoll et al. 2016; Pandit et al. 2013). Indeed, the splicing outcome for a particular exon depends on the position and sequence context of SR protein binding. When SR proteins bind to exons that flank an alternative exon, they repress its inclusion (Han et al. 2011). Similarly, SR proteins binding to introns repress splicing of the flanking exons (Erkelenz et al. 2013; Simard and Chabot 2002). Contributing to the complexity of AS regulation, individual SR proteins activate splice sites differently due to differences in their affinity for ESEs and cooperation with each other. Moreover, some atypical SR proteins, e.g. SRSF10 and SRSF12, act as inducible global splicing repressors (Shin and Manley 2002; Simard and Chabot 2002). SRSF10 is converted into a potent splicing repressor through dephosphorylation by PP1, which occurs during heat stress or M phase of the cell cycle, leading to an inhibition of the splicing of bound transcripts (Shin et al. 2004; Shin and Manley 2002). Finally, the activities of individual SR proteins can be antagonized by heterogeneous nuclear ribonucleoprotein (hnRNP) family proteins that bind to exonic splicing silencer (ESS) sequences within the same exons (Caceres and Kornblihtt 2002).

This illustrates the complex network of AS regulation, where SR proteins can have opposite effects on the splicing of a particular exon depending on the position and context of their binding sites, their phosphorylation state and the presence of synergizing or antagonizing factors. The integration of genome-wide splicing data and high-resolution binding maps as well as high-resolution structures of full-length SR proteins will help to better understand the prevalent modes and mechanisms of splicing regulation as well as the roles of PTMs (Rot et al. 2017).

## 3.7 mRNP Maturation and Remodelling

SR protein-containing mRNPs are extensively remodelled throughout the different stages of the splicing reaction. These rearrangements require adenosine triphosphate (ATP), suggesting the involvement of ATP-dependent helicases (Wongpalee et al. 2016). Once SR proteins bind to an exon, they recruit the early spliceosomal U1 and U2 snRNPs, which bind to 5′ss and 3′ss, respectively. Interactions via their RS domains bridge U1 and U2 snRNP proteins to form the exon definition complex (EDC). Further recruitment of SR proteins may also replace inhibitory hnRNP proteins (Fu and Ares 2014; Wongpalee et al. 2016).

During formation of the active spliceosome, SR proteins are dephosphorylated by the protein phosphatase PP1 and PP2A (Mermoud et al. 1994). Dephosphorylation is crucial for splicing catalysis, for the release of the splicing machinery from the mRNP and to gain nuclear export competency (Naro and Sette 2013). Although the exact mechanistic details of SR protein dephosphorylation are unknown, recent data suggest that prior to SRSF1 binding to the pre-mRNA, PP1 is already bound to SRSF1 (to a short RVxF motif within the RRM), but allosterically inhibited and unable to dephosphorylate SRSF1. Pre-mRNA binding by SRSF1 releases PP1 from the RRM, which is now free to dephosphorylate the RS domain of SRSF1 during the splicing reaction. Binding of CLK to the RS domain additionally protects it from PP1's activity until CLK is released by additional factors such as SRPK1 (Aubol et al. 2017; Ma et al. 2010) (Fig. 3.4). Clearly, the splicing activity of SR proteins is precisely controlled through the interplay between kinases and phosphatases.

After splicing, dephosphorylated SR proteins can follow one of two paths: one subset is removed from mature mRNPs and re-phosphorylated by CLK1 for further rounds of pre-mRNA splicing (Lin et al. 2005). In line with this, a recent CLIP-Seq study showed that many interactions of SR proteins with (pre)-mRNAs are compartment-specific (Brugiolo et al. 2017). The other subset remains associated with spliced mRNAs and regulates downstream steps of gene expression (Botti et al. 2017; Lai and Tarn 2004; Sanford et al. 2005). Indeed, all SR proteins tested to date have shown substantial crosslinking to spliced exon-exon junctions, indicating that a large fraction of SR proteins remain bound to mature mRNAs after splicing is completed (Botti et al. 2017; Müller-McNicoll et al. 2016; Pandit et al. 2013). Stable binding of SR proteins to spliced mRNPs may be assisted by exon-junction complexes (EJC), which are deposited ~20-24 nucleotides upstream of each exon–exon junction during splicing (Boehm and Gehring 2016). Depletion of the EJC subunit eIF4AIII reduced the RNA-binding activity of SRSF1 and SRSF3 (Singh et al. 2012). EJC proteins form higher-order complexes with hypo-phosphorylated SR proteins via their RS domains, which assist in the packaging and compaction of mRNPs for nuclear export (Singh et al. 2012). Consistent with this, CLIP-Seq of EJC RNA footprints revealed many additional EJC-binding sites that overlap with SR protein-binding sites (Sauliere et al. 2012).

## 3.8 3′End Processing and Alternative Polyadenylation

In addition to exons and introns, SR proteins also bind substantially to 3′UTRs (Änkö et al. 2012; Bradley et al. 2015; Müller-McNicoll et al. 2016; Pandit et al. 2013; Sanford et al. 2008). Depletion of individual SR proteins causes changes in the length and identity of 3′UTRs of specific target mRNAs, suggesting that some SR proteins regulate alternative polyadenylation (APA) (Bradley et al. 2015; Müller-McNicoll et al. 2016). For example, SRSF3 and SRSF7 often bind at a defined distance to proximal poly(A) sites and may regulate their usage in an opposite manner, as depletion of SRSF3 leads to an accumulation of transcripts with shorter 3′UTRs and depletion of SRSF7 causes 3′UTR extension (Müller-McNicoll et al. 2016). SR proteins could affect the length of 3′UTRs by various mechanisms. First, splicing and 3′end processing occur co-transcriptionally and compete kinetically with each other. Thus, by enhancing splicing, SR proteins may inhibit the usage of intronic poly(A) sites, which are frequent in last introns (Lou et al. 1998). Second, SR proteins regulate the alternative inclusion of last exons through AS and thereby determine the identity of 3′UTRs (Müller-McNicoll et al. 2016). Third, SR proteins assist in the recognition of last exons and enhance polyadenylation through interactions with cleavage and polyadenylation (CPA) factors and in turn may stabilize their binding at alternative poly(A) sites (Kaida 2016). Fourth, SR proteins may bind to ESEs located in close proximity to alternative poly(A) sites and selectively activate them by recruiting CPA factors via their RS domains (Hudson et al. 2016; Zhu et al. 2018). In line with this, it was shown that SRSF7 and the CPA factor CFIm68 interact via their RS domains and that SRSF7 is able to recruit the CPA machinery and enhance polyadenylation in Rous sarcoma virus transcripts (Dettwiler et al. 2004; Hudson et al. 2016). This suggests that SRSF7 may actively regulate 3′UTR length of selected cellular mRNAs (Müller-McNicoll et al. 2016). Fifth, it is possible that SR proteins affect APA only indirectly, e.g. by promoting the selective export of transcripts with long 3′UTRs or their stabilization in the cytoplasm. Finally, SR proteins may also affect the levels of CPA factors by splicing. The integration of global changes in poly(A) site usage and CPA factor binding after depletion of individual SR proteins will shed light on the impact and mechanisms of SR protein-regulated 3′end processing.

## 3.9 mRNP Export and Nucleo-Cytoplasmic Shuttling

Eukaryotic gene expression is compartmentalized into nuclear and cytoplasmic events, which are connected through shuttling RBPs. Most members of the SR protein family shuttle between the nucleus and the cytoplasm (Botti et al. 2017; Misteli et al. 1998; Sapra et al. 2009). Individual SR proteins exhibit differences in their shuttling activities, which correlate with the phosphorylation state of their RS domains and the extent of NXF1 recruitment to the mRNP (Botti et al. 2017; Caceres et al. 1998; Cazalla et al. 2002; Lin et al. 2005). It was proposed that dephosphorylation of SR proteins represents one mechanism for the selective export of spliced mRNAs (Huang and Steitz 2005). Indeed, most SR proteins serve as adaptors for NXF1 and their depletion affects the selective nuclear export of specific cellular mRNA isoforms (Müller-McNicoll et al. 2016). SR proteins bind NXF1 only in their hypo-phosphorylated state (Fig. 3.2). Two pairs of neighbouring Arg residues flanking a glycine-rich region in the linker domain are required for NXF1 interaction (Botti et al. 2017; Hargous et al. 2006; Huang and Steitz 2005; Lai and Tarn 2004; Tintaru et al. 2007). Interestingly, these Arg residues are differentially methylated in SRSF1 and SRSF5, which affects their NXF1 interaction and nucleo-cytoplasmic shuttling (Botti et al. 2017; Sinha et al. 2010).

Binding of SRSF3 and SRSF7 enhances the RNA-binding capacity of NXF1 in vitro and in vivo, suggesting that a structural change in NXF1 upon binding exposes its RNA-binding domain (RBD) as occurs upon interacting with other export adaptors such as ALYREF (Hautbergue et al. 2008; Müller-McNicoll et al. 2016; Viphakone et al. 2012). However, in contrast to ALYREF, which hands the mRNA over to NXF1, SR proteins and NXF1 appear to form a trimeric complex with the mRNA, as both proteins bind in close proximity to the same mRNAs and shuttle together to the cytoplasm (Botti et al. 2017; Müller-McNicoll et al. 2016). SRSF3 emerged as the most potent NXF1 adaptor and was shown to selectively export isoforms with long 3′UTRs, $m^6A$-containing mRNAs, intronless histone transcripts, *NANOG*, *PDCD4* and some viral transcripts (Escudero-Paunetto et al. 2010; Huang and Steitz 2001; Müller-McNicoll et al. 2016; Park and Jeong 2016; Ratnadiwakara et al. 2018; Roundtree et al. 2017). SRSF1 was shown to selectively export neurotoxic *C9orf72* mRNAs that contain G4C2 repeat expansions and thereby contribute to neurodegeneration (Hautbergue et al. 2017).

Nucleo-cytoplasmic shuttling of SRSF5 and SRSF2 is regulated depending on the cellular differentiation state through phosphorylation of SRSF2 (Botti et al. 2017). In HeLa cells and differentiated murine cells, SRSF2 and SRSF5 are confined to the nucleus (Botti et al. 2017; Caceres et al. 1998; Cazalla et al. 2002; Lin et al. 2005; Sapra et al. 2009). SRSF2 does not shuttle because its unusual RS domain renders it resistant to dephosphorylation by PP1 during splicing and, consequently, SRSF2 cannot recruit NXF1 (Cazalla et al. 2002; Lin et al. 2005). However, in undifferentiated cells, both proteins remain bound to mature mRNAs after splicing in a partially dephosphorylated state, recruit NXF1 to the mRNPs and shuttle to the cytoplasm, in this way likely contributing to selective mRNA export of pluripotency-specific transcripts (Botti et al. 2017; Hammarskjold and Rekosh 2017).

SR proteins may also prevent the recruitment of export factors to mature or partially spliced mRNPs and thereby cause their nuclear retention, possibly in nuclear speckles. This would provide the possibility to uncouple mRNA export from co-transcriptional splicing and allow a coordinated release of transcripts with related functions in response to external stimuli. For example, transcripts with retained introns are often sequestered in nuclear speckles and can be released in response to external stimuli, either through post-transcriptional splicing, through removal of retention factors or through acquisition of export factors (Wegener and Muller-McNicoll 2017; Wickramasinghe and Laskey 2015). SRSF1 was shown to cause retention of mRNAs related to inflammation in the nucleus of macrophages and their release upon stimulation with LPS (Zhou et al. 2017). LPS signalling causes migration of Interleukin-1 receptor-associated kinase-2 (IRAK2) to the nucleus and hyper-phosphorylation of SRSF1, which reduced its binding to target mRNAs and allows recruitment of the export adaptor ALYREF and NXF1 (Zhou et al. 2017). Given that SRSF1 normally recruits NXF1 efficiently to mRNAs and shuttles robustly between the nucleus and the cytoplasm (Botti et al. 2017; Caceres et al. 1998; Lai and Tarn 2004; Müller-McNicoll et al. 2016), it remains unclear why in this case SRSF1 binding prevents the export of specific mRNAs. One possibility is that these inflammation-specific transcripts have SRSF1-binding sites within their very long 3′UTRs, in which case SRSF1 would remain phosphorylated even after completion of splicing and thereby interfere with NXF1 recruitment. Altogether, selective mRNA export through SR proteins may constitute a novel layer of gene expression regulation, but the underlying mechanisms and the roles of individual SR proteins need to be further studied.

## 3.10 Nonsense-Mediated Decay

In the cytoplasm, SR protein-containing mRNPs can be either translated into proteins or degraded by nonsense-mediated decay (NMD). NMD is a cytoplasmic surveillance pathway that inspects mRNPs during the first round of translation for premature termination codons (PTCs) that can be introduced either through mutations or splicing errors. Functional mRNPs are remodelled for efficient bulk translation, whereas PTC-containing mRNAs are rapidly degraded (Nasif et al. 2017). SR proteins affect the efficiency of NMD in many different ways, both directly and indirectly. First, SR proteins promote the inclusion of poison cassette exons that contain PTCs into their own transcripts during splicing. This ‘unproductive splicing’ is widely used by RBPs to maintain homeostatic expression levels (Nasif et al. 2017). SRSF3 was shown to also cross-regulate other SR protein family members through unproductive splicing (Änkö et al. 2012). Second, SR proteins may promote the splicing of introns located within 3′UTRs, which turns normal stop codons into PTCs and renders these transcripts sensitive to NMD (Sureau et al. 2001). Third, through selective export and retention, SR proteins may affect the availability of mRNPs in the cytoplasm for NMD. Fourth, SR proteins may stabilize EJCs downstream of PTCs, which are required to trigger NMD, and thus may enhance NMD of specific isoforms (Singh et al. 2012; Zhang and Krainer 2004). Finally, SRSF1 was shown to directly promote NMD of bound PTC-containing reporter and endogenous mRNAs through a direct recruitment of the NMD factor UPF1 (Aznarez et al. 2018; Zhang and Krainer 2004). SRSF1 also promotes NMD by enhancing the translation of bound transcripts (see next section).

## 3.11 mRNP Translation

Although SRSF1 was only found in mRNPs containing the nuclear cap-binding protein CBP80, its overexpression shifted both CBP80- and eIF4E-bound mRNAs to heavier polysomes, suggesting that SRSF1 enhances both the pioneer and subsequent rounds of translation (Sato et al. 2008). The former is thought to occur through NXF1, which is recruited to the shuttling mRNP by SRSF1. NXF1 associates with translating ribosomes and was shown to enhance translation of viral RNAs (Jin et al. 2003; Sato et al. 2008). The stimulatory effect of SRSF1 on bulk translation depends on eIF4E. It was proposed that SRSF1 recruits the protein kinase mammalian target of rapamycin to a subset of ESE-containing reporter mRNAs, while inhibiting PP2A. This caused the hyper-phosphorylation of eukaryotic translation initiation factor 4E (4E-BP1), a competitive inhibitor of cap-dependent translation, its release from eIF4E and activation of translation (Michlewski et al. 2008; Sanford et al. 2004). Intriguingly, interaction of SRSF1 with PP2A and mTOR required the conserved SWQDLKD motif within the ΨRRM of SRSF1, which is also required for specific RNA binding (Clery et al. 2013; Michlewski et al. 2008). This implies three possible scenarios: (1) the specific activation of SRSF1-bound mRNAs causes the subsequent release of SRSF1 from the mRNA; (2) SRSF1 is not bound directly to the mRNA but held in place by other RNP components, such as NXF1 (Tintaru et al. 2007); and (3) translational targets of SRSF1 are only bound by the RRM1 of SRSF1. In favour of the first scenario, a genome-wide study identified 505 direct translational targets of SRSF1, which were bound directly by SRSF1 (CLIP-Seq) and contained an enriched purine-rich sequence motif similar to the ESE used in the reporter gene studies and structural studies with ΨRRM (Clery et al. 2013; Maslon et al. 2014).

Altogether, these data suggest that alternative splice isoforms may be subject to differential translation regulated by individual SR proteins. Indeed, more than 30% of alternative mRNA isoforms exhibit differential polysome association, and it was suggested that specific cellular functions, e.g. cell-cycle control, are subject to AS-dependent modulation of translation (Sterne-Weiler et al. 2013; Weatheritt et al. 2016). In line with this, shuttling SR proteins have been detected in polysomal fractions from mitotic cells (Aviner et al. 2017).

SRSF3, SRSF5 and SRSF7 may also regulate translation, since a fraction of them was found to associate with light polysomes. CLIP-Seq on polysomal fractions (piCLIP) revealed that SRSF3 and SRSF5 bind directly to mRNAs undergoing translation (Botti et al. 2017). SRSF3 was also shown to repress translation of the *PDCD4* mRNA (Kim et al. 2014), whereas SRSF7 enhances translation of unspliced viral RNAs that contain a constitutive transport element (CTE) bound by NXF1 (Swartz et al. 2007). Tethering SRSF1 and SRSF7 to reporter genes also supported their active role as translation activators (Mo et al. 2013). More studies are required to reveal the underlying mechanisms of translational regulation by individual SR proteins.

## 3.12 mRNP Remodelling and Re-import of SR Proteins

It is generally accepted that once in the cytoplasm, SR proteins are rapidly re-phosphorylated by SRPK1/2, which triggers their immediate dissociation from NXF1 and the exported mRNA cargo. SRPK1/2 phosphorylates the N-terminal half of the RS domain of SRSF1 and then dissociates from it (Ghosh and Adams 2011). The partially phosphorylated RS domain acts as a potent NLS and mediates the rapid nuclear re-import of SRSF1 via the SR-specific nuclear import receptor transportin-SR (TRN-SR) and its localization to nuclear speckles (Lai et al. 2001) (Fig. 3.2).

Despite extensive research on SRPK1, it is not clear how and where in the cytoplasm this phosphorylation-mediated disassembly occurs. RNA binding and SRPK1 binding are mutually exclusive events, since both occupy the same residues within the ΨRRM of SRSF1 (Clery et al. 2013; Ngo et al. 2008). In vitro structural studies suggested that SRSF1 is not directly bound to the mRNAs in shuttling mRNPs. Instead, the mRNA is handed over from SRSF1 to NXF1 during the formation of export-competent mRNPs (Tintaru et al. 2007); thus a free ΨRRM would allow SRPK1 to re-phosphorylate SRSF1. However, this model contrasts with CLIP-Seq studies, which showed that a proportion of SRSF1, SRSF3, SRSF5 and SRSF7 bind to the same sites on mRNAs whether they localize to the nucleoplasm, cytoplasm or polysomes and, in the case of SRSF1, regulate the translation of selected targets (Botti et al. 2017; Brugiolo et al. 2017; Maslon et al. 2014; Müller-McNicoll et al. 2016; Sanford et al. 2008). Moreover, NXF1 crosslinks are in close proximity to SRSF3-binding sites (Müller-McNicoll et al. 2016).

Thus, it appears more likely that SR proteins are rather re-phosphorylated by SRPK1/2 after they are removed from bound mRNAs during the pioneer round of translation. This would allow SR proteins to regulate the fate of bound isoforms based on where they bind in the transcript. Binding within the open-reading frame close to EJCs may enhance recognition of NMD targets, while binding in 5′UTRs may regulate translation initiation. In line with the latter possibility, SRSF3 binds to the 5′UTR of *PDCD4* and inhibits its translation. SRSF3 is also found in stress granules associated with stalled ribosomes (Jayabalan et al. 2016; Kim et al. 2014). Binding in the 3′UTRs would allow SR proteins to remain bound during several rounds of translation and thereby regulate mRNA stability, localization and/or bulk translation efficiency. Further investigations are needed to solve the discrepancies between in vitro and in vivo studies.

## 3.13 Concluding Remarks

SR proteins are extremely versatile and adjustable proteins that engage in a wide spectrum of mutually exclusive interactions with proteins and RNAs in different cellular compartments. Thus, SR proteins can be seen as molecular adaptors that connect gene expression and processing machineries, providing the unique possibility to investigate the roles of other RBPs in each step of the mRNP life cycle. Over the past 30 years, extensive research focussing on SRSF1 has provided a wealth of mechanistic insights into how SR proteins influence the assembly, maturation, function and turnover of mRNPs and has significantly contributed to deciphering the SR protein 'code'. However, many of the functions and mechanisms described herein are probably unique to SRSF1, either because the responsible protein domains or motifs are missing in other family members, because their RS domains are different or because each SR protein is uniquely modified. It has also become clear that individual SR proteins are present in distinct mRNPs and regulate distinct sets of genes and that they connect different steps of gene expression in different cell types. The composition of SR protein-containing mRNPs indicates that they often act in pairs, whereby one partner might provide binding specificity and the other, adjustability to the mRNP. The loss of one SR protein could cause the coordinated loss or the compensatory gain of another SR protein at bound exons. Obviously, we have barely scratched the surface, and much remains to be discovered by also studying the other members of this family of multitasking RBPs.

**Acknowledgements** We thank François McNicoll and Daniel Zenklusen for great proofreading and editing. We are grateful for funding from the Deutsche Forschungsgemeinschaft (CEF-MC and SFB902 to MMM).

## References

Adivarahan S, Livingston N, Nicholson B, Rahman S, Wu B, Rissland OS, Zenklusen D (2018) Spatial organization of single mRNPs at different stages of the gene expression pathway. Mol Cell 72:727–738.e725. https://doi.org/10.1016/j.molcel.2018.10.010

Änkö ML (2014) Regulation of gene expression programmes by serine-arginine rich splicing factors. Semin Cell Dev Biol 32:11–21. https://doi.org/10.1016/j.semcdb.2014.03.011

Änkö ML, Morales L, Henry I, Beyer A, Neugebauer KM (2010) Global analysis reveals SRp20- and SRp75-specific mRNPs in cycling and neural cells. Nat Struct Mol Biol 17:962–970. https://doi.org/10.1038/nsmb.1862

Änkö ML et al (2012) The RNA-binding landscapes of two SR proteins reveal unique functions and binding to diverse RNA classes. Genome Biol 13:R17. https://doi.org/10.1186/gb-2012-13-3-r17

Aubol BE et al (2014) N-terminus of the protein kinase CLK1 induces SR protein hyperphosphorylation. Biochem J 462:143–152. https://doi.org/10.1042/BJ20140494

Aubol BE et al (2016) Release of SR proteins from CLK1 by SRPK1: a symbiotic kinase system for phosphorylation control of pre-mRNA splicing. Mol Cell 63:218–228. https://doi.org/10.1016/j.molcel.2016.05.034

Aubol BE, Hailey KL, Fattet L, Jennings PA, Adams JA (2017) Redirecting SR protein nuclear trafficking through an allosteric platform. J Mol Biol 429:2178–2191. https://doi.org/10.1016/j.jmb.2017.05.022

Aviner R et al (2017) Proteomic analysis of polyribosomes identifies splicing factors as potential regulators of translation during mitosis. Nucleic Acids Res 45:5945–5957. https://doi.org/10.1093/nar/gkx326

Aznarez I, Nomakuchi TT, Tetenbaum-Novatt J, Rahman MA, Fregoso O, Rees H, Krainer AR (2018) Mechanism of nonsense-mediated mRNA decay stimulation by splicing factor SRSF1. Cell Rep 23:2186–2198. https://doi.org/10.1016/j.celrep.2018.04.039

Bjork P, Jin S, Zhao J, Singh OP, Persson JO, Hellman U, Wieslander L (2009) Specific combinations of SR proteins associate with single pre-messenger RNAs in vivo and contribute different functions. J Cell Biol 184:555–568. https://doi.org/10.1083/jcb.200806156

Blaustein M et al (2005) Concerted regulation of nuclear and cytoplasmic activities of SR proteins by AKT. Nat Struct Mol Biol 12:1037–1044. https://doi.org/10.1038/nsmb1020

Boehm V, Gehring NH (2016) Exon junction complexes: supervising the gene expression assembly line. Trends Genet 32:724–735. https://doi.org/10.1016/j.tig.2016.09.003

Botti V et al (2017) Cellular differentiation state modulates the mRNA export activity of SR proteins. J Cell Biol 216:1993–2009. https://doi.org/10.1083/jcb.201610051

Bradley T, Cook ME, Blanchette M (2015) SR proteins control a complex network of RNA-processing events. RNA 21:75–92. https://doi.org/10.1261/rna.043893.113

Bressan GC, Moraes EC, Manfiolli AO, Kuniyoshi TM, Passos DO, Gomes MD, Kobarg J (2009) Arginine methylation analysis of the splicing-associated SR protein SFRS9/SRP30C. Cell Mol Biol Lett 14:657–669. https://doi.org/10.2478/s11658-009-0024-2

Brugiolo M, Botti V, Liu N, Müller-McNicoll M, Neugebauer KM (2017) Fractionation iCLIP detects persistent SR protein binding to conserved, retained introns in chromatin, nucleoplasm and cytoplasm. Nucleic Acids Res 45:10452–10465. https://doi.org/10.1093/nar/gkx671

Busch A, Hertel KJ (2012) Evolution of SR protein and hnRNP splicing regulatory factors. Wiley Interdiscip Rev RNA 3:1–12. https://doi.org/10.1002/wrna.100

Caceres JF, Kornblihtt AR (2002) Alternative splicing: multiple control mechanisms and involvement in human disease. Trends Genet 18:186–193

Caceres JF, Misteli T, Screaton GR, Spector DL, Krainer AR (1997) Role of the modular domains of SR proteins in subnuclear localization and alternative splicing specificity. J Cell Biol 138:225–238

Caceres JF, Screaton GR, Krainer AR (1998) A specific subset of SR proteins shuttles continuously between the nucleus and the cytoplasm. Genes Dev 12:55–66

Cavaloc Y, Popielarz M, Fuchs JP, Gattoni R, Stevenin J (1994) Characterization and cloning of the human splicing factor 9G8: a novel 35 kDa factor of the serine/arginine protein family. EMBO J 13:2639–2649

Cavaloc Y, Bourgeois CF, Kister L, Stevenin J (1999) The splicing factors 9G8 and SRp20 transactivate splicing through different and specific enhancers. RNA 5:468–483

Cazalla D, Zhu J, Manche L, Huber E, Krainer AR, Caceres JF (2002) Nuclear export and retention signals in the RS domain of SR proteins. Mol Cell Biol 22:6871–6882

Chandradas S, Deikus G, Tardos JG, Bogdanov VY (2010) Antagonistic roles of four SR proteins in the biosynthesis of alternatively spliced tissue factor transcripts in monocytic cells. J Leukoc Biol 87:147–152. https://doi.org/10.1189/jlb.0409252

Chaudhary N, McMahon C, Blobel G (1991) Primary structure of a human arginine-rich nuclear protein that colocalizes with spliceosome components. Proc Natl Acad Sci U S A 88:8189–8193

Chen Y et al (2018) Mutually exclusive acetylation and ubiquitylation of the splicing factor SRSF5 control tumor growth. Nat Commun 9:2464. https://doi.org/10.1038/s41467-018-04815-3

Cho S, Hoang A, Chakrabarti S, Huynh N, Huang DB, Ghosh G (2011a) The SRSF1 linker induces semi-conservative ESE binding by cooperating with the RRMs. Nucleic Acids Res 39:9413–9421. https://doi.org/10.1093/nar/gkr663

Cho S, Hoang A, Sinha R, Zhong XY, Fu XD, Krainer AR, Ghosh G (2011b) Interaction between the RNA binding domains of Ser-Arg splicing factor 1 and U1-70K snRNP protein determines early spliceosome assembly. Proc Natl Acad Sci U S A 108:8233–8238. https://doi.org/10.1073/pnas.1017700108

Clery A et al (2013) Isolated pseudo-RNA-recognition motifs of SR proteins can regulate splicing using a noncanonical mode of RNA recognition. Proc Natl Acad Sci U S A 110:E2802–E2811. https://doi.org/10.1073/pnas.1303445110

Colwill K, Pawson T, Andrews B, Prasad J, Manley JL, Bell JC, Duncan PI (1996) The Clk/Sty protein kinase phosphorylates SR splicing factors and regulates their intranuclear distribution. EMBO J 15:265–275

Cowper AE, Caceres JF, Mayeda A, Screaton GR (2001) Serine-arginine (SR) protein-like factors that antagonize authentic SR proteins and regulate alternative splicing. J Biol Chem 276:48908–48914. https://doi.org/10.1074/jbc.M103967200

Daubner GM, Clery A, Jayne S, Stevenin J, Allain FH (2012) A syn-anti conformational difference allows SRSF2 to recognize guanines and cytosines equally well. EMBO J 31:162–174. https://doi.org/10.1038/emboj.2011.367

De Conti L, Baralle M, Buratti E (2013) Exon and intron definition in pre-mRNA splicing. Wiley Interdiscip Rev RNA 4:49–60. https://doi.org/10.1002/wrna.1140

de la Mata M, Kornblihtt AR (2006) RNA polymerase II C-terminal domain mediates regulation of alternative splicing by SRp20. Nat Struct Mol Biol 13:973–980. https://doi.org/10.1038/nsmb1155

de la Mata M et al (2003) A slow RNA polymerase II affects alternative splicing in vivo. Mol Cell 12:525–532

Dettwiler S, Aringhieri C, Cardinale S, Keller W, Barabino SM (2004) Distinct sequence motifs within the 68-kDa subunit of cleavage factor Im mediate RNA binding, protein-protein interactions, and subcellular localization. J Biol Chem 279:35788–35797. https://doi.org/10.1074/jbc.M403927200

Du L, Warren SL (1997) A functional interaction between the carboxy-terminal domain of RNA polymerase II and pre-mRNA splicing. J Cell Biol 136:5–18

Edmond V et al (2011) Acetylation and phosphorylation of SRSF2 control cell fate decision in response to cisplatin. EMBO J 30:510–523. https://doi.org/10.1038/emboj.2010.333

Erkelenz S, Mueller WF, Evans MS, Busch A, Schoneweis K, Hertel KJ, Schaal H (2013) Position-dependent splicing activation and repression by SR and hnRNP proteins rely on common mechanisms. RNA 19:96–102. https://doi.org/10.1261/rna.037044.112

Escudero-Paunetto L, Li L, Hernandez FP, Sandri-Goldin RM (2010) SR proteins SRp20 and 9G8 contribute to efficient export of herpes simplex virus 1 mRNAs. Virology 401:155–164. https://doi.org/10.1016/j.virol.2010.02.023

Fu XD, Ares M Jr (2014) Context-dependent control of alternative splicing by RNA-binding proteins. Nat Rev Genet 15:689–701. https://doi.org/10.1038/nrg3778

Fu XD, Maniatis T (1992) Isolation of a complementary DNA that encodes the mammalian splicing factor SC35. Science 256:535–538

Galganski L, Urbanek MO, Krzyzosiak WJ (2017) Nuclear speckles: molecular organization, biological function and role in disease. Nucleic Acids Res 45:10350–10368. https://doi.org/10.1093/nar/gkx759

Ge H, Manley JL (1990) A protein factor, ASF, controls cell-specific alternative splicing of SV40 early pre-mRNA in vitro. Cell 62:25–34

Ghosh G, Adams JA (2011) Phosphorylation mechanism and structure of serine-arginine protein kinases. FEBS J 278:587–597. https://doi.org/10.1111/j.1742-4658.2010.07992.x

Gui JF, Lane WS, Fu XD (1994) A serine kinase regulates intracellular localization of splicing factors in the cell cycle. Nature 369:678–682. https://doi.org/10.1038/369678a0

Hammarskjold ML, Rekosh D (2017) SR proteins: to shuttle or not to shuttle, that is the question. J Cell Biol 216:1875–1877. https://doi.org/10.1083/jcb.201705009

Han J, Ding JH, Byeon CW, Kim JH, Hertel KJ, Jeong S, Fu XD (2011) SR proteins induce alternative exon skipping through their activities on the flanking constitutive exons. Mol Cell Biol 31:793–802. https://doi.org/10.1128/MCB.01117-10

Hargous Y et al (2006) Molecular basis of RNA recognition and TAP binding by the SR proteins SRp20 and 9G8. EMBO J 25:5126–5137. https://doi.org/10.1038/sj.emboj.7601385

Hautbergue GM, Hung ML, Golovanov AP, Lian LY, Wilson SA (2008) Mutually exclusive interactions drive handover of mRNA from export adaptors to TAP. Proc Natl Acad Sci U S A 105:5154–5159. https://doi.org/10.1073/pnas.0709167105

Hautbergue GM et al (2017) SRSF1-dependent nuclear export inhibition of C9ORF72 repeat transcripts prevents neurodegeneration and associated motor deficits. Nat Commun 8:16063. https://doi.org/10.1038/ncomms16063

Hentze MW, Castello A, Schwarzl T, Preiss T (2018) A brave new world of RNA-binding proteins. Nat Rev Mol Cell Biol. https://doi.org/10.1038/nrm.2017.130

Hnilicova J, Hozeifi S, Duskova E, Icha J, Tomankova T, Stanek D (2011) Histone deacetylase activity modulates alternative splicing. PLoS One 6:e16727. https://doi.org/10.1371/journal.pone.0016727

Huang Y, Steitz JA (2001) Splicing factors SRp20 and 9G8 promote the nucleocytoplasmic export of mRNA. Mol Cell 7:899–905

Huang Y, Steitz JA (2005) SRprises along a messenger's journey. Mol Cell 17:613–615. https://doi.org/10.1016/j.molcel.2005.02.020

Hudson SW, McNally LM, McNally MT (2016) Evidence that a threshold of serine/arginine-rich (SR) proteins recruits CFIm to promote rous sarcoma virus mRNA 3′ end formation. Virology 498:181–191. https://doi.org/10.1016/j.virol.2016.08.021

Jankowsky E, Harris ME (2015) Specificity and nonspecificity in RNA-protein interactions. Nat Rev Mol Cell Biol 16:533–544. https://doi.org/10.1038/nrm4032

Jayabalan AK et al (2016) NEDDylation promotes stress granule assembly. Nat Commun 7:12125. https://doi.org/10.1038/ncomms12125

Ji X et al (2013) SR proteins collaborate with 7SK and promoter-associated nascent RNA to release paused polymerase. Cell 153:855–868. https://doi.org/10.1016/j.cell.2013.04.028

Jin L, Guzik BW, Bor YC, Rekosh D, Hammarskjold ML (2003) Tap and NXT promote translation of unspliced mRNA. Genes Dev 17:3075–3086. https://doi.org/10.1101/gad.1155703

Kaida D (2016) The reciprocal regulation between splicing and 3′-end processing. Wiley Interdiscip Rev RNA 7:499–511. https://doi.org/10.1002/wrna.1348

Keshwani MM et al (2015) Conserved proline-directed phosphorylation regulates SR protein conformation and splicing function. Biochem J 466:311–322. https://doi.org/10.1042/BJ20141373

Khong A, Parker R (2018) mRNP architecture in translating and stress conditions reveals an ordered pathway of mRNP compaction J Cell Biol. https://doi.org/10.1083/jcb.201806183

Kim E, Du L, Bregman DB, Warren SL (1997) Splicing factors associate with hyperphosphorylated RNA polymerase II in the absence of pre-mRNA. J Cell Biol 136:19–28

Kim J, Park RY, Chen JK, Kim J, Jeong S, Ohn T (2014) Splicing factor SRSF3 represses the translation of programmed cell death 4 mRNA by associating with the 5′-UTR region. Cell Death Differ 21:481–490. https://doi.org/10.1038/cdd.2013.171

Ko B, Gunderson SI (2002) Identification of new poly(A) polymerase-inhibitory proteins capable of regulating pre-mRNA polyadenylation. J Mol Biol 318:1189–1206

Krainer AR, Conway GC, Kozak D (1990) The essential pre-mRNA splicing factor SF2 influences 5′ splice site selection by activating proximal sites. Cell 62:35–42

Krchnakova Z, Thakur PK, Krausova M, Bieberstein N, Haberman N, Muller-McNicoll M, Stanek D (2018) Splicing of long non-coding RNAs primarily depends on polypyrimidine tract and 5′ splice-site sequences due to weak interactions with SR proteins. Nucleic Acids Res. https://doi.org/10.1093/nar/gky1147

Lai MC, Tarn WY (2004) Hypophosphorylated ASF/SF2 binds TAP and is present in messenger ribonucleoproteins. J Biol Chem 279:31745–31749. https://doi.org/10.1074/jbc.C400173200

Lai MC, Lin RI, Tarn WY (2001) Transportin-SR2 mediates nuclear import of phosphorylated SR proteins. Proc Natl Acad Sci U S A 98:10154–10159. https://doi.org/10.1073/pnas.181354098

Larsen SC et al (2016) Proteome-wide analysis of arginine monomethylation reveals widespread occurrence in human cells. Sci Signal 9:rs9. https://doi.org/10.1126/scisignal.aaf7329

Lee G et al (2017) Post-transcriptional regulation of de novo lipogenesis by mTORC1-S6K1-SRPK2 signaling. Cell 171:1545–1558.e1518. https://doi.org/10.1016/j.cell.2017.10.037

Lemaire R, Prasad J, Kashima T, Gustafson J, Manley JL, Lafyatis R (2002) Stability of a PKCI-1-related mRNA is controlled by the splicing factor ASF/SF2: a novel function for SR proteins. Genes Dev 16:594–607. https://doi.org/10.1101/gad.939502

Li X, Manley JL (2005) Inactivation of the SR protein splicing factor ASF/SF2 results in genomic instability. Cell 122:365–378. https://doi.org/10.1016/j.cell.2005.06.008

Lin S, Xiao R, Sun P, Xu X, Fu XD (2005) Dephosphorylation-dependent sorting of SR splicing factors during mRNP maturation. Mol Cell 20:413–425. https://doi.org/10.1016/j.molcel.2005.09.015

Lin S, Coutinho-Mansfield G, Wang D, Pandit S, Fu XD (2008) The splicing factor SC35 has an active role in transcriptional elongation. Nat Struct Mol Biol 15:819–826. https://doi.org/10.1038/nsmb.1461

Liu KJ, Harland RM (2005) Inhibition of neurogenesis by SRp38, a neuroD-regulated RNA-binding protein. Development 132:1511–1523. https://doi.org/10.1242/dev.01703

Long Y et al (2018) Distinct mechanisms govern the phosphorylation of different SR protein splicing factors. J Biol Chem. https://doi.org/10.1074/jbc.RA118.003392

Lou H, Neugebauer KM, Gagel RF, Berget SM (1998) Regulation of alternative polyadenylation by U1 snRNPs and SRp20. Mol Cell Biol 18:4977–4985

Ma CT, Ghosh G, Fu XD, Adams JA (2010) Mechanism of dephosphorylation of the SR protein ASF/SF2 by protein phosphatase 1. J Mol Biol 403:386–404. https://doi.org/10.1016/j.jmb.2010.08.024

Mabon SA, Misteli T (2005) Differential recruitment of pre-mRNA splicing factors to alternatively spliced transcripts in vivo. PLoS Biol 3:e374. https://doi.org/10.1371/journal.pbio.0030374

Malanga M, Czubaty A, Girstun A, Staron K, Althaus FR (2008) Poly(ADP-ribose) binds to the splicing factor ASF/SF2 and regulates its phosphorylation by DNA topoisomerase I. J Biol Chem 283:19991–19998. https://doi.org/10.1074/jbc.M709495200

Manley JL, Krainer AR (2010) A rational nomenclature for serine/arginine-rich protein splicing factors (SR proteins). Genes Dev 24:1073–1074. https://doi.org/10.1101/gad.1934910

Maris C, Dominguez C, Allain FH (2005) The RNA recognition motif, a plastic RNA-binding platform to regulate post-transcriptional gene expression. FEBS J 272:2118–2131. https://doi.org/10.1111/j.1742-4658.2005.04653.x

Maslon MM, Heras SR, Bellora N, Eyras E, Caceres JF (2014) The translational landscape of the splicing factor SRSF1 and its role in mitosis. Elife e02028. https://doi.org/10.7554/eLife.02028

Mermoud JE, Cohen PT, Lamond AI (1994) Regulation of mammalian spliceosome assembly by a protein phosphorylation mechanism. EMBO J 13:5679–5688

Metkar M, Ozadam H, Lajoie BR, Imakaev M, Mirny LA, Dekker J, Moore MJ (2018) Higher-order organization principles of pre-translational mRNPs. Mol Cell 72:715–726.e713. https://doi.org/10.1016/j.molcel.2018.09.012

Michlewski G, Sanford JR, Caceres JF (2008) The splicing factor SF2/ASF regulates translation initiation by enhancing phosphorylation of 4E-BP1. Mol Cell 30:179–189. https://doi.org/10.1016/j.molcel.2008.03.013

Misteli T, Spector DL (1996) Serine/threonine phosphatase 1 modulates the subnuclear distribution of pre-mRNA splicing factors. Mol Biol Cell 7:1559–1572

Misteli T, Spector DL (1999) RNA polymerase II targets pre-mRNA splicing factors to transcription sites in vivo. Mol Cell 3:697–705

Misteli T, Caceres JF, Clement JQ, Krainer AR, Wilkinson MF, Spector DL (1998) Serine phosphorylation of SR proteins is required for their recruitment to sites of transcription in vivo. J Cell Biol 143:297–307

Mo S, Ji X, Fu XD (2013) Unique role of SRSF2 in transcription activation and diverse functions of the SR and hnRNP proteins in gene expression regulation. Transcription 4:251–259

Müller-McNicoll M, Neugebauer KM (2013) How cells get the message: dynamic assembly and function of mRNA-protein complexes. Nat Rev Genet 14:275–287. https://doi.org/10.1038/nrg3434

Müller-McNicoll M et al (2016) SR proteins are NXF1 adaptors that link alternative RNA processing to mRNA export. Genes Dev 30:553–566. https://doi.org/10.1101/gad.276477.115

Naro C, Sette C (2013) Phosphorylation-mediated regulation of alternative splicing in cancer. Int J Cell Biol 2013:151839. https://doi.org/10.1155/2013/151839

Nasif S, Contu L, Muhlemann O (2017) Beyond quality control: the role of nonsense-mediated mRNA decay (NMD) in regulating gene expression. Semin Cell Dev Biol. https://doi.org/10.1016/j.semcdb.2017.08.053

Ngo JC et al (2008) A sliding docking interaction is essential for sequential and processive phosphorylation of an SR protein by SRPK1. Mol Cell 29:563–576. https://doi.org/10.1016/j.molcel.2007.12.017

Ninomiya K, Kataoka N, Hagiwara M (2011) Stress-responsive maturation of Clk1/4 pre-mRNAs promotes phosphorylation of SR splicing factor. J Cell Biol 195:27–40. https://doi.org/10.1083/jcb.201107093

Pan Q, Shai O, Lee LJ, Frey BJ, Blencowe BJ (2008) Deep surveying of alternative splicing complexity in the human transcriptome by high-throughput sequencing. Nat Genet 40:1413–1415. https://doi.org/10.1038/ng.259

Pandit S et al (2013) Genome-wide analysis reveals SR protein cooperation and competition in regulated splicing. Mol Cell 50:223–235. https://doi.org/10.1016/j.molcel.2013.03.001

Park SK, Jeong S (2016) SRSF3 represses the expression of PDCD4 protein by coordinated regulation of alternative splicing, export and translation. Biochem Biophys Res Commun 470:431–438. https://doi.org/10.1016/j.bbrc.2016.01.019

Patel NA et al (2005) Molecular and genetic studies imply Akt-mediated signaling promotes protein kinase CbetaII alternative splicing via phosphorylation of serine/arginine-rich splicing factor SRp40. J Biol Chem 280:14302–14309. https://doi.org/10.1074/jbc.M411485200

Paz S, Krainer AR, Caputi M (2014) HIV-1 transcription is regulated by splicing factor SRSF1. Nucleic Acids Res 42:13812–13823. https://doi.org/10.1093/nar/gku1170

Pelisch F et al (2010) The serine/arginine-rich protein SF2/ASF regulates protein sumoylation. Proc Natl Acad Sci U S A 107:16119–16124. https://doi.org/10.1073/pnas.1004653107

Phelan MM, Goult BT, Clayton JC, Hautbergue GM, Wilson SA, Lian LY (2012) The structure and selectivity of the SR protein SRSF2 RRM domain with RNA. Nucleic Acids Res 40:3232–3244. https://doi.org/10.1093/nar/gkr1164

Preussner M, Goldammer G, Neumann A, Haltenhof T, Rautenstrauch P, Müller-McNicoll M, Heyd F (2017) Body temperature cycles control rhythmic alternative splicing in mammals. Mol Cell 67:433–446.e434. https://doi.org/10.1016/j.molcel.2017.06.006

Ratnadiwakara M, Mohenska M, Anko ML (2017) Splicing factors as regulators of miRNA biogenesis – links to human disease. Semin Cell Dev Biol. https://doi.org/10.1016/j.semcdb.2017.10.008

Ratnadiwakara M et al (2018) SRSF3 promotes pluripotency through Nanog mRNA export and coordination of the pluripotency gene expression program. Elife 7. https://doi.org/10.7554/eLife.37419

Rot G et al (2017) High-resolution RNA maps suggest common principles of splicing and polyadenylation regulation by TDP-43. Cell Rep 19:1056–1067. https://doi.org/10.1016/j.celrep.2017.04.028

Roth MB, Zahler AM, Stolk JA (1991) A conserved family of nuclear phosphoproteins localized to sites of polymerase II transcription. J Cell Biol 115:587–596

Roundtree IA et al (2017) YTHDC1 mediates nuclear export of N(6)-methyladenosine methylated mRNAs. Elife 6. https://doi.org/10.7554/eLife.31311

Sanchez-Hernandez N et al (2017) Targeting proteins to RNA transcription and processing sites within the nucleus. Int J Biochem Cell Biol 91:194–202. https://doi.org/10.1016/j.biocel.2017.06.001

Sanford JR, Gray NK, Beckmann K, Caceres JF (2004) A novel role for shuttling SR proteins in mRNA translation. Genes Dev 18:755–768. https://doi.org/10.1101/gad.286404

Sanford JR, Ellis JD, Cazalla D, Caceres JF (2005) Reversible phosphorylation differentially affects nuclear and cytoplasmic functions of splicing factor 2/alternative splicing factor. Proc Natl Acad Sci U S A 102:15042–15047. https://doi.org/10.1073/pnas.0507827102

Sanford JR, Coutinho P, Hackett JA, Wang X, Ranahan W, Caceres JF (2008) Identification of nuclear and cytoplasmic mRNA targets for the shuttling protein SF2/ASF. PLoS One 3:e3369. https://doi.org/10.1371/journal.pone.0003369

Sanford JR et al (2009) Splicing factor SFRS1 recognizes a functionally diverse landscape of RNA transcripts. Genome Res 19:381–394. https://doi.org/10.1101/gr.082503.108

Sapra AK et al (2009) SR protein family members display diverse activities in the formation of nascent and mature mRNPs in vivo. Mol Cell 34:179–190. https://doi.org/10.1016/j.molcel.2009.02.031

Sato H, Hosoda N, Maquat LE (2008) Efficiency of the pioneer round of translation affects the cellular site of nonsense-mediated mRNA decay. Mol Cell 29:255–262. https://doi.org/10.1016/j.molcel.2007.12.009

Sauliere J et al (2012) CLIP-seq of eIF4AIII reveals transcriptome-wide mapping of the human exon junction complex. Nat Struct Mol Biol 19:1124–1131. https://doi.org/10.1038/nsmb.2420

Scheres SH, Nagai K (2017) CryoEM structures of spliceosomal complexes reveal the molecular mechanism of pre-mRNA splicing. Curr Opin Struct Biol 46:130–139. https://doi.org/10.1016/j.sbi.2017.08.001

Screaton GR et al (1995) Identification and characterization of three members of the human SR family of pre-mRNA splicing factors. EMBO J 14:4336–4349

Serrano P et al (2016) Directional phosphorylation and nuclear transport of the splicing factor SRSF1 is regulated by an RNA recognition motif. J Mol Biol 428:2430–2445. https://doi.org/10.1016/j.jmb.2016.04.009

Shen H, Green MR (2006) RS domains contact splicing signals and promote splicing by a common mechanism in yeast through humans. Genes Dev 20:1755–1765. https://doi.org/10.1101/gad.1422106

Shi J et al (2011) Cyclic AMP-dependent protein kinase regulates the alternative splicing of tau exon 10: a mechanism involved in tau pathology of Alzheimer disease. J Biol Chem 286:14639–14648. https://doi.org/10.1074/jbc.M110.204453

Shin C, Manley JL (2002) The SR protein SRp38 represses splicing in M phase cells. Cell 111:407–417

Shin C, Feng Y, Manley JL (2004) Dephosphorylated SRp38 acts as a splicing repressor in response to heat shock. Nature 427:553–558. https://doi.org/10.1038/nature02288

Simard MJ, Chabot B (2002) SRp30c is a repressor of 3′ splice site utilization. Mol Cell Biol 22:4001–4010

Singh G, Kucukural A, Cenik C, Leszyk JD, Shaffer SA, Weng Z, Moore MJ (2012) The cellular EJC interactome reveals higher-order mRNP structure and an EJC-SR protein nexus. Cell 151:750–764. https://doi.org/10.1016/j.cell.2012.10.007

Sinha R, Allemand E, Zhang Z, Karni R, Myers MP, Krainer AR (2010) Arginine methylation controls the subcellular localization and functions of the oncoprotein splicing factor SF2/ASF. Mol Cell Biol 30:2762–2774. https://doi.org/10.1128/MCB.01270-09

Soret J et al (1998) Characterization of SRp46, a novel human SR splicing factor encoded by a PR264/SC35 retropseudogene. Mol Cell Biol 18:4924–4934

Soret J, Gabut M, Dupon C, Kohlhagen G, Stevenin J, Pommier Y, Tazi J (2003) Altered serine/arginine-rich protein phosphorylation and exonic enhancer-dependent splicing in mammalian cells lacking topoisomerase I. Cancer Res 63:8203–8211

Sterne-Weiler T et al (2013) Frac-seq reveals isoform-specific recruitment to polyribosomes. Genome Res 23:1615–1623. https://doi.org/10.1101/gr.148585.112

Stoehr A et al (2016) Prolyl hydroxylation regulates protein degradation, synthesis, and splicing in human induced pluripotent stem cell-derived cardiomyocytes. Cardiovasc Res 110:346–358. https://doi.org/10.1093/cvr/cvw081

Sureau A, Gattoni R, Dooghe Y, Stevenin J, Soret J (2001) SC35 autoregulates its expression by promoting splicing events that destabilize its mRNAs. EMBO J 20:1785–1796. https://doi.org/10.1093/emboj/20.7.1785

Swartz JE, Bor YC, Misawa Y, Rekosh D, Hammarskjold ML (2007) The shuttling SR protein 9G8 plays a role in translation of unspliced mRNA containing a constitutive transport element. J Biol Chem 282:19844–19853. https://doi.org/10.1074/jbc.M701660200

Tacke R, Chen Y, Manley JL (1997) Sequence-specific RNA binding by an SR protein requires RS domain phosphorylation: creation of an SRp40-specific splicing enhancer. Proc Natl Acad Sci U S A 94:1148–1153

Tejedor JR, Papasaikas P, Valcarcel J (2015) Genome-wide identification of Fas/CD95 alternative splicing regulators reveals links with iron homeostasis. Mol Cell 57:23–38. https://doi.org/10.1016/j.molcel.2014.10.029

Tian B, Manley JL (2017) Alternative polyadenylation of mRNA precursors. Nat Rev Mol Cell Biol 18:18–30. https://doi.org/10.1038/nrm.2016.116

Tintaru AM, Hautbergue GM, Hounslow AM, Hung ML, Lian LY, Craven CJ, Wilson SA (2007) Structural and functional analysis of RNA and TAP binding to SF2/ASF. EMBO Rep 8:756–762. https://doi.org/10.1038/sj.embor.7401031

Tripathi V et al (2010) The nuclear-retained noncoding RNA MALAT1 regulates alternative splicing by modulating SR splicing factor phosphorylation. Mol Cell 39:925–938. https://doi.org/10.1016/j.molcel.2010.08.011

Tripathi V et al (2012) SRSF1 regulates the assembly of pre-mRNA processing factors in nuclear speckles. Mol Biol Cell 23:3694–3706. https://doi.org/10.1091/mbc.E12-03-0206

Van Nostrand EL et al (2016) Robust transcriptome-wide discovery of RNA-binding protein binding sites with enhanced CLIP (eCLIP). Nat Methods 13:508–514. https://doi.org/10.1038/nmeth.3810

Viphakone N et al (2012) TREX exposes the RNA-binding domain of Nxf1 to enable mRNA export. Nat Commun 3:1006. https://doi.org/10.1038/ncomms2005

Weatheritt RJ, Sterne-Weiler T, Blencowe BJ (2016) The ribosome-engaged landscape of alternative splicing. Nat Struct Mol Biol 23:1117–1123. https://doi.org/10.1038/nsmb.3317

Wegener M, Muller-McNicoll M (2017) Nuclear retention of mRNAs – quality control, gene regulation and human disease. Semin Cell Dev Biol. https://doi.org/10.1016/j.semcdb.2017.11.001

Wheeler EC, Van Nostrand EL, Yeo GW (2017) Advances and challenges in the detection of transcriptome-wide protein-RNA interactions. Wiley Interdiscip Rev RNA. https://doi.org/10.1002/wrna.1436

Wickramasinghe VO, Laskey RA (2015) Control of mammalian gene expression by selective mRNA export. Nat Rev Mol Cell Biol 16:431–442. https://doi.org/10.1038/nrm4010

Wongpalee SP, Vashisht A, Sharma S, Chui D, Wohlschlegel JA, Black DL (2016) Large-scale remodeling of a repressed exon ribonucleoprotein to an exon definition complex active for splicing. Elife 5. https://doi.org/10.7554/eLife.19743

Xiang S et al (2013) Phosphorylation drives a dynamic switch in serine/arginine-rich proteins. Structure 21:2162–2174. https://doi.org/10.1016/j.str.2013.09.014

Xiao W et al (2016) Nuclear m(6)A reader YTHDC1 regulates mRNA splicing. Mol Cell 61:507–519. https://doi.org/10.1016/j.molcel.2016.01.012

Yin X, Jiang X, Wang J, Qian S, Liu F, Qian W (2018) SIRT1 deacetylates SC35 and suppresses its function in Tau Exon 10 inclusion. J Alzheimers Dis 61:561–570. https://doi.org/10.3233/JAD-170418

Yuryev A, Patturajan M, Litingtung Y, Joshi RV, Gentile C, Gebara M, Corden JL (1996) The C-terminal domain of the largest subunit of RNA polymerase II interacts with a novel set of serine/arginine-rich proteins. Proc Natl Acad Sci U S A 93:6975–6980

Zahler AM, Lane WS, Stolk JA, Roth MB (1992) SR proteins: a conserved family of pre-mRNA splicing factors. Genes Dev 6:837–847

Zhang Z, Krainer AR (2004) Involvement of SR proteins in mRNA surveillance. Mol Cell 16:597–607. https://doi.org/10.1016/j.molcel.2004.10.031

Zhang WJ, Wu JY (1996) Functional properties of p54, a novel SR protein active in constitutive and alternative splicing. Mol Cell Biol 16:5400–5408

Zhou Z, Fu XD (2013) Regulation of splicing by SR proteins and SR protein-specific kinases. Chromosoma 122:191–207. https://doi.org/10.1007/s00412-013-0407-z

Zhou Z et al (2012) The Akt-SRPK-SR axis constitutes a major pathway in transducing EGF signaling to regulate alternative splicing in the nucleus. Mol Cell 47:422–433. https://doi.org/10.1016/j.molcel.2012.05.014

Zhou H et al (2017) IRAK2 directs stimulus-dependent nuclear export of inflammatory mRNAs. Elife 6. https://doi.org/10.7554/eLife.29630

Zhu Y et al (2018) Molecular mechanisms for CFIm-mediated regulation of mRNA alternative polyadenylation. Mol Cell 69:62–74.e64. https://doi.org/10.1016/j.molcel.2017.11.031

# Chapter 4
# The Nuclear RNA Exosome and Its Cofactors

**Manfred Schmid and Torben Heick Jensen**

**Abstract** The RNA exosome is a highly conserved ribonuclease endowed with 3′–5′ exonuclease and endonuclease activities. The multisubunit complex resides in both the nucleus and the cytoplasm, with varying compositions and activities between the two compartments. While the cytoplasmic exosome functions mostly in mRNA quality control pathways, the nuclear RNA exosome partakes in the 3′-end processing and complete decay of a wide variety of substrates, including virtually all types of noncoding (nc) RNAs. To handle these diverse tasks, the nuclear exosome engages with dedicated cofactors, some of which serve as activators by stimulating decay through oligoA addition and/or RNA helicase activities or, as adaptors, by recruiting RNA substrates through their RNA-binding capacities. Most nuclear exosome cofactors contain the essential RNA helicase Mtr4 (MTR4 in humans). However, apart from Mtr4, nuclear exosome cofactors have undergone significant evolutionary divergence. Here, we summarize biochemical and functional knowledge about the nuclear exosome and exemplify its cofactor variety by discussing the best understood model organisms—the budding yeast *Saccharomyces cerevisiae*, the fission yeast *Schizosaccharomyces pombe*, and human cells.

**Keywords** RNA exosome · Nuclear RNA decay · Exosome cofactors · Polyadenylation · TRAMP · NEXT · PAXT

## 4.1 The RNA Exosome

### *4.1.1 The Core*

The central core of the RNA exosome is barrel-shaped and composed of six RNase PH-like proteins that form a ring. This ring associates with 3 S1/KH RNA-binding

M. Schmid · T. H. Jensen (✉)
Department of Molecular Biology and Genetics, Aarhus University, Aarhus C, Denmark
e-mail: ms@mbg.au.dk; thj@mbg.au.dk

 
M. Oeffinger, D. Zenklusen (eds.), *The Biology of mRNA: Structure and Function*, Advances in Experimental Medicine and Biology 1203,
https://doi.org/10.1007/978-3-030-31434-7_4

domain containing proteins positioned on one end of the barrel, typically pictured as the "top" (Januszyk and Lima 2014; Liu et al. 2006; Lorentzen et al. 2007). The structure of the resulting 9 protein subunit core complex, termed Exo9, is very similar to eubacterial exonucleases RNase PH and PNPase, and the archaeal RNA exosome, which all have active phosphorolytic exonuclease sites positioned in the central cavity of a characteristic barrel-shaped structure (Januszyk and Lima 2014). In contrast, most eukaryotic Exo9 homologs have altered active site residues, which results in a catalytically inactive exosome core (Dziembowski et al. 2007; Liu et al. 2006). Notable exceptions are plants and some early-branching nonplant eukaryotes, where one of the RNase PH domains has retained phosphorolytic activity (Sikorska et al. 2017).

### *4.1.2 Catalytic Subunits*

To compensate for the widespread loss of activity within their cores, eukaryotic exosomes assemble with the processive 3′–5′ exonuclease and endonuclease Dis3 (often also referred to as Rrp44; DIS3 in humans) and the distributive 3′–5′ exonuclease Rrp6 (EXOSC10 in humans), which bind to the bottom and the top of the core exosome, respectively (Dziembowski et al. 2007; Makino et al. 2013; Mitchell et al. 1997; Wasmuth et al. 2014; Zinder et al. 2016). Exosome complexes comprising Dis3 or Dis3 plus Rrp6 are commonly referred to as Exo10 and Exo11, respectively. Dis3 receives RNAs that are threaded down through the central channel of the exosome core, whereas Rrp6 accesses RNA from the exosome top without threading through the core structure (Kowalinski et al. 2016; Liu et al. 2016; Makino et al. 2013; Wasmuth et al. 2014; Zinder et al. 2016). Even so, Rrp6 functions are intimately linked with the exosome core, and its position close to the entry site of the central channel is consistent with data, suggesting that Rrp6 may control RNA threading to Dis3 (Makino et al. 2015; Wasmuth et al. 2014). Conversely, core KH domain proteins contribute to the binding of RNAs processed by Rrp6 (Zinder et al. 2016). In addition, Rrp6 and its partner Rrp47 provide critical binding surfaces for exosome cofactors, such as Mtr4 (Fig. 4.1) (Falk et al. 2017a; Schuch et al. 2014).

Dis3 and Rrp6 association with the exosome core varies to some extent between organisms and subcellular compartments. Both budding and fission yeasts possess a single Dis3 and Rrp6 paralog, with Rrp6 being exclusively nuclear, while Dis3 is present on both nuclear and cytoplasmic exosomes (Allmang et al. 1999; Mitchell et al. 1997). The situation is more complex for higher eukaryotes; the human genome, for example, encodes three different Dis3 paralogs: DIS3, "DIS3 like" (DIS3L), and DIS3L2. While DIS3 and DIS3L inhabit nuclear and cytoplasmic exosomes, respectively (Staals et al. 2010; Tomecki et al. 2010), DIS3L2 exercises cytoplasmic 3′–5′ exonucleolytic activities independent of the core exosome (Chang et al. 2013; Lubas et al. 2013; Malecki et al. 2013). Moreover, even though the single human Rrp6 paralog EXOSC10 is primarily nuclear, some cytoplasmic presence has also been reported (Brouwer et al. 2001; Lejeune et al. 2003; Tomecki et al. 2010).

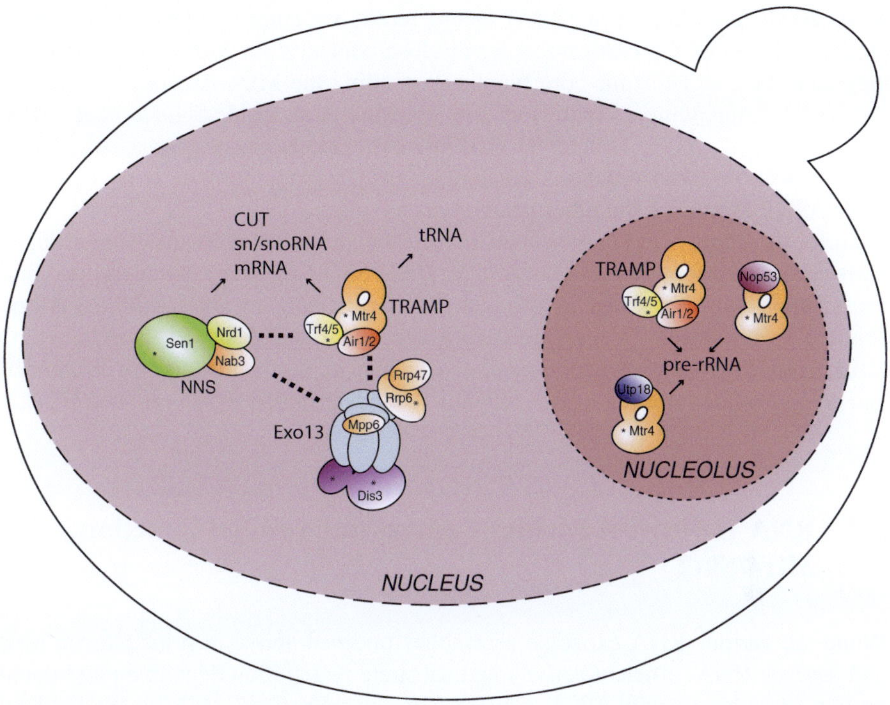


**Fig. 4.1** The exosome and its cofactors in the *S. cerevisiae* nucleus. The major exosome cofactor in *S. cerevisiae* nuclei is the TRAMP complex (Mtr4, Air1/Air2, and Trf4/Trf5), which acts in the nucleoplasm and nucleolus. In the nucleoplasm, the NNS adaptor complex (Nrd1, Nab3, Sen1) is important for exosome targeting of all major RNAPII transcript classes, such as CUTs, sn/snoRNAs, and mRNAs. In nucleoli, TRAMP and the Mtr4-interacting proteins Nop53 and Utp18 recruit the exosome for the processing of rRNA precursors and the decay of processing by-products. TRAMP also facilitates decay of hypomodified tRNA. Asterisks denote enzymatic activities. See text for more detail

The mechanisms determining the subcellular fractioning of exosomes and which nucleases they carry still remain to be elucidated.

### 4.1.3 Lrp1 and Mpp6

In addition to the core and catalytic components, the Lrp1 (often also referred to as Rrp47, C1D in humans) and Mpp6 (MPP6 in humans) proteins are considered constituents of the nuclear exosome, yielding Exo13 (Makino et al. 2015; Milligan et al. 2008; Mitchell et al. 2003; Schilders et al. 2005; Schuch et al. 2014; Wasmuth et al. 2017). Both Lrp1 and Mpp6 are nuclear restricted and were originally proposed to act as exosome adaptors by facilitating exosome access to specific substrates either by direct RNA binding or by contacting specific RNP components (Milligan et al. 2008; Schilders et al. 2005). More recent structural studies position both proteins on top of the exosome core, in close contact with Mtr4, suggesting that they contribute to exosome core function by aiding the Mtr4–exosome interaction and its RNA threading activity (Fig. 4.1) (Makino et al. 2015; Schuch et al. 2014; Wasmuth et al. 2017; Falk et al. 2017a). Lrp1 binds to, and stabilizes, Rrp6, wherefore its in vivo functions are largely overlapping those of Rrp6 (Mitchell et al. 2003). Mpp6, on the other hand, contacts other exosome core subunits, but it is curiously enough only associated in substoichiometric amounts, suggesting a more specialized function (Schilders et al. 2005; Shi et al. 2015). Recently, budding yeast Mpp6 was suggested to promote RNA threading to Dis3, whereas Mpp6 absence would result in the threading-independent decay by Rrp6 (Kim et al. 2016). Whether such an Mpp6-mediated switch in decay mechanism is general and conserved in other organisms remains to be determined.

## 4.2 RNA Helicase Activities Central to Exosome Function: Mtr4/Ski2

While the various RNA exosome assemblies outlined above in principle can bind and degrade RNA, efficient activity and substrate recognition depend on additional protein complexes, with RNA helicases of the Mtr4/Ski2 (MTR4 (SKIV2L2)/SKIV2L in humans) family playing central roles (Johnson and Jackson 2013; Zinder and Lima 2017). Binding to the top of the exosome, these proteins hand RNA substrates to the exosome core, possibly using the RNA helicase activity to inject the substrate for threading down to Dis3 or for presenting the RNA to Rrp6 (Falk et al. 2017a; Halbach et al. 2013; Zinder et al. 2016). Critically, Mtr4/Ski2 are also part of other complexes, containing so-called adaptor proteins, which serve to directly recognize exosome substrates (see below). Despite some commonalities, these complexes have diverged considerably within and between different eukaryotic species. Ski2 homologs are generally cytoplasmic, while Mtr4 homologs are nuclear (Zinder and Lima 2017). As this chapter focuses on nuclear exosome biology, the next sections will describe the different Mtr4-containing complexes in the three model organisms of choice.

## 4.3 *S. cerevisiae*

### 4.3.1 The TRAMP and NNS Complexes

The *S. cerevisiae* Trf4–Air2–Mtr4 polyadenylation (TRAMP) and Nrd1–Nab3–Sen1 (NNS) complexes were among the first nuclear exosome cofactors to be discovered (Fig. 4.1) (LaCava et al. 2005; Vanacova et al. 2005; Wyers et al.

2005). TRAMP consists of Mtr4, the poly(A) polymerase Trf4, and the RNA-binding protein Air2. Trf4 and Air2 can be replaced by their paralogs Trf5 and Air1, yielding different possible TRAMP compositions (LaCava et al. 2005; Vanacova et al. 2005; Wyers et al. 2005). Current evidence suggests that at least a subset of these different TRAMP complexes are present in vivo and serve partially nonoverlapping functions (San Paolo et al. 2009; Schmidt et al. 2012). The molecular contribution of TRAMP complexes to exosome activity is believed to involve the addition of short A-tails to RNA 3′ ends by the Trf4/5 enzymes (Schmidt and Butler 2013; Zinder and Lima 2017). Consistently, in wild-type cells, TRAMP targets can be found carrying short (~4 nt) oligo(A) tails, which are lost in Trf4/5 mutants but accumulate upon exosome inactivation (Jia et al. 2011; Tuck and Tollervey 2013; Wlotzka et al. 2011). These short unstructured tails are then suggested to facilitate the loading of RNA 3′ ends by Mtr4 to promote the exosomal threading of otherwise structured RNAs. The Zn-finger containing Air1/2 proteins may provide RNA-binding capacity to the TRAMP complex, while also promoting overall complex stability through cooperative binding with Trf4/5 to Mtr4 (Falk et al. 2014; Hamill et al. 2010).

In budding yeast, TRAMP engages in a wide variety of nuclear exosome functions, including the decay of aberrant tRNA, processing of rRNA precursors, and decay of processing by-products (Schmidt and Butler 2013). In these cases, the combined adenylation and helicase activities of TRAMP may allow for the decay of otherwise highly structured substrates resilient to exosomal attack. Moreover, TRAMP also targets RNA polymerase II (RNAPII) products, e.g., facilitating decay of the so-called cryptic unstable transcripts (CUTs; see below) and the 3′ trimming of snRNA and snoRNA precursors (Schmidt and Butler 2013). These latter exosome substrates are unlikely to be highly structured, reflecting that TRAMP might also serve as an RNA-binding adaptor, in addition to its role as enzymatic activator.

How does TRAMP get in contact with RNA? At least some targeting capacity is likely to be mediated by direct RNA contacts via the Air proteins (Holub et al. 2012; Schmidt and Butler 2013; Schmidt et al. 2012). However, in the case of RNAPII-produced substrates, target recognition is often mediated by the NNS complex through the sequence-specific RNA-binding domains of the Nrd1 and Nab3 proteins (Tudek et al. 2014; Vasiljeva and Buratowski 2006; Wlotzka et al. 2011). Nrd1 further contains a so-called C-terminal domain (CTD) interaction domain (CID), which specifically binds the Ser5P phosphorylated CTD of the largest subunit of RNAPII, while also directly binding to the TRAMP complex component Trf4 (Gudipati et al. 2008; Tudek et al. 2014; Vasiljeva et al. 2008). Hence, the NNS complex associates with both, early elongating RNAPII, and the TRAMP and exosome complexes (Fig. 4.1). In doing so, it serves two functions: (1) promoting transcription termination of RNAPII from short transcription units (TUs), and (2) channeling 3′ ends derived from such termination events for TRAMP/exosome-mediated trimming or decay (Arigo et al. 2006; Gudipati et al. 2008; Schulz et al. 2013; Steinmetz et al. 2006; Thiebaut et al. 2006; Tudek et al. 2014). The contemporary view suggests that NNS function relies on the binding of Nrd1 and Nab3 to their respective RNA recognition sites during early transcription (Porrua and Libri 2015). This likely involves the interaction of Nrd1 with RNAPII, since the CTD Ser5-P modification is most prominent at TU 5′ ends. Since the helicase activity of Sen1 can promote the disassembly of RNAPII transcription complexes in vitro (Porrua and Libri 2013), this explains the termination function of the NNS complex. After transcription termination, NNS supposedly "hands" the resulting transcript to the RNA exosome via the Nrd1–Trf4 interaction (Tudek et al. 2014). However, disruption of this interaction causes only a moderate stabilization of NNS targets (Tudek et al. 2014). Moreover, Nrd1/Nab3 can directly contact the exosome components Mpp6 and Rrp6 independent of TRAMP (Fasken et al. 2015; Kim et al. 2016). Thus, TRAMP appears to not be strictly required for exosome association with NNS targets but may rather serve to promote degradation of transcripts that are not directly amenable to exosomal decay.

Exosome removal of NNS-targeted transcripts is highly efficient, and typically, these RNAs are only revealed in NNS-, TRAMP-, or exosome-depleted cells, hence their nomenclature as CUTs (Neil et al. 2009; Wyers et al. 2005; Xu et al. 2009). The RNA sequence motifs recognized by Nrd1 and Nab3 are short and abundantly present in the *S. cerevisiae* genome but conspicuously absent from the coding strand of protein-coding genes (Cakiroglu et al. 2016; Schulz et al. 2013). This explains how the NNS complex discriminates the numerous RNAs produced by spurious transcription, either bidirectionally from gene promoters or antisense to mRNAs, from protein-coding transcripts. At sn/snoRNA TUs, NNS activity facilitates the production of short stable RNAs. This is presumably due to the highly structured and protein-bound nature of mature sn/snoRNAs, which stops RNA exosome progress after its initial removal (processing) of the unstructured 3′ extensions (Coy et al. 2013). CUTs do not assemble stable structures and are thus completely decayed.

Although the NNS and TRAMP complexes were long believed to target only ncRNAs, recent data revealed NNS and Mtr4 interaction with a host of mRNAs whose expression changes in response to glucose depletion (Bresson et al. 2017). This suggests the interesting possibility that mRNAs can be targeted for nuclear decay and that this can be regulated in a stimulus-specific manner. Such potential re-purposing of the NNS complex from ncRNA to mRNA targeting is consistent with an earlier study, showing that Nrd1 is dephosphorylated during nutrient depletion and that this influences nutrient-dependent protein-coding gene expression (Darby et al. 2012). However, whether Nrd1/Mtr4 mRNA targeting elicits decay and if so, how such regulation may occur remains to be determined.

### *4.3.2 Nucleolar Exosome Cofactors*

In addition to its role in TRAMP, *S. cerevisiae* Mtr4 also interacts directly with the nucleolar proteins Nop53 and Utp18 (Fig. 4.1, "NUCLEOLUS"). This occurs through the so-called arch domain of Mtr4, which binds a conserved short sequence motif, the arch interaction motif (AIM) (Falk et al. 2017b; Thoms et al. 2015).

Nop53 is a component of nuclear ribosomal pre-60S particles, which contain 5.8S rRNA precursors, and its interaction with Mtr4 is required for the exosomal trimming of 3′ extensions of 5.8S pre-rRNAs. Utp18, instead, is part of ribosomal pre-90S particles and takes part in the release, and Mtr4-dependent decay, of the nonfunctional 5′ external transcribed spacer (5′ETS) (Thoms et al. 2015). Interestingly, TRAMP is also implicated in 5′ETS removal (Houseley and Tollervey 2006), but the functional relationship between Mtr4's action in the context of TRAMP and together with Utp18 has not been disentangled. Assembly of the TRAMP complex does not depend on the Mtr4 arch domain, and it is therefore possible that Utp18 recruits Mtr4 as part of the TRAMP complex for 5′ETS decay (Falk et al. 2017b; Thoms et al. 2015). At the same time, there are nonessential direct contacts between the Mtr4 arch domain and Air2 within TRAMP (Falk et al. 2017b), suggesting that Air2 and Utp18 interactions with Mtr4p influence each other to control 5′ETS decay.

## 4.4 *S. pombe*

### *4.4.1 TRAMP*

The composition of the fission yeast TRAMP complex is overall similar to its budding yeast paralog with subunits Cid14 (homologous to Trf4/5), Air1, and Mtr4 (Fig. 4.2, "NUCLEOLUS") (Keller et al. 2010). Compared to *S. cerevisiae*, *S. pombe* TRAMP appears to be a more specialized exosome cofactor, still implicated in the processing or decay of nucleolar substrates but with a less general role in the nucleoplasm (Larochelle et al. 2012; Win et al. 2006). Consistently, a functional analog of the *S. cerevisiae* NNS complex has not been identified (Lemay et al. 2016; Wittmann et al. 2017). Instead, Cid14 and the exosome subunit Rrp6 were shown to be involved in RNAi-independent heterochromatin formation processes, pointing toward a still ill-defined link between decay of heterochromatin-derived transcripts and the deposition of chromatin marks (Buhler et al. 2007; Keller et al. 2010; Reyes-Turcu et al. 2011; Wang et al. 2008).

### *4.4.2 MTREC*

To engage in nuclear activities outside of nucleoli, the fission-yeast specific nucleoplasmic-residing Mtr4 paralog, called Mtr4-like 1 (Mtl1), forms a tight complex with the Zn-finger protein Red1 (Fig. 4.2). This dimer then interacts with numerous other proteins to form higher-order complexes termed MTREC (Mtl1–Red1 core) or NURS (nuclear RNA silencing) (Egan et al. 2014; Lee et al. 2013; Zhou et al. 2015). Red1 is required for MTREC's association with the *S. pombe* exosome, supposedly compensating for Mtl1's loss of a specific N-terminal domain required for Mtr4:Rrp6/Lsd1 interaction (Schuch et al. 2014; Zhou et al. 2015). The

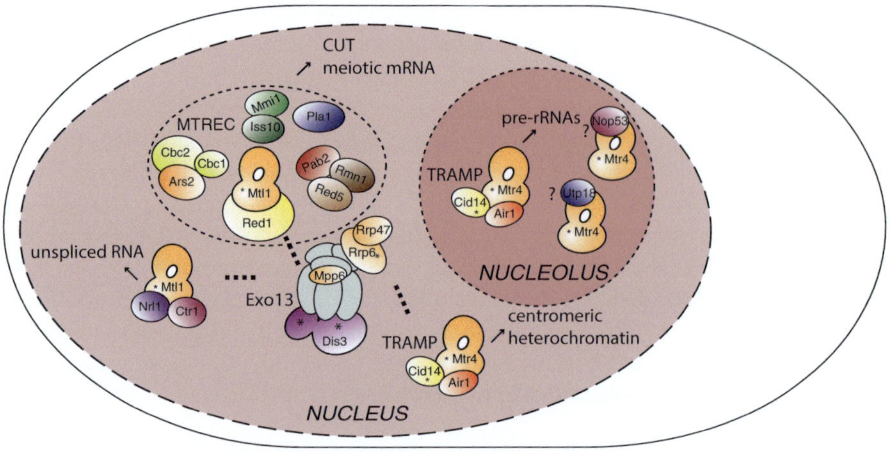


**Fig. 4.2** The exosome and its cofactors in the *S. pombe* nucleus. Mtl1 and Red1 form the core of the major exosome cofactors in the *S. pombe* nucleoplasm. MTREC comprises Mtl1–Red1 and a number of other protein complexes, including nCBC (Cbc1, Cbc2, Ars2); a subcomplex comprising Mmi1 and Iss10; a subcomplex of Pab2, Rmn1, and Red5; and the poly(A) polymerase Pla1. MTREC targets include meiosis-specific RNAs during vegetative growth but also other RNAPII-derived transcripts, such as CUTs. In addition to MTREC, Mtl1 is part of a complex including Ctr1 and Nrl1, which also binds the exosome and is involved in decay of unspliced RNAs. The *S. pombe* TRAMP complex functions in the nucleoplasm where it targets centromeric heterochromatin and in the nucleolus where it is involved in rRNA processing. Utp18 and Nop53 homologs contain AIM domains and are conserved, and therefore also likely to act as exosome cofactors via Mtr4. Asterisks denote enzymatic activities. Question marks are used to symbolize that roles of Utp18 and Nop53 have not been demonstrated in *S. pombe*. See text for more detail

MTREC complex comprises a number of other proteins, including Red5, Iss10, Mmi1, poly(A)-binding protein Pab2, poly(A) polymerase Pla1, and the nuclear mRNA 5′ cap-binding complex (nCBC) proteins (Egan et al. 2014; Lee et al. 2013; Zhou et al. 2015) (Fig. 4.2). The presence of Pla1 and Pab2 in MTREC might provide a means to add and recognize A-tails of MTREC targets, independent of TRAMP. At the same time, Pla1 (and probably Pab2) are also required for the production of regular mRNAs that normally need to avoid nuclear decay. How this distinction is achieved is presently under intense investigation (see below).

MTREC should not be seen as a single well-defined complex, but rather comprises a number of functionally distinct subcomplexes associating around the Mtl1–Red1 core (Fig. 4.2). Mtl1 also engages in a Red1-independent complex with the Nrl1 and Ctr1 proteins, which interact with the splicing machinery and seem to be specifically involved in exosomal decay of unspliced transcripts, including those containing so-called cryptic introns (Lee et al. 2013; Zhou et al. 2015).

Interestingly, homologs of many MTREC components are also cofactors of the human RNA exosome (Table 4.1) and several harbor RNA-binding domains, which might contribute to target recognition. An example is the sequence-specific RNA-binding protein Mmi1, which serves a highly specific role in the targeting of

**Table 4.1** Exosome components and cofactors

| Complex | *S. cerevisiae* | *S. pombe* | Human | Domains |
|---|---|---|---|---|
| Exo13 | Csl4 | Csl4 | EXOSC1 | S1 |
| | Rrp4 | Rrp4 | EXOSC2 | S1/KH |
| | Rrp40 | Rrp40 | EXOSC3 | S1/KH |
| | Rrp41 | Rrp41 | EXOSC4 | RNase PH |
| | Rrp46 | Rrp46 | EXOSC5 | RNase PH |
| | Mtr3 | Mtr3 | EXOSC6 | RNase PH |
| | Rrp42 | Rrp42 | EXOSC7 | RNase PH |
| | Rrp43 | Rrp43 | EXOSC8 | RNase PH |
| | Rrp45 | Rrp45 | EXOSC9 | RNase PH |
| | Rrp6 | Rrp6 | EXOSC10 | 3′–5′ exonuclease (RNase D) |
| | Dis3 (Rrp44) | Dis3 | DIS3, DIS3L | 3′–5′ exonuclease (RNase II), PIN endonuclease domain |
| | Lrp1 (Rrp47) | Cti1 | C1D (LRP1) | C1D |
| | Mpp6 | Mpp6 | MPP6 | – |
| Mtr4 | Mtr4 | Mtr4 Mtl1 | MTREX (MTR4, SKIV2L2) | ATP-dependent RNA helicase |
| TRAMP | Trf4, Trf5 | Cid14 | PAPD5 (TRF4-2) | poly(A) polymerase |
| | Air1, Air2 | Air1 | ZCCHC7 (AIR1) | Zn-knuckle |
| NNS | Nrd1 | Seb1[#] | ? SCAF4, SCAF8 | RRM, CID |
| | Nab3 | Nab3[#] | ? RALY | RRM |
| | Sen1 | Sen1[#] | SETX (ALS4, AOA2)[#] | ATP-dependent RNA helicase |
| NEXT | | | RBM7 | RRM |
| | | | ZCCHC8 | Zn-finger |
| nCBC | ? Cbc2 (Cbp20) | Cbc2 | NCBP2 (CBP20) | RRM |
| | ? Sto1 (Cbp80) | Cbc1 | NCBP1 (CBP80) | – |
| | | Pir2 | SRRT (ARS2) | Zn-finger |
| | | | ZC3H18 | Zn-finger |
| MTREC/ PAXT | | Red1 | ZFC3H1 | Zn-finger |
| | | Red5 | ? ZC3H3 | Zn-finger |
| | | Mmi1 | ? YTHDF1/2/3 | YTH domain |
| | | Iss10 | ZFC3H1 N terminus | – |
| | | Rmn1 | ? RBM26, RBM27 | RRM |
| | Pap1 | Pla1 | ? PAPOLA, PAPOLG | Poly(A) polymerase |
| | | Pab2 | PABPN1 | RRM (poly(A) binding) |
| – | | Ctr1 | ? CCDC174 | – |
| | | Nrl1 | ? NRDE2 | – |

(continued)

**Table 4.1** (continued)

| Complex | *S. cerevisiae* | *S. pombe* | Human | Domains |
|---|---|---|---|---|
| – | Utp18 | ? Utp18 | ? UTP18 (WDR50) | WDR40, AIM |
| | Nop53 | ? Rrp16 | ? NOP53 (GLTSCR2) | AIM |
| – | ? Rix7 | ? Rix7 | NVL (NVL2) | AAA ATPase |

List of exosome components and cofactors from *S. cerevisiae*, *S. pombe*, and human cells. Listed are standard gene names from the *S. cerevisiae* and *S. pombe* genome databases (www.yeastgenome.org and www.pombase.org) as well as approved symbols for human genes from the "HUGO" Gene Nomenclature Committee (www.genenames.org). Alternative, commonly used names are in parenthesis. Sequence homologs which are not proposed to have a functional connection with the nuclear RNA exosome are marked with "#," and sequence homologs where a functional connection to the exosome is possible but not yet demonstrated are marked with "?"

meiosis-specific mRNAs during *S. pombe* vegetative growth (Harigaya et al. 2006). This silencing is partly achieved by the posttranscriptional decay of these transcripts mediated by Mmi1–MTREC binding to cognate sites in the target RNAs and their subsequent handover to the nuclear exosome (Chen et al. 2011; Harigaya et al. 2006; Yamashita et al. 2012). In addition to posttranscriptional decay, silencing of meiosis-specific genes involves the formation of heterochromatic islands around affected loci, and the MTREC complex is involved in the deposition of repressive chromatin marks (Egan et al. 2014; Lee et al. 2013). This activity is independent on Cid14 and therefore provides a unique link between RNA decay and heterochromatin formation in *S. pombe* that has not been reported in other organisms.

## 4.5 Human

### *4.5.1 TRAMP*

A TRAMP-like complex, although still poorly characterized, also exists in human cells and is composed of MTR4, the poly(A) polymerase PAPD5, and the Zn-finger protein ZCCHC7 (Lubas et al. 2011). This complex, hTRAMP, localizes to nucleoli, and its depletion mainly results in phenotypes affecting nucleolar substrates (Lubas et al. 2011), which is consistent with the presence of distinct adaptor complexes serving exosome functions in the nucleoplasm. This yields a conceptually similar setup as for *S. pombe* (Fig. 4.3). MTR4 also interacts with other nucleolar proteins, such as NVL, which promotes pre-rRNA processing, and the human homologs of *S. cerevisiae* Nop53 and Utp18, suggesting that these proteins are also exosome cofactors in human rRNA metabolism (Lubas et al. 2011; Yoshikatsu et al. 2015).

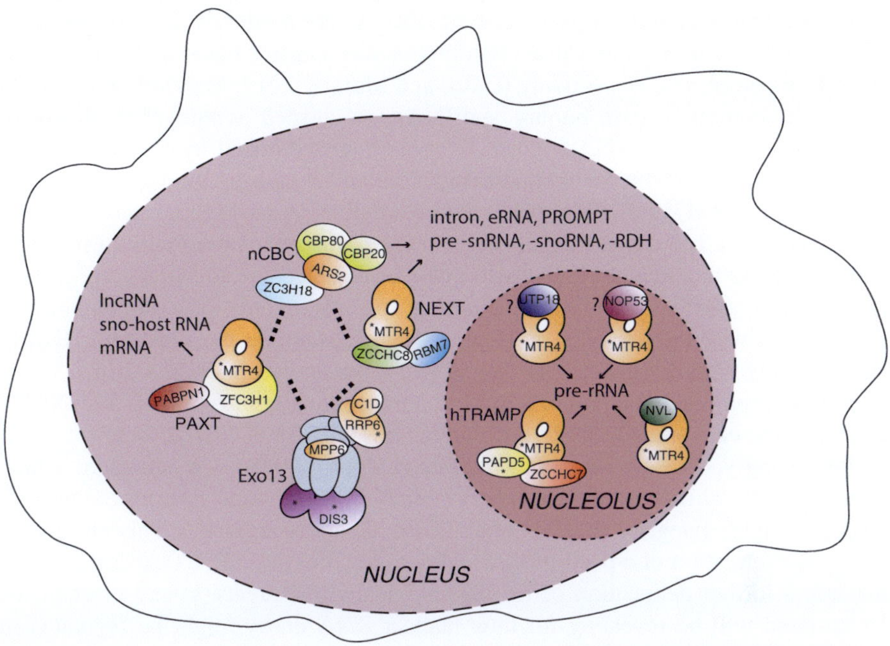


**Fig. 4.3** The exosome and its cofactors in the human cell nucleus. Human exosome cofactors NEXT (MTR4, ZCCHC8, and RBM7) and PAXT (MTR4, ZFC3H1, and PABPN1) are present in the nucleoplasm. PAXT targets poly(A)$^+$ lncRNAs, including the subclass hosting intronic snoRNAs ("sno-host RNA"), and mRNAs. NEXT facilitates decay of numerous unstable transcripts (i.e., some PROMPTs, eRNAs, and spliced-out introns) and the 3′ processing of pre-snRNAs, pre-snoRNAs, and replication-dependent histone (RDH)-encoding mRNAs. Both PAXT and NEXT bind to the CBCA complex (CBC80, CBC20, ARS2) via the protein ZC3H18, but a functional role of CBCA in exosomal decay has primarily been shown for NEXT targets. Cofactors in human nucleoli include hTRAMP (MTR4, ZCCHC7, PAPD5) and the MTR4-interacting protein NVL. In addition, human MTR4 associates with NOP53 and UTP18, suggesting a conserved role of these proteins, even though the AIM domain is only conserved in NOP53. Nucleolar cofactors facilitate pre-rRNA processing and decay of processing by-products. Asterisks denote enzymatic activities. Question marks are used to symbolize that roles of UTP18 and NOP53 have not been demonstrated in humans. See text for more detail

## *4.5.2 NEXT*

The nuclear exosome targeting (NEXT) complex is presently the best-characterized human exosome adaptor complex. It consists of the RNA-binding protein RBM7, linked to MTR4 by the Zn-finger protein ZCCHC8 (Lubas et al. 2011) (Fig. 4.3). NEXT facilitates the exosomal decay of many promoter-upstream transcripts (PROMPTs, also called upstream antisense (ua)RNAs) and other labile ncRNAs, like enhancer RNAs (eRNAs) (Lubas et al. 2011, 2015; Meola et al. 2016). Moreover, it mediates the exosomal trimming of 3′-end extensions of snRNAs, snoRNAs, and histone-encoding mRNAs (Lubas et al. 2011, 2015).

Human snRNAs and histone-encoding mRNAs are produced from autonomous TUs using specialized transcription termination mechanisms based on the Integrator and CPSF complexes, respectively (Guiro and Murphy 2017; Marzluff and Koreski 2017). In contrast, most human snoRNAs are hosted within the introns of pre-mRNAs and pre-ncRNAs, where from they are produced by the trimming of excised intron 5′ ends by the exonuclease XRN2 and 3′ ends by the exosome (Valen et al. 2011). RBM7 Individual-nucleotide resolution Cross-Linking and Immuno-Precipitation (iCLIP) experiments demonstrated that the protein promiscuously contacts RNAs in a manner unlikely to involve sequence-specific target recognition (Lubas et al. 2015). Thus, RBM7 binding appears to only be consequential in combination with the presence of an unprotected 3′-end. At the same time, RBM7 also interacts with the splicing factor SF3B2 (also termed SAP145), which likely underlies the enriched binding of RBM7 to intronic 3′ ends and explains how NEXT facilitates the exosomal decay of intronic regions (Falk et al. 2016).

Interestingly, RBM7 gets phosphorylated upon cellular UV damage, which debilitates the ability of the protein to bind RNA without otherwise affecting NEXT complex integrity (Blasius et al. 2014; Tiedje et al. 2015). This provides a first characterization of a posttranslational modification of an RNA exosome cofactor, and a further delineation of its physiological consequence(s) and mechanistic background will be revealing for how nuclear RNA decay might be regulated in response to external stimuli.

### *4.5.3 PAXT*

A third human MTR4-containing complex assembles around the stable MTR4–ZFC3H1 dimer (Meola et al. 2016). ZFC3H1 and MTR4 depletions both lead to the accumulation of the mature products of some snoRNA host genes as well as numerous other nuclear transcripts (Meola et al. 2016; Ogami et al. 2017). Many ZFC3H1-specific targets are also stabilized upon depletion of the nuclear poly(A)-binding protein PABPN1, and exosomal decay depends on their polyadenylation by the canonical poly(A) polymerase PAP prompting the idea of a so-called PAP-mediated RNA decay (PPD) pathway (Beaulieu et al. 2012; Bresson and Conrad 2013; Bresson et al. 2015; Meola et al. 2016). The PPD pathway was originally suggested to primarily affect ncRNAs, but transcriptome-wide analysis of ZFC3H1 and PABPN1 inactivation indicated that mRNAs are also frequently targeted (Meola et al. 2016; Silla et al. 2018). Taken all evidence together, the emerging picture suggests that PABPN1 binding to RNA poly(A) tails will lead to recruitment of the exosome via MTR–ZFC3H1 unless the RNA manages to escape the nucleus (Meola and Jensen 2017). Consistently, PABPN1 associates with MTR4 in a ZFC3H1-dependent manner (Meola et al. 2016), yet, this interaction is less robust than that of the core MTR4–ZFC3H1 dimer, inspiring the proposition of a "poly(A) exosome targeting" (PAXT) connection, comprising MTR4, ZFC3H1, and PABPN1 (Meola et al. 2016) (Fig. 4.3). The term "connection" recognizes that this is not a stable complex and the suboptimal binding of PABPN1 may indeed help explain how stable polyadenylated transcripts evade decay (see below).

ZFC3H1 and PABPN1 are the human homologs of the *S. pombe* Red1 and Pab2 MTREC components, suggesting an overall conserved function between PAXT and MTREC in the decay of polyadenylated nuclear RNA, including mRNA.

### *4.5.4 The Nuclear RNA Cap-Binding Complex*

Both NEXT and PAXT components can be physically bridged to the nuclear 5′ cap-binding complex (nCBC) and nCBC proteins also co-IP the nuclear exosome (Andersen et al. 2013; Lubas et al. 2011; Meola et al. 2016). The link between NEXT/PAXT and the nCBC is mediated by the ZC3H18 protein, which further binds to the nCBC proteins Cbp20 and Cbp80 (also termed NCBP1 and NCBP2) via the protein ARS2 (also termed SRRT) (Giacometti et al. 2017) (Fig. 4.3). Individual depletion of all of these proteins leads to the stabilization of some nuclear exosome substrates, suggesting that nCBC in some instances contribute to exosome recruitment (Andersen et al. 2013; Iasillo et al. 2017). In addition, nCBC and ARS2, but neither ZC3H18 nor NEXT, are required for the efficient termination of RNAPII transcription at PROMPT, snRNA, and histone mRNA TUs, which suggests an active coupling between transcription termination and decay (Andersen et al. 2013; Iasillo et al. 2017). This is reminiscent of budding yeast NNS activity, which also promotes transcription termination before offering substrates to the exosome for decay. nCBC components are also part of *S. pombe* MTREC (Egan et al. 2014; Lee et al. 2013; Zhou et al. 2015), suggesting an omnipresent role of the RNA 5′ cap in facilitating nuclear 3′–5′ decay. While this at first sight seems counterproductive due to the unwanted targeting of capped RNAs with functional roles in the cell, it may indeed provide an important connection, enabling the quality control of capped transcripts.

## 4.6 Nuclear Decay vs. RNA Export

An emerging concept in RNA biology suggests that nuclear RNA decay, as described above, is in competition with RNA nuclear export, to prevent the cytoplasmic appearance of too many nonfunctional molecules. In line with this notion, there are clear indications that nuclear exosome cofactors impact RNA export. This is perhaps best exemplified by the nCBC and ARS2 (forming the CBCA complex), which also actively facilitates RNA export by interacting with the "phosphorylated adaptor for RNA export" (PHAX) protein (Giacometti et al. 2017; Hallais et al. 2013). Interestingly, this interaction is mutually exclusive with the binding of ZC3H18 to the CBC, which would otherwise bridge the CBCA complex to the exosome adaptors NEXT and PAXT (Giacometti et al. 2017). Moreover, the nuclear mRNA export factor ALYREF binds at transcript 5′- and 3′-ends via interactions with the nCBC and PABPN1, respectively (Fan et al. 2017; Shi et al. 2017). As described above, both nCBC and PABPN1 also interact with PAXT and/or NEXT, indicating that RNA association with exosome cofactors is generally mutually exclusive with binding of export factors. Consistent with this idea, the PAXT component ZFC3H1 appears to be capable of retaining RNA exosome substrates in the nucleus, as upon depletion of ZFC3H1, numerous PAXT targets are now found in the cytoplasm where they may even engage in translation (Ogami et al. 2017; Silla et al. 2018). This ability of ZFC3H1 to counter untimely RNA export appears to reach beyond simply preventing the binding of export factors as exosome substrates accumulating in exosome-depleted cells concentrate in ZFC3H1-dependent subnuclear aggregates (Silla et al. 2018). ZFC3H1 contains long low complexity regions, suggesting a direct role of ZFC3H1 in forming such foci, which probably reflect RNP complexes formed to prevent their unsolicited export from the nucleus.

But what then decides how RNAs are chosen for decay or export? Targeting of polyadenylated RNAs for exosomal decay mediated by PAXT or MTREC involves PABP recruitment, which occurs not only on exosome targets but also on to stable mRNAs. This conundrum has inspired the so-called nuclear timer model, where PABPs serve to initially protect poly(A) tailed RNA only later to elicit decay of transcripts that remain nuclear (Libri 2010; Meola and Jensen 2017). Mechanistically, this could be achieved through transient interactions of PABPN1/Pab2 with MTR4–ZFC3H1/Mtl1–Red and the exosome, leading to the slow assembly of a decay-promoting complex and avoiding decay of timely exported mRNAs (Meola et al. 2016).

## 4.7 Concluding Remarks

The nuclear RNA exosome partakes in the processing and/or decay of virtually all types of transcripts. Being able to handle such diverse tasks depends on exosome interaction with a number of adapter proteins as described above. This places the RNA exosome as a central player in cellular RNA metabolism. It may therefore come as no surprise that the RNA exosome and some of its cofactors have been linked to various disease states. For example, the exosome subunit DIS3 is recurrently found mutated in multiple myeloma, and mutations in several exosome core subunits are linked to inherited neurodegenerative diseases (Morton et al. 2017; Robinson et al. 2015). In addition, the RNA exosome and its cofactors also figure in the arms races occurring between viruses and their hosts. This is exemplified by influenza viruses, which, on one hand, have been shown to actively hijack the nuclear RNA exosome to produce RNA fragments required for priming transcription of their genomes (Rialdi et al. 2017), while, on the other hand, the cellular defense against other types of RNA viruses involves export of hTRAMP proteins to the cytoplasm to aid in the decay of viral RNA (Molleston et al. 2016). Although still immature, these examples provide a glimpse of the central role of the nuclear RNA exosome in cell biology. While the composition and function of basic exosome machinery is now reasonably understood, much still remains to be learned about the regulation and cellular function of the various exosome cofactors and their relation to cell physiology in different systems.

## References

Allmang C, Petfalski E, Podtelejnikov A, Mann M, Tollervey D, Mitchell P (1999) The yeast exosome and human PM-Scl are related complexes of $3' \rightarrow 5'$ exonucleases. Genes Dev 13:2148–2158

Andersen PR, Domanski M, Kristiansen MS, Storvall H, Ntini E, Verheggen C, Schein A, Bunkenborg J, Poser I, Hallais M et al (2013) The human cap-binding complex is functionally connected to the nuclear RNA exosome. Nat Struct Mol Biol 20:1367–1376

Arigo JT, Eyler DE, Carroll KL, Corden JL (2006) Termination of cryptic unstable transcripts is directed by yeast RNA-binding proteins Nrd1 and Nab3. Mol Cell 23:841–851

Beaulieu YB, Kleinman CL, Landry-Voyer AM, Majewski J, Bachand F (2012) Polyadenylation-dependent control of long noncoding RNA expression by the poly(A)-binding protein nuclear 1. PLoS Genet 8:e1003078

Blasius M, Wagner SA, Choudhary C, Bartek J, Jackson SP (2014) A quantitative 14-3-3 interaction screen connects the nuclear exosome targeting complex to the DNA damage response. Genes Dev 28:1977–1982

Bresson SM, Conrad NK (2013) The human nuclear poly(a)-binding protein promotes RNA hyperadenylation and decay. PLoS Genet 9:e1003893

Bresson SM, Hunter OV, Hunter AC, Conrad NK (2015) Canonical poly(A) polymerase activity promotes the decay of a wide variety of mammalian nuclear RNAs. PLoS Genet 11:e1005610

Bresson S, Tuck A, Staneva D, Tollervey D (2017) Nuclear RNA decay pathways aid rapid remodeling of gene expression in yeast. Mol Cell 65(787–800):e785

Brouwer R, Allmang C, Raijmakers R, van Aarssen Y, Egberts WV, Petfalski E, van Venrooij WJ, Tollervey D, Pruijn GJ (2001) Three novel components of the human exosome. J Biol Chem 276:6177–6184

Buhler M, Haas W, Gygi SP, Moazed D (2007) RNAi-dependent and -independent RNA turnover mechanisms contribute to heterochromatic gene silencing. Cell 129:707–721

Cakiroglu SA, Zaugg JB, Luscombe NM (2016) Backmasking in the yeast genome: encoding overlapping information for protein-coding and RNA degradation. Nucleic Acids Res 44:8065–8072

Chang HM, Triboulet R, Thornton JE, Gregory RI (2013) A role for the Perlman syndrome exonuclease Dis3l2 in the Lin28-let-7 pathway. Nature 497:244–248

Chen HM, Futcher B, Leatherwood J (2011) The fission yeast RNA binding protein Mmi1 regulates meiotic genes by controlling intron specific splicing and polyadenylation coupled RNA turnover. PLoS One 6:e26804

Coy S, Volanakis A, Shah S, Vasiljeva L (2013) The Sm complex is required for the processing of non-coding RNAs by the exosome. PLoS One 8:e65606

Darby MM, Serebreni L, Pan X, Boeke JD, Corden JL (2012) The Saccharomyces cerevisiae Nrd1-Nab3 transcription termination pathway acts in opposition to Ras signaling and mediates response to nutrient depletion. Mol Cell Biol 32:1762–1775

Dziembowski A, Lorentzen E, Conti E, Seraphin B (2007) A single subunit, Dis3, is essentially responsible for yeast exosome core activity. Nat Struct Mol Biol 14:15–22

Egan ED, Braun CR, Gygi SP, Moazed D (2014) Post-transcriptional regulation of meiotic genes by a nuclear RNA silencing complex. RNA 20:867–881

Falk S, Weir JR, Hentschel J, Reichelt P, Bonneau F, Conti E (2014) The molecular architecture of the TRAMP complex reveals the organization and interplay of its two catalytic activities. Mol Cell 55:856–867

Falk S, Finogenova K, Melko M, Benda C, Lykke-Andersen S, Jensen TH, Conti E (2016) Structure of the RBM7-ZCCHC8 core of the NEXT complex reveals connections to splicing factors. Nat Commun 7:13573

Falk S, Bonneau F, Ebert J, Kogel A, Conti E (2017a) Mpp6 incorporation in the nuclear exosome contributes to RNA channeling through the Mtr4 helicase. Cell Rep 20:2279–2286

Falk S, Tants JN, Basquin J, Thoms M, Hurt E, Sattler M, Conti E (2017b) Structural insights into the interaction of the nuclear exosome helicase Mtr4 with the preribosomal protein Nop53. RNA 23:1780–1787

Fan J, Kuai B, Wu G, Wu X, Chi B, Wang L, Wang K, Shi Z, Zhang H, Chen S et al (2017) Exosome cofactor hMTR4 competes with export adaptor ALYREF to ensure balanced nuclear RNA pools for degradation and export. EMBO J 36:2870–2886

Fasken MB, Laribee RN, Corbett AH (2015) Nab3 facilitates the function of the TRAMP complex in RNA processing via recruitment of Rrp6 independent of Nrd1. PLoS Genet 11:e1005044

Giacometti S, Benbahouche NEH, Domanski M, Robert MC, Meola N, Lubas M, Bukenborg J, Andersen JS, Schulze WM, Verheggen C et al (2017) Mutually exclusive CBC-containing complexes contribute to RNA fate. Cell Rep 18:2635–2650

Gudipati RK, Villa T, Boulay J, Libri D (2008) Phosphorylation of the RNA polymerase II C-terminal domain dictates transcription termination choice. Nat Struct Mol Biol 15:786–794

Guiro J, Murphy S (2017) Regulation of expression of human RNA polymerase II-transcribed snRNA genes. Open Biol 7:170073

Halbach F, Reichelt P, Rode M, Conti E (2013) The yeast ski complex: crystal structure and RNA channeling to the exosome complex. Cell 154:814–826

Hallais M, Pontvianne F, Andersen PR, Clerici M, Lener D, Benbahouche Nel H, Gostan T, Vandermoere F, Robert MC, Cusack S et al (2013) CBC-ARS2 stimulates 3′-end maturation of multiple RNA families and favors cap-proximal processing. Nat Struct Mol Biol 20:1358–1366

Hamill S, Wolin SL, Reinisch KM (2010) Structure and function of the polymerase core of TRAMP, a RNA surveillance complex. Proc Natl Acad Sci U S A 107:15045–15050

Harigaya Y, Tanaka H, Yamanaka S, Tanaka K, Watanabe Y, Tsutsumi C, Chikashige Y, Hiraoka Y, Yamashita A, Yamamoto M (2006) Selective elimination of messenger RNA prevents an incidence of untimely meiosis. Nature 442:45–50

Holub P, Lalakova J, Cerna H, Pasulka J, Sarazova M, Hrazdilova K, Arce MS, Hobor F, Stefl R, Vanacova S (2012) Air2p is critical for the assembly and RNA-binding of the TRAMP complex and the KOW domain of Mtr4p is crucial for exosome activation. Nucleic Acids Res 40:5679–5693

Houseley J, Tollervey D (2006) Yeast Trf5p is a nuclear poly(A) polymerase. EMBO Rep 7:205–211

Iasillo C, Schmid M, Yahia Y, Maqbool MA, Descostes N, Karadoulama E, Bertrand E, Andrau JC, Jensen TH (2017) ARS2 is a general suppressor of pervasive transcription. Nucleic Acids Res 45:10229–10241

Januszyk K, Lima CD (2014) The eukaryotic RNA exosome. Curr Opin Struct Biol 24:132–140

Jia H, Wang X, Liu F, Guenther UP, Srinivasan S, Anderson JT, Jankowsky E (2011) The RNA helicase Mtr4p modulates polyadenylation in the TRAMP complex. Cell 145:890–901

Johnson SJ, Jackson RN (2013) Ski2-like RNA helicase structures: common themes and complex assemblies. RNA Biol 10:33–43

Keller C, Woolcock K, Hess D, Buhler M (2010) Proteomic and functional analysis of the noncanonical poly(A) polymerase Cid14. RNA 16:1124–1129

Kim K, Heo DH, Kim I, Suh JY, Kim M (2016) Exosome cofactors connect transcription termination to RNA processing by guiding terminated transcripts to the appropriate exonuclease within the nuclear exosome. J Biol Chem 291:13229–13242

Kowalinski E, Kogel A, Ebert J, Reichelt P, Stegmann E, Habermann B, Conti E (2016) Structure of a cytoplasmic 11-subunit RNA exosome complex. Mol Cell 63:125–134

LaCava J, Houseley J, Saveanu C, Petfalski E, Thompson E, Jacquier A, Tollervey D (2005) RNA degradation by the exosome is promoted by a nuclear polyadenylation complex. Cell 121:713–724

Larochelle M, Lemay JF, Bachand F (2012) The THO complex cooperates with the nuclear RNA surveillance machinery to control small nucleolar RNA expression. Nucleic Acids Res 40:10240–10253

Lee NN, Chalamcharla VR, Reyes-Turcu F, Mehta S, Zofall M, Balachandran V, Dhakshnamoorthy J, Taneja N, Yamanaka S, Zhou M et al (2013) Mtr4-like protein coordinates nuclear RNA processing for heterochromatin assembly and for telomere maintenance. Cell 155:1061–1074

Lejeune F, Li X, Maquat LE (2003) Nonsense-mediated mRNA decay in mammalian cells involves decapping, deadenylating, and exonucleolytic activities. Mol Cell 12:675–687

Lemay JF, Marguerat S, Larochelle M, Liu X, van Nues R, Hunyadkurti J, Hoque M, Tian B, Granneman S, Bahler J et al (2016) The Nrd1-like protein Seb1 coordinates cotranscriptional 3′ end processing and polyadenylation site selection. Genes Dev 30:1558–1572

Libri D (2010) Nuclear poly(a)-binding proteins and nuclear degradation: take the mRNA and run? Mol Cell 37:3–5

Liu Q, Greimann JC, Lima CD (2006) Reconstitution, activities, and structure of the eukaryotic RNA exosome. Cell 127:1223–1237

Liu JJ, Niu CY, Wu Y, Tan D, Wang Y, Ye MD, Liu Y, Zhao W, Zhou K, Liu QS et al (2016) CryoEM structure of yeast cytoplasmic exosome complex. Cell Res 26:822–837

Lorentzen E, Dziembowski A, Lindner D, Seraphin B, Conti E (2007) RNA channelling by the archaeal exosome. EMBO Rep 8:470–476

Lubas M, Christensen MS, Kristiansen MS, Domanski M, Falkenby LG, Lykke-Andersen S, Andersen JS, Dziembowski A, Jensen TH (2011) Interaction profiling identifies the human nuclear exosome targeting complex. Mol Cell 43:624–637

Lubas M, Damgaard CK, Tomecki R, Cysewski D, Jensen TH, Dziembowski A (2013) Exonuclease hDIS3L2 specifies an exosome-independent 3′-5′ degradation pathway of human cytoplasmic mRNA. EMBO J 32:1855–1868

Lubas M, Andersen PR, Schein A, Dziembowski A, Kudla G, Jensen TH (2015) The human nuclear exosome targeting complex is loaded onto newly synthesized RNA to direct early ribonucleolysis. Cell Rep 10:178–192

Makino DL, Baumgartner M, Conti E (2013) Crystal structure of an RNA-bound 11-subunit eukaryotic exosome complex. Nature 495:70–75

Makino DL, Schuch B, Stegmann E, Baumgartner M, Basquin C, Conti E (2015) RNA degradation paths in a 12-subunit nuclear exosome complex. Nature 524:54–58

Malecki M, Viegas SC, Carneiro T, Golik P, Dressaire C, Ferreira MG, Arraiano CM (2013) The exoribonuclease Dis3L2 defines a novel eukaryotic RNA degradation pathway. EMBO J 32:1842–1854

Marzluff WF, Koreski KP (2017) Birth and death of histone mRNAs. Trends Genet 33:745–759

Meola N, Jensen TH (2017) Targeting the nuclear RNA exosome: poly(A) binding proteins enter the stage. RNA Biol 14:820–826

Meola N, Domanski M, Karadoulama E, Chen Y, Gentil C, Pultz D, Vitting-Seerup K, Lykke-Andersen S, Andersen JS, Sandelin A et al (2016) Identification of a nuclear exosome decay pathway for processed transcripts. Mol Cell 64:520–533

Milligan L, Decourty L, Saveanu C, Rappsilber J, Ceulemans H, Jacquier A, Tollervey D (2008) A yeast exosome cofactor, Mpp6, functions in RNA surveillance and in the degradation of noncoding RNA transcripts. Mol Cell Biol 28:5446–5457

Mitchell P, Petfalski E, Shevchenko A, Mann M, Tollervey D (1997) The exosome: a conserved eukaryotic RNA processing complex containing multiple $3' \rightarrow 5'$ exoribonucleases. Cell 91:457–466

Mitchell P, Petfalski E, Houalla R, Podtelejnikov A, Mann M, Tollervey D (2003) Rrp47p is an exosome-associated protein required for the 3′ processing of stable RNAs. Mol Cell Biol 23:6982–6992

Molleston JM, Sabin LR, Moy RH, Menghani SV, Rausch K, Gordesky-Gold B, Hopkins KC, Zhou R, Jensen TH, Wilusz JE et al (2016) A conserved virus-induced cytoplasmic TRAMP-like complex recruits the exosome to target viral RNA for degradation. Genes Dev 30:1658–1670

Morton DJ, Kuiper EG, Jones SK, Leung SW, Corbett AH, Fasken MB (2017) The RNA exosome and RNA exosome-linked disease. RNA 24(2):127–142

Neil H, Malabat C, d'Aubenton-Carafa Y, Xu Z, Steinmetz LM, Jacquier A (2009) Widespread bidirectional promoters are the major source of cryptic transcripts in yeast. Nature 457:1038–1042

Ogami K, Richard P, Chen Y, Hoque M, Li W, Moresco JJ, Yates JR III, Tian B, Manley JL (2017) An Mtr4/ZFC3H1 complex facilitates turnover of unstable nuclear RNAs to prevent their cytoplasmic transport and global translational repression. Genes Dev 31(12):1257–1271

Porrua O, Libri D (2013) A bacterial-like mechanism for transcription termination by the Sen1p helicase in budding yeast. Nat Struct Mol Biol 20:884–891

Porrua O, Libri D (2015) Transcription termination and the control of the transcriptome: why, where and how to stop. Nat Rev Mol Cell Biol 16:190–202

Reyes-Turcu FE, Zhang K, Zofall M, Chen E, Grewal SI (2011) Defects in RNA quality control factors reveal RNAi-independent nucleation of heterochromatin. Nat Struct Mol Biol 18:1132–1138

Rialdi A, Hultquist J, Jimenez-Morales D, Peralta Z, Campisi L, Fenouil R, Moshkina N, Wang ZZ, Laffleur B, Kaake RM et al (2017) The RNA exosome syncs IAV-RNAPII transcription to promote viral Ribogenesis and infectivity. Cell 169(679-692):e614

Robinson SR, Oliver AW, Chevassut TJ, Newbury SF (2015) The 3′ to 5′ Exoribonuclease DIS3: from structure and mechanisms to biological functions and role in human disease. Biomol Ther 5:1515–1539

San Paolo S, Vanacova S, Schenk L, Scherrer T, Blank D, Keller W, Gerber AP (2009) Distinct roles of non-canonical poly(A) polymerases in RNA metabolism. PLoS Genet 5:e1000555

Schilders G, Raijmakers R, Raats JM, Pruijn GJ (2005) MPP6 is an exosome-associated RNA-binding protein involved in 5.8S rRNA maturation. Nucleic Acids Res 33:6795–6804

Schmidt K, Butler JS (2013) Nuclear RNA surveillance: role of TRAMP in controlling exosome specificity. Wiley Interdiscip Rev RNA 4:217–231

Schmidt K, Xu Z, Mathews DH, Butler JS (2012) Air proteins control differential TRAMP substrate specificity for nuclear RNA surveillance. RNA 18:1934–1945

Schuch B, Feigenbutz M, Makino DL, Falk S, Basquin C, Mitchell P, Conti E (2014) The exosome-binding factors Rrp6 and Rrp47 form a composite surface for recruiting the Mtr4 helicase. EMBO J 33:2829–2846

Schulz D, Schwalb B, Kiesel A, Baejen C, Torkler P, Gagneur J, Soeding J, Cramer P (2013) Transcriptome surveillance by selective termination of noncoding RNA synthesis. Cell 155:1075–1087

Shi Y, Pellarin R, Fridy PC, Fernandez-Martinez J, Thompson MK, Li Y, Wang QJ, Sali A, Rout MP, Chait BT (2015) A strategy for dissecting the architectures of native macromolecular assemblies. Nat Methods 12:1135–1138

Shi M, Zhang H, Wu X, He Z, Wang L, Yin S, Tian B, Li G, Cheng H (2017) ALYREF mainly binds to the 5′ and the 3′ regions of the mRNA in vivo. Nucleic Acids Res 45:9640–9653

Sikorska N, Zuber H, Gobert A, Lange H, Gagliardi D (2017) RNA degradation by the plant RNA exosome involves both phosphorolytic and hydrolytic activities. Nat Commun 8:2162

Silla T, Karadoulama E, Mąkosa D, Lubas M, Jensen TH (2018) The RNA exosome adaptor ZFC3H1 functionally competes with nuclear export activity to retain target transcripts. Cell Rep 23(7):2199–2210

Staals RH, Bronkhorst AW, Schilders G, Slomovic S, Schuster G, Heck AJ, Raijmakers R, Pruijn GJ (2010) Dis3-like 1: a novel exoribonuclease associated with the human exosome. EMBO J 29:2358–2367

Steinmetz EJ, Warren CL, Kuehner JN, Panbehi B, Ansari AZ, Brow DA (2006) Genome-wide distribution of yeast RNA polymerase II and its control by Sen1 helicase. Mol Cell 24:735–746

Thiebaut M, Kisseleva-Romanova E, Rougemaille M, Boulay J, Libri D (2006) Transcription termination and nuclear degradation of cryptic unstable transcripts: a role for the nrd1-nab3 pathway in genome surveillance. Mol Cell 23:853–864

Thoms M, Thomson E, Bassler J, Gnadig M, Griesel S, Hurt E (2015) The exosome is recruited to RNA substrates through specific adaptor proteins. Cell 162:1029–1038

Tiedje C, Lubas M, Tehrani M, Menon MB, Ronkina N, Rousseau S, Cohen P, Kotlyarov A, Gaestel M (2015) p38MAPK/MK2-mediated phosphorylation of RBM7 regulates the human nuclear exosome targeting complex. RNA 21:262–278

Tomecki R, Kristiansen MS, Lykke-Andersen S, Chlebowski A, Larsen KM, Szczesny RJ, Drazkowska K, Pastula A, Andersen JS, Stepien PP et al (2010) The human core exosome interacts with differentially localized processive RNases: hDIS3 and hDIS3L. EMBO J 29:2342–2357

Tuck AC, Tollervey D (2013) A transcriptome-wide atlas of RNP composition reveals diverse classes of mRNAs and lncRNAs. Cell 154:996–1009

Tudek A, Porrua O, Kabzinski T, Lidschreiber M, Kubicek K, Fortova A, Lacroute F, Vanacova S, Cramer P, Stefl R et al (2014) Molecular basis for coordinating transcription termination with noncoding RNA degradation. Mol Cell 55:467–481

Valen E, Preker P, Andersen PR, Zhao X, Chen Y, Ender C, Dueck A, Meister G, Sandelin A, Jensen TH (2011) Biogenic mechanisms and utilization of small RNAs derived from human protein-coding genes. Nat Struct Mol Biol 18:1075–1082

Vanacova S, Wolf J, Martin G, Blank D, Dettwiler S, Friedlein A, Langen H, Keith G, Keller W (2005) A new yeast poly(A) polymerase complex involved in RNA quality control. PLoS Biol 3:e189

Vasiljeva L, Buratowski S (2006) Nrd1 interacts with the nuclear exosome for 3′ processing of RNA polymerase II transcripts. Mol Cell 21:239–248

Vasiljeva L, Kim M, Mutschler H, Buratowski S, Meinhart A (2008) The Nrd1-Nab3-Sen1 termination complex interacts with the Ser5-phosphorylated RNA polymerase II C-terminal domain. Nat Struct Mol Biol 15:795–804

Wang SW, Stevenson AL, Kearsey SE, Watt S, Bahler J (2008) Global role for polyadenylation-assisted nuclear RNA degradation in posttranscriptional gene silencing. Mol Cell Biol 28:656–665

Wasmuth EV, Januszyk K, Lima CD (2014) Structure of an Rrp6-RNA exosome complex bound to poly(A) RNA. Nature 511:435–439

Wasmuth EV, Zinder JC, Zattas D, Das M, Lima CD (2017) Structure and reconstitution of yeast Mpp6-nuclear exosome complexes reveals that Mpp6 stimulates RNA decay and recruits the Mtr4 helicase. Elife 6:e29062

Win TZ, Draper S, Read RL, Pearce J, Norbury CJ, Wang SW (2006) Requirement of fission yeast Cid14 in polyadenylation of rRNAs. Mol Cell Biol 26:1710–1721

Wittmann S, Renner M, Watts BR, Adams O, Huseyin M, Baejen C, El Omari K, Kilchert C, Heo DH, Kecman T et al (2017) The conserved protein Seb1 drives transcription termination by binding RNA polymerase II and nascent RNA. Nat Commun 8:14861

Wlotzka W, Kudla G, Granneman S, Tollervey D (2011) The nuclear RNA polymerase II surveillance system targets polymerase III transcripts. EMBO J 30:1790–1803

Wyers F, Rougemaille M, Badis G, Rousselle JC, Dufour ME, Boulay J, Regnault B, Devaux F, Namane A, Seraphin B et al (2005) Cryptic pol II transcripts are degraded by a nuclear quality control pathway involving a new poly(A) polymerase. Cell 121:725–737

Xu Z, Wei W, Gagneur J, Perocchi F, Clauder-Munster S, Camblong J, Guffanti E, Stutz F, Huber W, Steinmetz LM (2009) Bidirectional promoters generate pervasive transcription in yeast. Nature 457:1033–1037

Yamashita A, Shichino Y, Tanaka H, Hiriart E, Touat-Todeschini L, Vavasseur A, Ding DQ, Hiraoka Y, Verdel A, Yamamoto M (2012) Hexanucleotide motifs mediate recruitment of the RNA elimination machinery to silent meiotic genes. Open Biol 2:120014

Yoshikatsu Y, Ishida Y, Sudo H, Yuasa K, Tsuji A, Nagahama M (2015) NVL2, a nucleolar AAA-ATPase, is associated with the nuclear exosome and is involved in pre-rRNA processing. Biochem Biophys Res Commun 464:780–786

Zhou Y, Zhu J, Schermann G, Ohle C, Bendrin K, Sugioka-Sugiyama R, Sugiyama T, Fischer T (2015) The fission yeast MTREC complex targets CUTs and unspliced pre-mRNAs to the nuclear exosome. Nat Commun 6:7050

Zinder JC, Lima CD (2017) Targeting RNA for processing or destruction by the eukaryotic RNA exosome and its cofactors. Genes Dev 31:88–100

Zinder JC, Wasmuth EV, Lima CD (2016) Nuclear RNA exosome at 3.1 A reveals substrate specificities, RNA paths, and allosteric inhibition of Rrp44/Dis3. Mol Cell 64:734–745

# Chapter 5
# $3'$-UTRs and the Control of Protein Expression in Space and Time

**Traude H. Beilharz, Michael M. See, and Peter R. Boag**

**Abstract** The noncoding elements of an mRNA influence multiple aspects of its fate. For example, $3'$-UTRs serve as physical and sequence-based information hubs that direct the time, place, and level of translation of the protein encoded in *cis*, but often also have additional roles in *trans*. Understanding the information content of $3'$-UTRs has been a challenge. Bioinformatic searches for motifs, such as those that encode the polyadenylation signal or microRNA seed regions, are simple enough, but rarely do these inferred positions in genomes correlate well with the actual sites chosen by the relevant nanomachines in living cells. This is almost certainly due to three-dimensional complexity of RNA, the physical states of which are recognized by RNA-binding proteins that serve to read and interpret the information content. Here, we follow the $3'$-UTR-mediated posttranscriptional metabolism of mRNA in the germline of the nematode worm *Caenorhabditis elegans*. While many areas still require the clarification only detailed fundamental research can provide, this model system can serve as a basis of $3'$-mediated regulatory control for elaboration in more complex metazoan systems.

**Keywords** $3'$-UTR · Alternative polyadenylation · Poly(A)-tail · Germ granules · Regulation of germline translation · RNA-seq

---

T. H. Beilharz (✉) · P. R. Boag
Department of Biochemistry and Molecular Biology, Development and Stem Cells Program, Monash Biomedicine Discovery Institute, Monash University, Melbourne, VIC, Australia
e-mail: traude.beilharz@monash.edu; peter.boag@monash.edu

M. M. See
Department of Biochemistry and Molecular Biology, Development and Stem Cells Program, Monash Biomedicine Discovery Institute, Monash University, Melbourne, VIC, Australia

Monash Bioinformatics Platform, Monash University, Melbourne, VIC, Australia



M. Oeffinger, D. Zenklusen (eds.), *The Biology of mRNA: Structure and Function*, Advances in Experimental Medicine and Biology 1203,
https://doi.org/10.1007/978-3-030-31434-7_5

## 5.1 Introduction

The genome is brought to life by expression of RNA which, either directly or indirectly, controls all aspects of a living organism. The maturation of nascent protein-coding RNA in eukaryotes concludes with cleavage and polyadenylation after a segment of noncoding RNA termed the 3′-untranslated region (3′-UTR), the length of which scales with organismal complexity (Mayr 2017). The machinery for this terminal modification travels with the RNA polymerase and cuts the mRNA shortly after recognition of the polyadenylation signal (PAS). The canonical nuclear poly(A) polymerase then extends a non-templated poly(A)-tail to a length that is typical for the species in question. In the nematode worm *Caenorhabditis elegans*, this tail is extended to approximately 250 bases (Janicke et al. 2012). However, this maximal length is rapidly trimmed, such that the steady-state poly(A)-tail for any given transcript is typically distributed between ~20 and 100 adenosine residues (Nousch et al. 2013). When more than one position for cleavage and adenylation is available, transcript isoforms with 3′-UTR of different lengths can be generated. At least 60% of the worm transcriptome has been reported to undergo such alternative polyadenylation (APA) (Blazie et al. 2017; Mangone et al. 2010). The mechanism that decides switching between PAS sites is still under active investigation. However, it likely includes the combined influence of epigenetic marks, the transcriptional rate, and the effective concentration of the locally available cleavage and polyadenylation machinery (Gruber et al. 2014; Tian and Manley 2017; Turner et al. 2018). The result of such switching is the inclusion/exclusion of additional regulatory information (Fig. 5.1).

The information content of 3′-UTRs is generally considered to be centered around the stability, localization, and translational control of its associated protein-coding open reading frame (Turner et al. 2018). For example, AU-rich elements and microRNA-binding sites are well-documented meditators of both translation and stability (Chen and Shyu 1995; Filipowicz et al. 2008). Moreover, the role of the 3′-UTR in transcript localization has long been recognized (Heym and Niessing 2012; Holt and Bullock 2009; Jung et al. 2012; Lecuyer et al. 2007). However, recent research also points toward a more integrative role for 3′-UTR in the organization of cellular gene expression. For example, pioneering work from Christine Mayr's laboratory shows that 3′-UTR can function as assembly platforms to chaperone interactions between proteins and, by extension, that APA can influence participation of the encoded protein into higher order protein complexes (Berkovits and Mayr 2015; Mayr 2017). Alternatively, 3′-UTRs might also operate *in trans* to sponge microRNA (Ebert and Sharp 2010; Thomson and Dinger 2016) and possibly other regulatory elements and thereby alter the cellular regulatory paradigm.

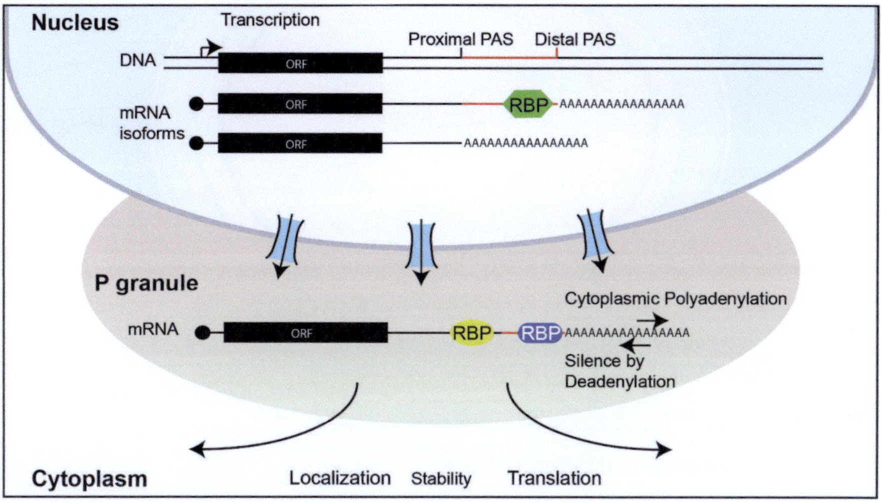


**Fig. 5.1** Schematic illustration of mRNA 3′-end processing and subsequent sorting. Alternative polyadenylation signals (PAS) encoded in the genome create the possibility of mRNA with 3′-UTR length isoforms. This creates scope for alternative regulatory fates for the encoded protein upon export to the cytoplasm. In the *C. elegans* germline, nuclear pore complexes direct their cargo into P granules that directly about the nuclear membrane. These are thought to warehouse mRNA for distribution and translation at later time points in development

## 5.2 The *C. elegans* System as a Tool for Investigating Spatiotemporal Control of Translation

The free-living, nonparasitic soil nematode *C. elegans* has served as an excellent model for understanding many facets of germ cell biology, including its RNA biology, with much of the focus being on the hermaphrodite germline that produces sperm during the fourth larval stage and then switches to producing oocytes during adulthood. The hermaphrodite germline has a linear organization, in which Notch signaling by the somatic distal tip cell maintains the germline stem cell population (Kimble and Simpson 1997). Once beyond the influence of the Notch signaling, the germ cells enter mitotic prophase 1 and progress through spatially separated cell cycle stages before arresting in diakinesis at the proximal end of the gonad (Fig. 5.2). This simple organization has provided a unique window within which to observe the progression of germ cell development and visualize changes in signaling pathways, transcription levels, and localization of mRNA and proteins. Collectively, this has led to a highly detailed understanding of how germ cell development is controlled by changes in gene expression and mRNA dynamics, processes that are likely to be conserved among sexually reproducing species.

A common feature of oogenesis across species is the broad suppression of transcription during meiosis. Although the exact timing of the transcriptional quiescence varies between species, it most often occurs early in meiosis (Lesch and Page 2012) and is thought to permit the reorganization of the oocyte genome in preparation for union with the sperm genome (Eckersley-Maslin et al. 2018). This

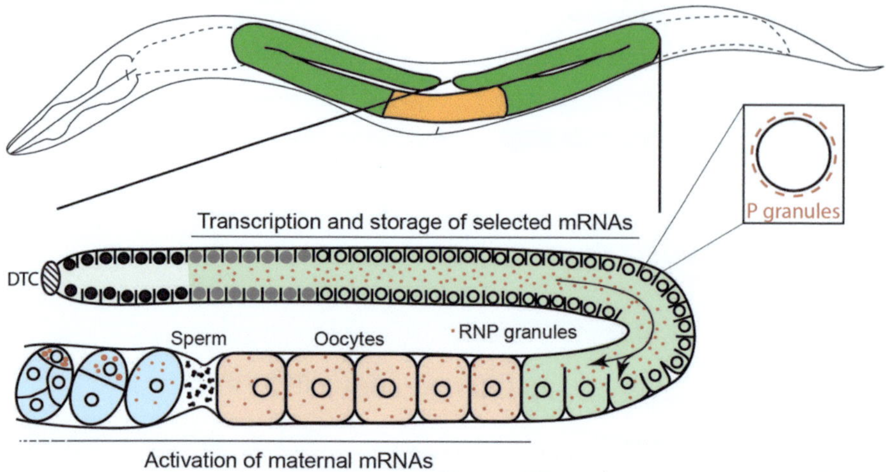


**Fig. 5.2** Illustration depicting the organization of the adult hermaphrodite *C. elegans* gonads. The adult worm has two U-shaped gonad arms (green) that share a common uterus (blue). Each gonad arm consists of ~1000 germ cells organized in a linear developmental pattern. The gonad initially consists of a syncytium, in which germ cells are only partially enclosed in a membrane and share a common cytoplasmic core. At the distal end of the gonad, a single somatic cell, the distal tip cell (DTC), maintains the germline stem cell population (solid black nuclei). Once cells have moved ~20 cell diameters from the DTC, the germ cells enter mitotic prophase I (solid gray nuclei) in the "transition zone." As the cells move further proximally, they enter the pachytene stage of mitotic prophase I and exit this stage as they progress through the gonad "loop" to develop into completely cellularized oocytes arrested in diakinesis. Transcription levels are low in the stem cell population (light green) and then dramatically increase as germ cells enter mitotic prophase I (green). Transcription is essentially inhibited in diakinesis-stage oocytes awaiting fertilization. Zygotic transcription does not broadly commence until the 4-cell stage of embryogenesis. The maternal RNAs stored in P granules (pink) are selectively released and translated in the maturing oocytes and early embryo (blue). At the first cell division, P granules coalesce and distribute mainly to the anterior (P0) blastomere and become further asymmetrically segregated at the second cell division toward the germ cell lineage

quiescence provides a significant challenge for germ cell development, as new protein products are still required to drive key temporally defined developmental changes (Huelgas-Morales and Greenstein 2017). To overcome this challenge, germ cells produce what can be considered as two distinct transcriptomes. The first transcriptome is used shortly after it is produced and encodes proteins required for building the germline, as well as those required for chromosome synapsis and the repair of meiotic double-stranded DNA breaks. The second transcriptome is translationally repressed and activated at specific points later in development when transcriptional quiescence prevails. The activation of repressed mRNAs shows exquisite specificity and fidelity, with different mRNA species differentially translated between neighboring oocytes or in specific cell lineages in the early embryo (Merritt et al. 2008). Translational repression is largely mediated through the 3′-UTR (Merritt et al. 2008). Indeed, both repression and activation of mRNAs rely upon a complex interplay between 3′-UTR binding of diverse array of RNA-binding proteins (RBPs) that combine to form ribonucleoprotein (RNP) granules.

## 5.3 *C. elegans* Germline RNP Granules

A conserved feature of germ cells is the presence of large RNA-protein-rich granules, often called nuage, in close proximity to the nucleus. In *C. elegans*, the most prominent germ granule is the "P granule," and in the early stages of germ cell development, these are physically associated with the cytoplasmic face of the nuclear pores (Fig. 5.1). This association has led to the idea that they extend the nuclear pore environment (Andralojc et al. 2017), providing a space where RNAs can be "sorted" and bound by their appropriate RBP(s). Interestingly, diverse RNA pathways appear to share the P granule environment. For example, many core components of the multiple endo-siRNA pathways localize to P granules, as well as proteins whose functions appear to be more closely involved in translational control pathways (Boag et al. 2005; Ciosk et al. 2006; Sengupta et al. 2013). The sequestration of RNP into P granules may help preserve the totipotency of germ cells by providing an environment that enables suppression of the somatic differentiation program until the onset of zygotic development (Ciosk et al. 2006; Updike et al. 2014). Defects in P granule formation or organization often result in sterility, highlighting their importance in maintaining correct gene expression, while inappropriate early translation of embryonic mRNA can result in a tumorous germline (Francis et al. 1995).

The existence of P granules in the germline requires active transcription. It has been experimentally shown that inhibition of transcription leads to P granule loss, suggesting that these RNA hubs are highly dynamic structures that rapidly turnover their RNA and protein complement (Sheth et al. 2010). Moreover, the maintenance of P granule integrity requires many different RNA pathways, including small RNA and silent maternal mRNA. Recent evidence suggests that perinuclear P granules detach from the nuclear pores by shear force stress (Brangwynne et al. 2009) and are subsequently distributed among the enlarging oocytes by actin-dependent cytoplasmic streaming. In cellularized oocytes where transcription is inhibited, P granules provide a reservoir of translationally repressed mRNAs for use later in development. Upon fertilization, P granules rapidly coalesce and localize to germline blastomeres in the first embryonic cell division (Fig. 5.2). This complexity of P granule assembly and localization is driven by an intricate interplay of protein-protein and protein-RNA events that are only now beginning to be understood. Mutagenic and RNAi screens have identified over 100 proteins that result in defects in P granule formation and/or organization (Updike and Strome 2009; Wood et al. 2016). Many of the P granule-resident proteins contain prion-like, low complexity, or intrinsically disordered regions. Accordingly, like other RNA-protein granules, P granules display liquid-like behaviors (Brangwynne et al. 2009) and RNA-induced phase separation. These properties allow self-assembly of RNPs in a highly dynamic subcellular environment.

The complexity of the described RNP granules within germ cells continues to grow. Several decades after P granules were first identified, a new RNP granule type, designated as Mutator foci, was described and is defined by proteins that are required for the efficient amplification of the RNAi gene silencing pathways (Phillips et al. 2012). Although Mutator foci are found physically adjacent to P granules, their formation appears to be independent of core P granule components, suggesting that they are a unique subcellular component required for small RNA amplification and mRNA silencing (Phillips et al. 2012). The most recent addition to the family, Z granule, appears sandwiched between P granules and Mutator foci in adult germ cells (Wan et al. 2018). Z granules are marked by the helicase ZNFX-1 and the Argonaute WAGO-4, both of which are required for small RNA-mediated epigenetic inheritance (Ishidate et al. 2018; Wan et al. 2018). Interestingly, ZNFX-1 and WAGO-4 localize to P granules during early embryogenesis but then separate from them during germ cell development. How Z granules form, and their functional relationship to P granules and Mutator foci, remains unclear; however, the distinct spatial arrangement of these liquid-like RNP granules in adult germ cells suggests that they are highly specialized structures that are unique, but possibly also interrelated, in their mechanisms of formation and function. Unravelling how these RNP granules form and interact and, most importantly, how their diverse cargo is released, will provide an exciting new layer of complexity in how germline transcripts are spatiotemporally regulated.

## 5.4 Translational Activation of mRNAs During the Oocyte-to-Embryo Transition

As a final step in oocyte development, the trigger for oocyte maturation is usually a signaling hormone that initiates a cascade of events that lead to significant changes in both nuclear and cytoplasmic organization (Von Stetina and Orr-Weaver 2011). Integral to these changes is the cytoplasmic reorganization of repressive RNP complexes that had blocked translation of stored mRNAs, but whose cargo now need to become activated for translation. Key players in this process are two antagonistic proteins, the TRIM-NHL RNA-binding protein LIN-41 and the redundant TIS11 zinc-finger RNA-binding proteins OMA-1 and OMA-2. LIN-41 promotes oocyte growth by inhibiting M-phase entry via translational repression of the key cell-cycle activating proteins (Spike et al. 2014). Interestingly, LIN-41 and the OMA proteins appear to be components of a large RNP complex present in transcriptionally silent late-stage germ cells. Further research found that this RNP complex also contained the conserved cytoplasmic poly(A) polymerase GLD-2 and its accessory proteins GLD-3 and RNP-8, the CCR4-NOT deadenylase complex, and translation initiation factors (Tsukamoto et al. 2017). Analysis of transcripts that associate with LIN-41 or the OMA proteins revealed many shared mRNA, but also distinct subsets. Interestingly, many of the transcripts that encode components of this complex are also enriched within the complex, suggesting there may be autoregulatory feedback mechanisms to control protein levels. Many of the LIN-41-associated mRNAs were also GLD-2 target mRNAs, suggesting that re-adenylation of LIN-41 target mRNAs enhances the translation of these transcripts. It is intriguing that the RNP granules contain proteins with antagonistic function; LIN-41 is a repressor (e.g., of *spn-4* and *meg-1*), whereas the OMA proteins activate translation; GLD-2 is a poly(A) polymerase, whereas CCR4-NOT is a deadenylase. Co-localization of these opposing functions suggests that there may not be a simple binary switch between states. Instead, the relative activities of the antagonistic pathways may prove a delicate rheostat in which translation can be tightly regulated at a subcellular level.

The complexity of posttranscriptional control is revealed by the number of different RBPs that can act on a single mRNA in different cells of the embryo. For example, the *zif-1* 3′-UTR can be bound by multiple RBPs which result in different outcomes in different cell types. In oocytes, the redundant OMA-1/2 pair represses its translation (Guven-Ozkan et al. 2010), while in the 1-cell embryo, the translational repression of *zif*-1 requires both the KH domain containing MEX-3 and RRM domain containing SPN-4 (Oldenbroek et al. 2012). In germline blastomeres, the Zn-finger POS-1 is sufficient to repress translation (Tabara et al. 1999). On the other hand, activation of translation of *zif-1* in the somatic blastomeres is promoted by the redundant Zn-finger proteins MEX-5/6 (Oldenbroek et al. 2012). Establishment of RNP gradients by asymmetric localization and degradation appears to be a critical process, and the integration of cell-type-specific regulatory networks ensures robust repression/activation of maternal RNA. Further work is required to understand the molecular details of these complexes and possibly competitive interactions that refine mRNA fate.

## 5.5 Cytoplasmic Polyadenylation Is Key to Activation of Silent mRNA

One way to dissect the combinatorial influence of the RNA-binding proteins on translation is to measure their influence over the adenylation state of the transcriptome. The steady-state poly(A)-tail length distribution of any transcriptome is a sum of multiple different activities. Nascent transcripts are fully adenylated at synthesis, but rapidly trimmed by the sequential activities of the PAN2 and CCR4/CAF1 deadenylases (Harrison et al. 2015; Janicke et al. 2012). In metazoans, noncanonical cytoplasmic polyadenylation further complicates understanding of what steady-state adenylation status means. The gene encoding the cytoplasmic poly(A) polymerase, GLD-2, was identified in worms in a genetic screen for germline-defective (GLD) mutants (Francis et al. 1995; Wang et al. 2002). Loss of GLD-2 results in major germline dysfunction, failure of oocyte-germline transition, and ultimately inviable embryos. Since transcription is globally shutdown for the final stages of oocyte maturation, any subsequent development is largely driven by new proteins synthesized from stored maternal mRNAs. How these stored mRNAs are activated at specific points in oocyte maturation and in early embryonic development is one of the major mysteries of germ cell biology. One aspect that is clear is that lengthening of the poly(A)-tail by GLD-2 is involved (Charlesworth et al. 2013). Early work showed that in vitro the GLD-2 enzyme required only a tether to recruit its poly(A) polymerase function to its targets (Kwak et al. 2004), and in vivo this function is achieved by a series of 3′-UTR-binding proteins.

We recently investigated 3′-UTR-mediated control of cytoplasmic polyadenylation in the *C. elegans* germline and early embryo as a surrogate for activation of mRNA for translation in a spatiotemporal manner (Boag et al. 2018). Our research is built upon pioneering work in the *Xenopus* oocyte (Hake and Richter 1994), contemporary work in the fly (Lim et al. 2016), and a body of research surrounding the *gld-2* mutant worm (Janicke et al. 2012; Nousch et al. 2017; Sengupta et al. 2013; Wang et al. 2002). Specifically, we wondered whether the embryonic lethality of worms without functional GLD-2 activity was due to a failure to activate a small number of regulatory mRNA or reflected a widespread dependence on this process within the sequential activation of mRNA at different developmental times within the germline and early embryo.

To probe the landscape of cytoplasmic polyadenylation in *C. elegans*, we used the PAT-seq approach to analyze 3′-end dynamics (Harrison et al. 2015). This method identifies changes in poly(A)-tail length distribution between transcriptomes, in addition to the gene expression and 3′-end idenification of more traditional 3′-end focused sequencing approaches. We discovered that more than 1000 transcripts depend on GLD-2-mediated polyadenylation. These transcripts contained overall longer 3′-UTR than nontargeted transcripts (Fig. 5.3), consistent with the idea that length scales with the complexity of regulation. Moreover, while the 3′-UTR of GLD-2-associated transcripts tended to be more cytosine rich and the PAS tended to be noncanonical (AAUGAA), there were no specific linear sequence motifs that associated particularly with the requirement for cytoplasmic polyadenylation (Boag et al. 2018). This contrasted strongly with the literature in both *Xenopus* and mammalian cells, where a U-rich cytoplasmic polyadenylation element (CPE) upstream of the canonical (AAUAAA) PAS was suggested as the anchor recognized by the CPE-binding protein (CPEB), which, in turn, tethered the GLD-2 activity to 3′-UTR (Hake and Richter 1994; Pique et al. 2008). The explanation for this is likely that CPEB is just one of many proteins that tether GLD-2 to its targets. Thus, like the *zif-1* example discussed above (see also Fig. 5.3c), the native adenylation state is likely the integrated outcome of multiple opposing activities in the mixed cell-types of the whole animal.

Our search was for mRNA that depends on GLD-2 enzymatic activity for normal, wild-type poly(A)-tail length distribution in whole animals and early embryos (Boag et al. 2018). A similar study compared the adenylation state of the fly embryo +/− Whisp, the *Drosophila* GLD-2 homolog (Lim et al. 2016). Neither study showed enrichment of a U-rich CPE or alternative global explanatory motif. However, by reanalysis of previous data testing the adenylation state in early embryos +/− the 3′-UTR tether for GLD-2, GLD-1, or either of the two 3′-regulatory RBPs, MEX-5 and POS-1 (Elewa et al. 2015), we could demonstrate that, similar to the germ cell to embryo transition, the wild-type adenylation state of developing early embryos is sculpted by multiple activities to restrict protein expression. For example, in the

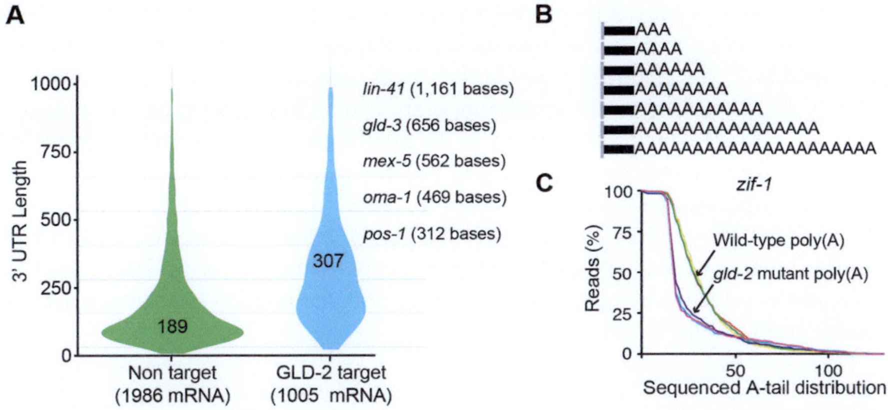


**Fig. 5.3** (**a**) The 3′-UTRs of GLD-2 target mRNA are overall longer than nontarget mRNA, consistent with increased regulatory information required for the posttranscriptional fate of such transcripts. In green are the length distributions of the 3′-UTR of 1986 mRNA that did not change adenylation state in our study. The width of the violin plots reflects the relative transcript abundance at the indicated 3′-UTR length. The GLD-2 target transcripts included those that were significantly shorter in their polyadenylation state in the adult *gld-2* mutant worm ($p < 0.01$) than in the wild type. The average length of nontarget 3′-UTR is 189 bases, whereas the GLD-2 targets have an average 3′-UTR length of 307 bases. The indicated RNA-binding proteins are each encoded by transcripts that themselves have long 3′-UTRs consistent with their complex autoregulation. (**b**) The poly(A)-tail length distribution recorded by PAT-seq and other next-generation sequencing approaches can be visualized by cumulative distribution plots as shown in the schematic. (**c**) The cumulative distribution of the poly(A)-tail associated with *zif-1* transcripts in either the adult wild-type worm or the *gld-2* mutant worm (triplicate of each). The length of poly(A)-tails sequenced in the *gld-2* mutant was significantly distributed toward shorter. In total, 1064 transcripts were identified having a significant difference in poly(A)-length distribution (Boag et al. 2018). The changes in poly(A)-distribution between conditions serve a surrogate read-out for regulatory control. For example, changes to the adenylation state of *zif-1* after knockdown of specific RBP (see main text) or after ablation of 3′-UTR sequences might help to define the regulatory circuitry of this developmental regulator

worm, cell fate in the early embryo is entirely posttranscriptionally programed (Evans and Hunter 2005). In the 2-cell embryo, one cell is destined to specify the endomesoderm linage, while the other will asymmetrically divide to produce mesodermal muscles, intestinal endoderm, and germline lineages (Rose and Gonczy 2014). By the second embryonic cell division, the major body-plan polarity axes are established. RNA-binding proteins are integral to these asymmetric divisions. By a complex autoregulatory mechanism, MEX-5 is expressed predominantly in the AB-cell that specifies anterior fate, whereas POS-1 is active in the P1 cells that will become the germline and posterior tissues such as muscles and the intestine. Our research showed that it is possible to distinguish a maternal mRNA destined for expression in one or other cell type based on its dependence on MEX-5 or POS-1 for a native poly(A)-length distribution. One example of this is NEG-1, which is essential for correct anterior specification (Elewa et al. 2015; Osborne Nishimura et al. 2015). Failure to curtail cytoplasmic adenylation in the P1 linage due to a loss of POS-1 results in overall longer poly(A)-tails (than wild-type) and aberrant protein expression (Elewa et al. 2015). Both these phenotypes require the 3′-UTR anchoring protein GLD-3. We identified many further transcripts whose asymmetric translation is likely regulated in a similar spatiotemporal manner, to restrict protein expression to a specific cell-fate lineage (Boag et al. 2018). By combinatorial depletion of RNA-binding proteins, we predict it will become possible to dissect the regulatory circuitry of transcripts whose expression is regulated in space and in time. Beyond its role within the germline, we predict that many other asymmetric divisions in biology, such as those in muscle progenitor cells (Gurevich et al. 2016), will be governed by selective cytoplasmic polyadenylation, recruited to target transcripts by similar mechanisms.

## 5.6 Scaling 3′-UTR Regulatory Capacity with Complexity

The extent of noncoding sequence in genomes is broadly correlated to evolutionary complexity (Taft et al. 2007). The noncoding regions of the transcriptome, especially 3′-UTR, are even more tightly correlated to complexity (Chen et al. 2012). Whereas prokaryotic, and their derivative mitochondrial, mRNA, often have minimal or no 3′-regulatory sequence, the 3′-UTR of eukaryotes globally increased in length with higher evolutionary complexity (Fig. 5.4). Hence, the small, tightly packed genome of *S. cerevisiae* has quite short 3′-UTRs (139 bases, on average), the *C. elegans* transcriptome displays an intermediate length (226 bases, on average), and the mouse 3′-end repertoire is even longer (1049 bases, on average). Note that the transcriptome of both worms (Fig. 5.3) and mouse (Fig. 5.4) contains 3′-UTR with different length distributions based on the regulatory complexity of the encoded protein. In the case of the worm, the transcripts subject to cytoplasmic polyadenylation tend to be longer (as discussed above). Similarly, transcripts encoding house-keeping functions such as those annotated with the GO term "structural constituent of the ribosome," in the mouse tend to have shorter 3′-UTR than transcripts whose expression is more highly regulated such as those associated with the GO term "transcription coactivator activity" or "integral component of the endoplasmic reticulum membrane" (Fig. 5.4). Specific secretory and tissue-restricted proteins have also previously been shown to be prone to longer more complex 3′-UTR (Berkovits and Mayr 2015; Mayr 2016, 2017). The transcriptome analyzed in Fig. 5.4 was obtained from murine bone marrow-derived macrophages differentiated in vitro (Tucey et al. 2018). We expect that analysis of a brain transcriptome would reveal an even more biased length distribution for those mRNAs that are distributed along dendrites for localized translation at the stimulated synapse. For example, in mice, the brain-derived neurotrophic factor (BDNF) expresses two 3′-UTR isoforms, with the shorter 3′-UTR resulting in mRNA localization to cell body and the longer 3′-UTR leading to localization to dendrites (An et al. 2008). Neuronal cells might represent the most extreme regulatory complexity of 3′-UTR

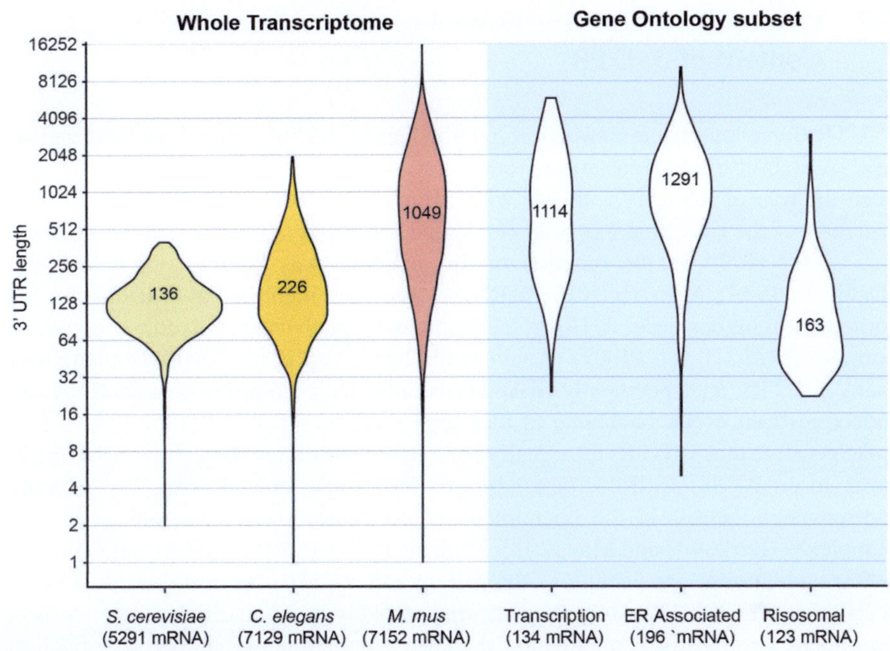


**Fig. 5.4** 3′-UTR length scales with morphological complexity and regulatory capacity. The 3′-UTR length distribution of baker's yeast *Saccharomyces cerevisiae* (Verma-Gaur et al. 2015), the *C. elegans* worm (Boag et al. 2018), or untreated murine bone marrow-derived macrophages (Tucey et al. 2018) are shown in the violin plot. The average length is given within the center of each. The global 3′-UTR distribution can be subset into different Gene Ontology categories having different length distributions. Shown are the *M. mus* data subset to display GO:0003713, GO:0010256, and GO:0003735 which correspond to "transcription coactivator activity," "integral component of the endoplasmic reticulum membrane," and "structural constituent of ribosome," respectively. *Note*: For these bioinformatic analyses, the 3′-UTR length represents the length of the major 3′-UTR isoform detected in the experiment analyzed. Where multiple 3′-UTRs were detected, the isoform having the most reads (under the condition tested) was utilized, irrespective of annotated transcript coordinates

use and will likely make an excellent model for studies into the functional consequence of 3′-UTR isoform selection and/or switching.

A propensity to scale 3′-UTR regulatory capacity with evolution is paralleled by the number and complexity of described 3′-UTR *cis* regulatory factors. In *S. cerevisiae*, the major 3′-regulatory factors are RNA-binding proteins (Hogan et al. 2008). With increasing organismal complexity, the repertoire of such proteins is much expanded and to it are added microRNAs and further RNA-mediated regulatory mechanism such as the parallel evolution of retro-transposon-derived sequences into mammalian 3′-UTRs to stimulate Staufen-mediated decay (Lucas et al. 2018). As a final regulatory layer, the growing list of modified nucleotides such as methyl-6-adenosine, collectively termed the epitranscriptome (Saletore et al. 2012), are likely to refine the regulatory landscape of 3′-UTR even further.

## 5.7 Future Directions for Dissection of the Information Content of 3′-UTR

3′-UTRs have long been viewed through the lens of mRNA subcellular localization, mRNA stability, and translational efficiency. However, hidden in plain sight has been the role of RNA as scaffolds with complex three-dimensional topology. The structure of the ribosome has been elucidated for nearly 20 years, and the essential role of the rRNA in the function of this machine was known well before that (Dahlberg 1989). More recently, lncRNAs have been identified as scaffolds for the formation of higher-order RNA-protein complexes, such as the polycomb repressive complexes (PRC1 and PRC2) (Khalil et al. 2009; Yap et al. 2010). The length of many 3′-UTRs, and specifically GLD-2 target mRNA, is comparable to lncRNA and indeed also the RNA backbone of the small ribosomal subunit (18S rRNA). It is quite possible that 3′-UTRs are equally complex in their three-dimensional folding in order to create the scaffold for binding of RBPs and small RNAs. The recently rediscovered interest in the phenomena of co-translational formation of protein complexes (Berkovits and Mayr 2015; Duncan and Mata 2011; Fulton and L'Ecuyer 1993) adds a new perspective to the role of the 3′-UTR. Assembly at the site of translation, via 3′-UTR assembly platforms, is likely to be an efficient way to form multimeric complexes, compared to the chance association of their parts, by random diffusion in the cytoplasm (posttranslational assembly). The use of alternative 3′-UTR would allow for different three-dimensional surface topologies (for the same protein) and thereby binding of distinct cohorts of interacting proteins. Further, the complex topology achievable by RNA folding could rapidly be reshaped by RNA helicases and the binding of additional RBPs providing a swift response to changes in the local cellular environment. An exciting opportunity now exists to effectively examine RNA three-dimensional surfaces using the expanded technologies of chemical mapping coupled with RNA sequencing, for example, *s*elective 2′-*h*ydroxyl *a*cylation analyzed by *p*rimer *e*xtension (SHAPE) (Merino et al. 2005; Poulsen et al. 2015) and its related next-generation sequencing approaches (Smola and Weeks 2018; Watters et al. 2016) that are now being applied to in vitro and in vivo contexts. Although these approaches still hold their challenges, including questions regarding their accuracy, they will continue to be refined and will provide important insights into the complex three-dimensional topology of 3′-UTRs.

Finally, the emerging single-cell RNA-sequencing technology (scRNA-seq) has the potential to further enlighten our understanding of the regulatory landscape of the early embryo. Recently, each cell from the *C. elegans* 1-cell zygote to the 16-cell stage embryo was individually analyzed (Tintori et al. 2016). It will be fascinating to use similar approaches to dissect the role of specific RBPs in the coordinated localization and translational activation of mRNAs across development.

**Acknowledgments** We thank members of the Boag and Beilharz laboratories for critical feedback on the manuscript and acknowledge support from the National Medical Health and Research Council (NHMRC project grant APP1042848). THB was supported by a Biomedicine Discovery

Fellowship from Monash University and acknowledges support from the Australian Research Council (ARC Discovery Project DP170100569 and Future Fellowship FT180100049). We acknowledge the Monash Bioinformatics Platform for their long-standing support. All figures are originals that were prepared specifically for this chapter.

## References

An JJ, Gharami K, Liao GY, Woo NH, Lau AG, Vanevski F, Torre ER, Jones KR, Feng Y, Lu B et al (2008) Distinct role of long 3′ UTR BDNF mRNA in spine morphology and synaptic plasticity in hippocampal neurons. Cell 134:175–187

Andralojc KM, Campbell AC, Kelly AL, Terrey M, Tanner PC, Gans IM, Senter-Zapata MJ, Khokhar ES, Updike DL (2017) ELLI-1, a novel germline protein, modulates RNAi activity and P-granule accumulation in *Caenorhabditis elegans*. PLoS Genet 13:e1006611

Berkovits BD, Mayr C (2015) Alternative 3′ UTRs act as scaffolds to regulate membrane protein localization. Nature 522:363–367

Blazie SM, Geissel HC, Wilky H, Joshi R, Newbern J, Mangone M (2017) Alternative Polyadenylation directs tissue-specific miRNA targeting in *Caenorhabditis elegans* somatic tissues. Genetics 206:757–774

Boag PR, Nakamura A, Blackwell TK (2005) A conserved RNA-protein complex component involved in physiological germline apoptosis regulation in *C. elegans*. Development 132:4975–4986

Boag PR, Harrison PF, Barugahare AA, Pattison AD, Swaminathan A, Monk S, Davis GM, Heinz E, Powell DR, Beilharz TH (2018) Widespread cytoplasmic polyadenylation programs asymmetry in the germline and early embryo. bioRxiv. https://doi.org/10.1101/428540

Brangwynne CP, Eckmann CR, Courson DS, Rybarska A, Hoege C, Gharakhani J, Julicher F, Hyman AA (2009) Germline P granules are liquid droplets that localize by controlled dissolution/condensation. Science 324:1729–1732

Charlesworth A, Meijer HA, de Moor CH (2013) Specificity factors in cytoplasmic polyadenylation. Wiley Interdiscip Rev RNA 4:437–461

Chen CY, Shyu AB (1995) AU-rich elements: characterization and importance in mRNA degradation. Trends Biochem Sci 20:465–470

Chen CY, Chen ST, Juan HF, Huang HC (2012) Lengthening of 3′UTR increases with morphological complexity in animal evolution. Bioinformatics 28:3178–3181

Ciosk R, DePalma M, Priess JR (2006) Translational regulators maintain totipotency in the *Caenorhabditis elegans* germline. Science 311:851–853

Dahlberg AE (1989) The functional role of ribosomal RNA in protein synthesis. Cell 57:525–529

Duncan CD, Mata J (2011) Widespread cotranslational formation of protein complexes. PLoS Genet 7:e1002398

Ebert MS, Sharp PA (2010) MicroRNA sponges: progress and possibilities. RNA 16:2043–2050

Eckersley-Maslin MA, Alda-Catalinas C, Reik W (2018) Dynamics of the epigenetic landscape during the maternal-to-zygotic transition. Nat Rev Mol Cell Biol 19:436–450

Elewa A, Shirayama M, Kaymak E, Harrison PF, Powell DR, Du Z, Chute CD, Woolf H, Yi D, Ishidate T et al (2015) POS-1 promotes endo-mesoderm development by inhibiting the cytoplasmic polyadenylation of neg-1 mRNA. Dev Cell 34:108–118

Evans TC, Hunter CP (2005) Translational control of maternal RNAs. WormBook 10:1–11

Filipowicz W, Bhattacharyya SN, Sonenberg N (2008) Mechanisms of post-transcriptional regulation by microRNAs: are the answers in sight? Nat Rev Genet 9:102–114

Francis R, Barton MK, Kimble J, Schedl T (1995) Gld-1, a tumor suppressor gene required for oocyte development in *Caenorhabditis elegans*. Genetics 139:579–606

Fulton AB, L'Ecuyer T (1993) Cotranslational assembly of some cytoskeletal proteins: implications and prospects. J Cell Sci 105(Pt 4):867–871
Gruber AR, Martin G, Keller W, Zavolan M (2014) Means to an end: mechanisms of alternative polyadenylation of messenger RNA precursors. Wiley Interdiscip Rev RNA 5:183–196
Gurevich DB, Nguyen PD, Siegel AL, Ehrlich OV, Sonntag C, Phan JM, Berger S, Ratnayake D, Hersey L, Berger J et al (2016) Asymmetric division of clonal muscle stem cells coordinates muscle regeneration in vivo. Science 353:aad9969
Guven-Ozkan T, Robertson SM, Nishi Y, Lin R (2010) Zif-1 translational repression defines a second, mutually exclusive OMA function in germline transcriptional repression. Development 137:3373–3382
Hake LE, Richter JD (1994) CPEB is a specificity factor that mediates cytoplasmic polyadenylation during Xenopus oocyte maturation. Cell 79:617–627
Harrison PF, Powell DR, Clancy JL, Preiss T, Boag PR, Traven A, Seemann T, Beilharz TH (2015) PAT-seq: a method to study the integration of 3′-UTR dynamics with gene expression in the eukaryotic transcriptome. RNA 21:1502–1510
Heym RG, Niessing D (2012) Principles of mRNA transport in yeast. Cell Mol Life Sci 69:1843–1853
Hogan DJ, Riordan DP, Gerber AP, Herschlag D, Brown PO (2008) Diverse RNA-binding proteins interact with functionally related sets of RNAs, suggesting an extensive regulatory system. PLoS Biol 6:e255
Holt CE, Bullock SL (2009) Subcellular mRNA localization in animal cells and why it matters. Science 326:1212–1216
Huelgas-Morales G, Greenstein D (2017) Control of oocyte meiotic maturation in *C. elegans*. Semin Cell Dev Biol 757:277–320
Ishidate T, Ozturk AR, Durning DJ, Sharma R, Shen EZ, Chen H, Seth M, Shirayama M, Mello CC (2018) ZNFX-1 functions within Perinuclear Nuage to balance epigenetic signals. Mol Cell 70:639–649 e636
Janicke A, Vancuylenberg J, Boag PR, Traven A, Beilharz TH (2012) ePAT: a simple method to tag adenylated RNA to measure poly(a)-tail length and other 3′ RACE applications. RNA 18:1289–1295
Jung H, Yoon BC, Holt CE (2012) Axonal mRNA localization and local protein synthesis in nervous system assembly, maintenance and repair. Nat Rev Neurosci 13:308–324
Khalil AM, Guttman M, Huarte M, Garber M, Raj A, Rivea Morales D, Thomas K, Presser A, Bernstein BE, van Oudenaarden A et al (2009) Many human large intergenic noncoding RNAs associate with chromatin-modifying complexes and affect gene expression. Proc Natl Acad Sci U S A 106:11667–11672
Kimble J, Simpson P (1997) The LIN-12/notch signaling pathway and its regulation. Annu Rev Cell Dev Biol 13:333–361
Kwak JE, Wang L, Ballantyne S, Kimble J, Wickens M (2004) Mammalian GLD-2 homologs are poly(a) polymerases. Proc Natl Acad Sci U S A 101:4407–4412
Lecuyer E, Yoshida H, Parthasarathy N, Alm C, Babak T, Cerovina T, Hughes TR, Tomancak P, Krause HM (2007) Global analysis of mRNA localization reveals a prominent role in organizing cellular architecture and function. Cell 131:174–187
Lesch BJ, Page DC (2012) Genetics of germ cell development. Nat Rev Genet 13:781–794
Lim J, Lee M, Son A, Chang H, Kim VN (2016) mTAIL-seq reveals dynamic poly(a) tail regulation in oocyte-to-embryo development. Genes Dev 30:1671–1682
Lucas BA, Lavi E, Shiue L, Cho H, Katzman S, Miyoshi K, Siomi MC, Carmel L, Ares M Jr, Maquat LE (2018) Evidence for convergent evolution of SINE-directed Staufen-mediated mRNA decay. Proc Natl Acad Sci U S A 115:968–973
Mangone M, Manoharan AP, Thierry-Mieg D, Thierry-Mieg J, Han T, Mackowiak SD, Mis E, Zegar C, Gutwein MR, Khivansara V et al (2010) The landscape of *C. elegans* 3′UTRs. Science 329:432–435
Mayr C (2016) Evolution and biological roles of alternative 3′UTRs. Trends Cell Biol 26:227–237

Mayr C (2017) Regulation by 3′-Untranslated regions. Annu Rev Genet 51:171–194

Merino EJ, Wilkinson KA, Coughlan JL, Weeks KM (2005) RNA structure analysis at single nucleotide resolution by selective 2′-hydroxyl acylation and primer extension (SHAPE). J Am Chem Soc 127:4223–4231

Merritt C, Rasoloson D, Ko D, Seydoux G (2008) 3′ UTRs are the primary regulators of gene expression in the *C. elegans* germline. Curr Biol 18:1476–1482

Nousch M, Techritz N, Hampel D, Millonigg S, Eckmann CR (2013) The Ccr4-not deadenylase complex constitutes the main poly(a) removal activity in *C. elegans*. J Cell Sci 126:4274–4285

Nousch M, Minasaki R, Eckmann CR (2017) Polyadenylation is the key aspect of GLD-2 function in *C. elegans*. RNA 23:1180–1187

Oldenbroek M, Robertson SM, Guven-Ozkan T, Gore S, Nishi Y, Lin R (2012) Multiple RNA-binding proteins function combinatorially to control the soma-restricted expression pattern of the E3 ligase subunit ZIF-1. Dev Biol 363:388–398

Osborne Nishimura E, Zhang JC, Werts AD, Goldstein B, Lieb JD (2015) Asymmetric transcript discovery by RNA-seq in *C. elegans* blastomeres identifies neg-1, a gene important for anterior morphogenesis. PLoS Genet 11:e1005117

Phillips CM, Montgomery TA, Breen PC, Ruvkun G (2012) MUT-16 promotes formation of perinuclear mutator foci required for RNA silencing in the *C. elegans* germline. Genes Dev 26:1433–1444

Pique M, Lopez JM, Foissac S, Guigo R, Mendez R (2008) A combinatorial code for CPE-mediated translational control. Cell 132:434–448

Poulsen LD, Kielpinski LJ, Salama SR, Krogh A, Vinther J (2015) SHAPE selection (SHAPES) enrich for RNA structure signal in SHAPE sequencing-based probing data. RNA 21:1042–1052

Rose L, Gonczy P (2014) Polarity establishment, asymmetric division and segregation of fate determinants in early *C. elegans* embryos. WormBook 30:1–43

Saletore Y, Meyer K, Korlach J, Vilfan ID, Jaffrey S, Mason CE (2012) The birth of the Epitranscriptome: deciphering the function of RNA modifications. Genome Biol 13:175

Sengupta MS, Low WY, Patterson JR, Kim HM, Traven A, Beilharz TH, Colaiacovo MP, Schisa JA, Boag PR (2013) Ifet-1 is a broad-scale translational repressor required for normal P granule formation in *C. elegans*. J Cell Sci 126:850–859

Sheth U, Pitt J, Dennis S, Priess JR (2010) Perinuclear P granules are the principal sites of mRNA export in adult *C. elegans* germ cells. Development 137:1305–1314

Smola MJ, Weeks KM (2018) In-cell RNA structure probing with SHAPE-MaP. Nat Protoc 13:1181–1195

Spike CA, Coetzee D, Eichten C, Wang X, Hansen D, Greenstein D (2014) The TRIM-NHL protein LIN-41 and the OMA RNA-binding proteins antagonistically control the prophase-to-metaphase transition and growth of *Caenorhabditis elegans* oocytes. Genetics 198:1535–1558

Tabara H, Hill RJ, Mello CC, Priess JR, Kohara Y (1999) Pos-1 encodes a cytoplasmic zinc-finger protein essential for germline specification in C. elegans. Development 126:1–11

Taft RJ, Pheasant M, Mattick JS (2007) The relationship between non-protein-coding DNA and eukaryotic complexity. Bioessays 29:288–299

Thomson DW, Dinger ME (2016) Endogenous microRNA sponges: evidence and controversy. Nat Rev Genet 17:272–283

Tian B, Manley JL (2017) Alternative polyadenylation of mRNA precursors. Nat Rev Mol Cell Biol 18:18–30

Tintori SC, Osborne Nishimura E, Golden P, Lieb JD, Goldstein B (2016) A transcriptional lineage of the Early *C. elegans* embryo. Dev Cell 38:430–444

Tsukamoto T, Gearhart MD, Spike CA, Huelgas-Morales G, Mews M, Boag PR, Beilharz TH, Greenstein D (2017) LIN-41 and OMA Ribonucleoprotein complexes mediate a translational repression-to-activation switch controlling oocyte meiotic maturation and the oocyte-to-embryo transition in *Caenorhabditis elegans*. Genetics 206:2007–2039

Tucey TM, Verma J, Harrison PF, Snelgrove SL, Lo TL, Scherer AK, Barugahare AA, Powell DR, Wheeler RT, Hickey MJ et al (2018) Glucose homeostasis is important for immune cell viability during Candida challenge and host survival of systemic fungal infection. Cell Metab 27:988–1006 e1007

Turner RE, Pattison AD, Beilharz TH (2018) Alternative polyadenylation in the regulation and dysregulation of gene expression. Semin Cell Dev Biol 75:61–69

Updike DL, Strome S (2009) A genomewide RNAi screen for genes that affect the stability, distribution and function of P granules in *Caenorhabditis elegans*. Genetics 183:1397–1419

Updike DL, Knutson AK, Egelhofer TA, Campbell AC, Strome S (2014) Germ-granule components prevent somatic development in the *C. elegans* germline. Curr Biol 24:970–975

Verma-Gaur J, Qu Y, Harrison PF, Lo TL, Quenault T, Dagley MJ, Bellousoff M, Powell DR, Beilharz TH, Traven A (2015) Integration of posttranscriptional gene networks into metabolic adaptation and biofilm maturation in *Candida albicans*. PLoS Genet 11:e1005590

Von Stetina JR, Orr-Weaver TL (2011) Developmental control of oocyte maturation and egg activation in metazoan models. Cold Spring Harb Perspect Biol 3:a005553

Wan G, Fields BD, Spracklin G, Shukla A, Phillips CM, Kennedy S (2018) Spatiotemporal regulation of liquid-like condensates in epigenetic inheritance. Nature 557:679–683

Wang L, Eckmann CR, Kadyk LC, Wickens M, Kimble J (2002) A regulatory cytoplasmic poly (a) polymerase in *Caenorhabditis elegans*. Nature 419:312–316

Watters KE, Yu AM, Strobel EJ, Settle AH, Lucks JB (2016) Characterizing RNA structures in vitro and in vivo with selective 2′-hydroxyl acylation analyzed by primer extension sequencing (SHAPE-Seq). Methods 103:34–48

Wood MP, Hollis A, Severance AL, Karrick ML, Schisa JA (2016) RNAi screen identifies novel regulators of RNP granules in the *Caenorhabditis elegans* germ line. G3 (Bethesda) 6:2643–2654

Yap KL, Li S, Munoz-Cabello AM, Raguz S, Zeng L, Mujtaba S, Gil J, Walsh MJ, Zhou MM (2010) Molecular interplay of the noncoding RNA ANRIL and methylated histone H3 lysine 27 by polycomb CBX7 in transcriptional silencing of INK4a. Mol Cell 38:662–674

# Chapter 6
# Communication Is Key: 5′–3′ Interactions that Regulate mRNA Translation and Turnover

**Hana Fakim and Marc R. Fabian**

**Abstract** Most eukaryotic mRNAs maintain a 5′ cap structure and 3′ poly(A) tail, cis-acting elements that are often separated by thousands of nucleotides. Nevertheless, multiple paradigms exist where mRNA 5′ and 3′ termini interact with each other in order to regulate mRNA translation and turnover. mRNAs recruit translation initiation factors to their termini, which in turn physically interact with each other. This physical bridging of the mRNA termini is known as the "closed loop" model, with years of genetic and biochemical evidence supporting the functional synergy between the 5′ cap and 3′ poly (A) tail to enhance mRNA translation initiation. However, a number of examples exist of "non-canonical" 5′–3′ communication for cellular and viral RNAs that lack 5′ cap structures and/or poly(A) tails. Moreover, in several contexts, mRNA 5′–3′ communication can function to repress translation. Overall, we detail how various mRNA 5′–3′ interactions play important roles in posttranscriptional regulation, wherein depending on the protein factors involved can result in translational stimulation or repression.

**Keywords** mRNA translation · mRNA decay · RNA-binding proteins · Posttranscriptional control · Protein–protein interactions

## 6.1 The Genesis of the Closed-Loop Model

The idea that the termini of eukaryotic mRNAs functionally interact in order to regulate protein synthesis is not a new hypothesis. Primarily based on electron micrograph images of polysome-bound mRNAs, it was proposed in the mid

H. Fakim
Jewish General Hospital, Lady Davis Institute, Montreal, Canada

M. R. Fabian (✉)
Jewish General Hospital, Lady Davis Institute, Montreal, Canada

Department of Oncology, McGill University, Montreal, Canada
e-mail: marc.fabian@mcgill.ca

 
M. Oeffinger, D. Zenklusen (eds.), *The Biology of mRNA: Structure and Function*, Advances in Experimental Medicine and Biology 1203,
https://doi.org/10.1007/978-3-030-31434-7_6

twentieth century that eukaryotic mRNAs translate as circular complexes rather than as linear molecules (Mathias et al. 1964; Philipps 1965; Baglioni et al. 1969). This was posited to allow for terminating ribosomes to be "recycled" rather than falling off the mRNA, thus enhancing mRNA translation. While enticing, this model was nevertheless proposed without any genetic or biochemical evidence to support it. Studies in the following decades identified the mRNA 5′ cap structure and 3′ poly(A) tail, as well as the translation factors that bind these cis-acting elements and stimulate translation, including the 5′ cap-bound eIF4F complex and the poly(A)-binding protein (PABP) (Tarun et al. 1997; Imataka et al. 1998). Importantly, it was shown that eIF4F physically binds PABP to stimulate protein synthesis and lent credence to a model where this interaction helps to bridge the mRNA termini. This model became commonly referred to as the "closed loop" model for translational control (Gallie 1991; Amrani et al. 2008). Moreover, there now exist multiple examples where alternative 5′–3′ interactions between protein and RNA elements at the mRNA termini are utilized to stimulate the translation of select cellular and viral mRNAs that lack either a 5′ cap and/or poly(A) tail. Finally, just as 5′–3′ mRNA interactions promote translation, a number of examples exist where the remodeling of mRNA circularization plays an important role in repressing the translation of specific mRNAs. In this chapter, we provide a broad overview of the different modes of communication between mRNA termini and how they stimulate or inhibit mRNA translation and decay.

## 6.2 5′ Cap- and 3′ Poly(A) Tail-Dependent Translation

The majority of eukaryotic mRNAs maintain a 5′ cap structure ($m^7$GpppN) and a 3′ poly(A) tail. Early experiments in cells and cell-free systems established that the cap and poly(A) tail elements stabilize mRNAs in an additive manner, but synergistically stimulate mRNA translation (Gallie 1991; Iizuka et al. 1994; Tarun et al. 1997; Preiss and Hentze 1998). This interdependency between the cap and poly(A) tail led to the hypotehsis that these elements must be directly communicating to engender optimal mRNA translation (Gallie 1991). Data that supported a physical interaction between these terminal elements came with the discovery of translation initiation factors that bind the cap and poly(A) tail structures. The 5′ cap is bound by the eIF4F complex, which consists of eIF4E, eIF4A, and eIF4G (Merrick and Pavitt 2018). eIF4E physically contacts the 5′ cap and binds to eIF4G, a scaffold protein that also interacts with a number of translation factors including eIF4A, an ATP-dependent RNA helicase, and eIF3. Ultimately, these translation initiation factors function to recruit the 40S ribosomal subunit as part of the 43S pre-initiation complex. Following scanning of the 5′ UTR and the identification of a proper start codon, the 60S ribosomal subunit joins the 40S subunit to form a functional 80S ribosome that can initiate translation. Regardless of being at the opposite end of the mRNA, the 3′ poly (A) tail stimulates translation by recruiting the poly(A)-binding protein (PABP), which serves as a bona fide translation initiation factor (Kahvejian et al. 2005).

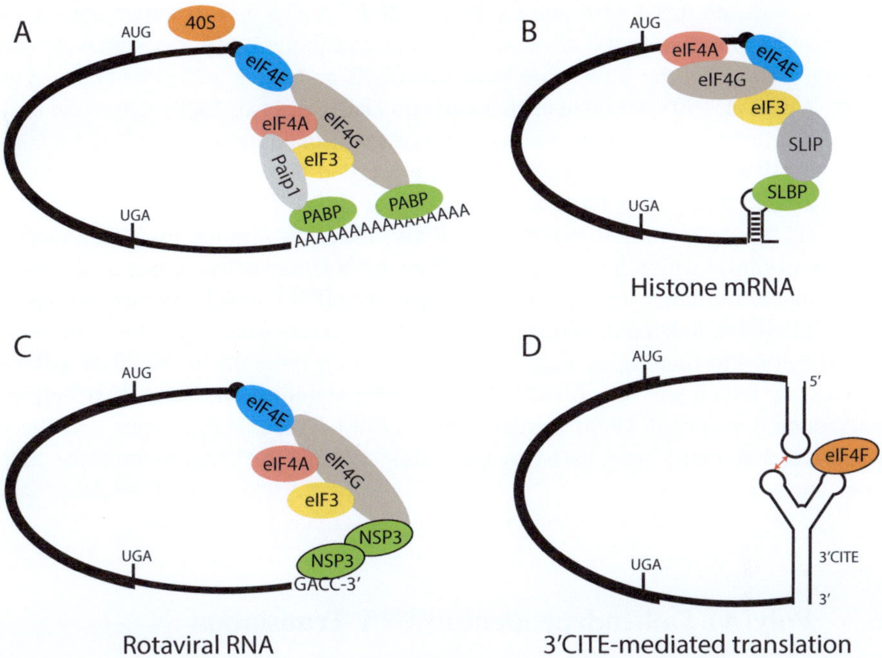


**Fig. 6.1** Schematic diagram of mRNA translation initiation mechanisms. (**a**) Canonical cap- and poly(A) tail-dependent translation. eIF4E interacts with the 5′ cap structure and forms the eIF4F complex by binding eIF4A and eIF4G. eIF4G mediates mRNA circularization by simultaneously binding to both eIF4E and PABP on the 3′ poly(A) tail. Paip1 may also assist in mRNA circularization by simultaneously interacting with eIF4A, eIF3, and PABP. (**b**) Histone mRNA translation. Histone mRNAs maintain a 3′ stemloop (SL) that recruits the SL-binding protein (SLBP). SLBP in turn interacts with the SLBP-interacting protein (SLIP) that binds eIF3 in order to promote translation initiation. (**c**) Rotaviral RNA translation. Rotaviral mRNAs maintain a 3′ terminal sequence (GACC-3′) that recruits NSP3, which dimerizes and interacts with eIF4G to promote mRNA translation. (**d**) 3′ cap-independent translational enhancer (3′ CITE)-mediated translation. The 3′ CITE is located in the 3′ UTR of the viral RNA, where it physically interacts with eIF4F. eIF4F then is brought into proximity with the 5′ end of the RNA by a long-distance RNA–RNA interaction between the 3′ CITE and an RNA structure in the 5′ UTR. Base pairing between the 3′ CITE and the 5′ UTR is denoted by a red arrow

PABP stimulates translation at least in part by physically contacting eIF4G, an interaction that is conserved from yeast to humans as well as in plants (Tarun and Sachs 1995; Le et al. 1997; Gray et al. 2000; Wakiyama et al. 2000; Kahvejian et al. 2005). As PABP and eIF4G are bound at the 5′ and 3′ termini of mRNAs, it is postulated that their interaction helps to form a "closed loop" by bridging the two ends of the mRNA (Fig. 6.1a). This model was further reinforced by mRNAs forming closed-loop structures, as observed by atomic force microscopy, in the presence of yeast eIF4G, eIF4E and PABP in vitro (Wells et al. 1998). What exactly is the biochemical mechanism by which PABP-eIF4G contact stimulates translation initiation? Several lines of evidence, both in yeast and in cell-free in vitro translation systems, indicate that PABP and the PABP-eIF4G interaction stimulate translation initiation by promoting 40S ribosomal subunit recruitment, 60S ribosomal subunit joining, as well as the interaction between eIF4E and the 5′ cap (Ptushkina et al. 1998; Wei et al. 1998; Borman et al. 2000; von Der Haar et al. 2000; Kahvejian et al. 2005). Furthermore, experiments with recombinant mammalian PABP and a fragment of eIF4G that binds PABP have demonstrated that eIF4G binding to PABP increases its affinity to poly(A) RNAs (Safaee et al. 2012).

In addition to binding to eIF4G, PABP also interacts with the PABP-interacting protein 1 (Paip1) which functions to stimulate mRNA translation (Craig et al. 1998). Paip1 shares similarity with the middle domain of eIF4G, which interacts with both eIF3 and eIF4A. In keeping with this, Paip1 also interacts with both eIF4A and eIF3. Specifically, the binding of Paip1 to eIF3 has been reported to stimulate mRNA translation, which was suggested to be due to the stabilization of the PABP-eIF4G interaction (Craig et al. 1998; Martineau et al. 2008). Based on these data, it has been proposed that Paip1 may assist in generating circular mRNAs to promote their translation (Fig. 6.1a).

## 6.3 Poly(A) Tail-Independent mRNA Translation

As mentioned above, most eukaryotic mRNAs maintain both a 5′ cap and 3′ poly (A) tail, elements that stimulate their translation. However, several examples exist of cellular and viral RNAs that do not possess a poly(A) tail. Nevertheless, these mRNAs have adopted alternative PABP-independent mechanisms to stimulate their translation that still rely upon contact between their 5′ and 3′ termini. Two key examples of this are mRNAs that code for replication-dependent histone proteins and rotaviral mRNAs.

**Histone mRNA Translation** Histones are evolutionarily conserved amongst eukaryotes and maintain pivotal roles in packaging genetic material into chromatin and regulating transcription. There are two types of histones: replication-independent and -dependent histones. Replication-independent histones are expressed throughout the cell cycle and act to modulate chromatin state in a locus-specific manner (Talbert and Henikoff 2017). Replication-dependent histones (referred from hereon as histones) include core histones (H2A, H2B, H3 and H4) that form nucleosomes, the structural unit of chromatin, and H1 linker histones that are found between nucleosomes (Marzluff et al. 2008). Transcription of core histones-encoding genes increases at the beginning of S-phase to accommodate DNA replication, but these transcripts are rapidly degraded at the end of S-phase given that an imbalance between histone and DNA abundance is detrimental, whereby it has been shown to cause chromosome loss and genomic instability (Singh et al. 2010).

Like all eukaryotic mRNAs, histone-encoding mRNAs possess a 5′ cap structure. However, histone-encoding mRNAs are unique in that they are the only cellular mRNA species to lack a 3′ poly(A) tail. Nonetheless, despite the lack of a poly

(A) tail and the consequential absence of PABP association, the 5′ and 3′ termini of these histone transcripts interface in order to efficiently recruit the pre-initiation complex (Fig. 6.1b). Instead of a poly(A) tail, histone mRNAs possess a conserved 25–26 nucleotide 3′ terminal stem loop (SL). This terminal structure functions to stimulate histone mRNA translation by interacting with the histone stem-loop-binding protein (SLBP), a protein that plays a key role in regulating histone mRNA maturation, degradation, and translation (Wang et al. 1996; Tan et al. 2013; Marzluff and Koreski 2017). Just as PABP stimulates mRNA translation by interacting with eIF4G to circularize canonical mRNAs, SLBP also interacts with the 5′ cap-associated translation machinery on histone mRNAs. However, unlike PABP, SLBP does not directly bind to eIF4G. Instead, SLBP recruits the SLBP-interacting protein (SLIP1), a middle domain of initiation factor 4G (MIF4G)-like protein, which simultaneously interacts with both SLBP and eIF3 to circularize the transcript and promote efficient translation (Neusiedler et al. 2012; von Moeller et al. 2013) (Fig. 6.1b). In keeping with the cell-cycle dependent regulation of histone mRNAs, SLBP levels increase during the G1/S phase to stimulate the production of histones, and is rapidly degraded by the proteasome by the end of the S-phase (Marzluff and Koreski 2017).

**Rotaviral PABP-Independent Translation** Rotaviral mRNAs maintain a 5′ cap but lack a poly(A) tail and instead terminate with a 3′ GACC sequence (Vende et al. 2000) (Fig. 6.1c). However, rotaviral mRNAs are efficiently translated due to the recruitment of the viral nonstructural protein 3 (NSP3) to this 3′ terminal element. NSP3 enhances viral translation by simultaneously interacting with the 3′ GACC viral element and with eIF4G in a manner similar to PABP (Vende et al. 2000; Groft and Burley 2002; Gratia et al. 2015). Interestingly, NSP3 has been reported to interact with eIF4G as a dimer, which binds eIF4G with a tenfold higher affinity as compared to PABP (Deo et al. 2002). In addition to directly enhancing rotaviral mRNA translation initiation, NSP3 binding to eIF4G is proposed to assist rotavirus infection by displacing PABP from eIF4G and leading to PABP nuclear localization, thereby shutting down host protein synthesis (Harb et al. 2008). Thus, NSP3 provides an alternative mode of circularizing viral mRNAs and selectively enhancing viral translation in the absence of a 3′ poly(A) tail (Fig. 6.1c).

## 6.4 Long-Distance RNA-RNA Interactions that Support Cap- and Poly(A) Tail-Independent Translation

A number of positive-strand plant RNA viruses lack both cap and poly(A) tail structures but have adopted unique modes of mRNA circularization in order to stimulate the production of viral proteins (Miller and White 2006; Nicholson and White 2011, 2014). These include viruses from the *Tombusviridae* and *Luteoviridae* families, which maintain highly structured RNA elements in their 3′ UTRs that are termed cap-independent translational enhancer (CITE) elements (Simon and Miller

2013). In general, 3′ CITEs function by physically recruiting the eIF4F translation initiation complex to the 3′ UTR of viral RNAs (Gazo et al. 2004; Treder et al. 2008; Wang et al. 2009; Nicholson et al. 2010, 2013). However, some viral 3′ CITE elements directly bind ribosomal subunits independently of eIF4F to stimulate viral translation (Stupina et al. 2008; Gao et al. 2012). While 3′ CITEs are necessary to promote viral translation, they must communicate with the viral 5′ UTR. This is mediated by long-distance RNA–RNA interactions between RNA stem loop structures in the 5′ UTR and the 3′ CITE, thus generating RNA–RNA-based closed-loop interactions (Fig. 6.1d). Site-directed mutagenesis experiments have demonstrated the functional significance of these long-distance RNA–RNA interactions in stimulating viral translation. Viral translation was inhibited when 5′ UTR/3′ CITE base pairing was disrupted, however, compensatory mutations that reestablished these long-distance interactions efficiently rescued translation (Guo et al. 2001; Fabian and White 2004, 2006; Nicholson and White 2008; Nicholson et al. 2010, 2013). Thus, it has been proposed that 3′ CITE elements recruit translation factors or ribosomal subunits to viral 3′ UTRs, which are then brought into proximity with the 5′ UTR in order to facilitate translation initiation.

## 6.5 5′–3′ Interactions that Repress mRNA Translation

Just as 5′–3′ interactions are critical for stimulating eukaryotic translation, a number of repressive mechanisms exist that rely upon contact between the mRNA termini in order to inhibit mRNA translation. An overarching theme for these regulatory mechanisms is the tethering of translational repressor proteins to specific cis-acting elements in the 3′ UTRs of select mRNAs. These, in turn, interface with the mRNA 5′ terminus and shut down protein synthesis. In general, these translational repressors fall into two classes: 5′ cap-binding proteins, such as the eIF4E homolog protein 4EHP, or eIF4E-binding proteins, such 4E-T, CUP, Maskin and Neuroguidin.

**4EHP *and 4E-T*** The eIF4E homolog protein (4EHP) represents a translational repressor that is similar to eIF4E, in that it binds to the mRNA 5′ cap structure (Rom et al. 1998). However, unlike eIF4E, 4EHP does not interact with eIF4G and therefore acts to repress mRNA translation initiation. In flies, *Drosophila* 4EHP (d4EHP) targets select mRNAs during embryogenesis, including the caudal and hunchback encoding mRNAs (Cho et al. 2005, 2006; Lasko 2011). d4EHP is recruited to the caudal mRNA via Bicoid, an RNA-binding protein that interacts with the Bicoid-binding region (BBR) in the caudal mRNA 3′ UTR (Cho et al. 2005). d4EHP then interacts with the caudal mRNA 5′ cap structure, thus circularizing the mRNA, displacing eIF4E and shutting down caudal protein synthesis (Fig. 6.2a). d4EHP uses a similar mechanism to inhibit hunchback mRNA translation by simultaneously interacting with the 5′ cap and a 3′ UTR-bound RNA protein complex. However, instead of binding Bicoid, d4EHP interacts with a complex of three proteins, Nanos (NOS), Pumilio (PUM), and brain tumor protein (BRAT),

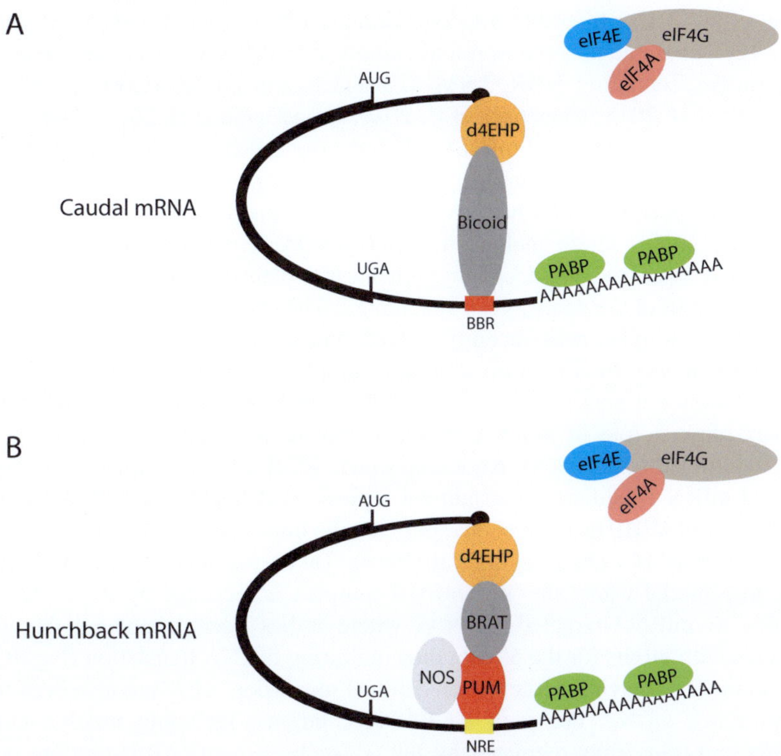


**Fig. 6.2** Schematic diagram of drosophila 4EHP (d4EHP)-mediated translational repression. (**a**) Caudal mRNA is translationally repressed by the d4EHP/Bicoid complex, which simultaneously binds to the Bicoid-binding region (BBR) in the caudal mRNA 3′ UTR and the 5′ cap structure. (**b**) Huncback mRNA is translationally repressed Pumilio (PUM), which binds to the Nanos response element (NRE). There, it interacts with Nanos (NOS) and Brain Tumor (BRAT). BRAT binds d4EHP, which interacts with the 5′ cap structure to inhibit mRNA translation

which are recruited to the Nanos-responsive element (NRE) in the hunchback 3′ UTR (Cho et al. 2006) (Fig. 6.2b). d4EHP therefore plays an important role in the development of the Drosophila embryo by making sure that Caudal and Hunchback proteins are produced in the proper locations within the embryo.

Proteomic and structural analyses of mammalian 4EHP have determined that it has two major binding partners: the RNA-binding protein GIGYF2 and the eIF4E-binding protein 4E-T (Morita et al. 2012; Chapat et al. 2017; Peter et al. 2017; Amaya Ramirez et al. 2018). Although less is currently known regarding the function of the GIGYF2/4EHP complex, several groups have implicated 4E-T in the translational repression and turnover of microRNA-targeted mRNAs (Kamenska et al. 2014a, 2016; Nishimura et al. 2015; Ozgur et al. 2015; Chapat et al. 2017; Duchaine and Fabian 2019). Like other 4E-BPs, 4E-T competes with eIF4G for binding to eIF4E (Dostie et al. 2000). However, in contrast to eIF4E-binding proteins such as eIF4G, 4E-T also has the ability to bind to 4EHP (Kubacka et al. 2013;

Chapat et al. 2017). In addition to containing an N-terminal eIF4E/4EHP-binding motif, 4E-T also interacts with proteins involved in mRNA translational repression and turnover, including UNR, LSM14, PATL1, and DDX6 (Dostie et al. 2000; Nishimura et al. 2015; Kamenska et al. 2016; Brandmann et al. 2018). How is 4E-T recruited to miRNA-targeted mRNAs? Briefly, the miRNA-induced silencing complex (miRISC) recruits a number of factors to targeted mRNAs that engender translational repression and mRNA decay (Jonas and Izaurralde 2015; Duchaine and Fabian 2019). These include the CCR4-NOT deadenylase complex and the translational repressor and decapping enhancer protein DDX6, which binds to the CNOT1 subunit of the deadenylase machinery. The crystal structure of the CNOT1/DDX6/4E-T complex was recently solved and demonstrates that DDX6, when directly bound to CNOT1, forms a unique complex with 4E-T (Ozgur et al. 2015). From a functional standpoint, several studies have reported that the 4E-T/4EHP complex plays a role in miRNA-mediated translational repression (Chapat et al. 2017; Jafarnejad et al. 2018). Knocking down 4EHP in mammalian cells partially impaired miRNA-mediated translational repression (Chapat et al. 2017; Chen and Gao 2017) and 4EHP has also been reported to be important for silencing the DUSP6 mRNA by miR-145 (Jafarnejad et al. 2018). Taken together, these data lend credence to a model where the 4E-T/4EHP complex is recruited by the CCR4-NOT complex to miRNA-targeted mRNAs where it has been postulated that 4EHP competes with eIF4E for the 5′ cap, thus inhibiting mRNA translation (Fig. 6.3a).

In addition to playing a role in translational repression, 4E-T has also been linked to enhancing mRNA decay of CCR4-NOT targets, including miRNA-targeted mRNAs and transcripts regulated by the AU-rich element (ARE)-binding protein tristetraprolin (TTP) (Ferraiuolo et al. 2005; Nishimura et al. 2015). Complementation experiments in HeLa cells demonstrated that a 4E-T mutant that cannot bind to eIF4E (or 4EHP) was unable to efficiently bring about the destabilization of miRNA- and TTP-targeted mRNAs. It was this suggested that 4E-T acts to enhance mRNA decay by bringing its interaction partners, that include the decapping factors LSM14, PATL1 and DDX6, into proximity with the 5′ terminus by binding to 4EHP (Fig. 6.3a) or eIF4E (Fig. 6.3b) (Nishimura et al. 2015).

**CUP** The Drosophila protein CUP is a well-characterized eIF4E-binding protein (4E-BP) that functions in the spatial and temporal regulation of specific mRNAs during oogenesis and embryogenesis. 4E-BPs [reviewed in (Kamenska et al. 2014b)] represent a class of translational regulators that compete with eIF4G bound on eIF4E to inhibit translation initiation. 4E-BPs include, but are not limited to, 4E-T, 4E-BP1, and Thor, which possess a canonical (C) (YXXXXLΦ, where X is any residue and Φ is hydrophobic) and a noncanonical (NC) eIF4E-binding site (Igreja et al. 2014). CUP was initially identified as a cytoplasmic protein present in *Drosophila* oocytes that functions in the translational repression and localization of oskar mRNA (Keyes and Spradling 1997; Wilhelm et al. 2003). Importantly, the proper localization of oskar mRNA is critical for posterior patterning of the embryo and germ line establishment (Ephrussi et al. 1991; Kimha et al. 1991). Instead of acting as a general translational repressor, CUP represses specific mRNAs (oskar and nanos)

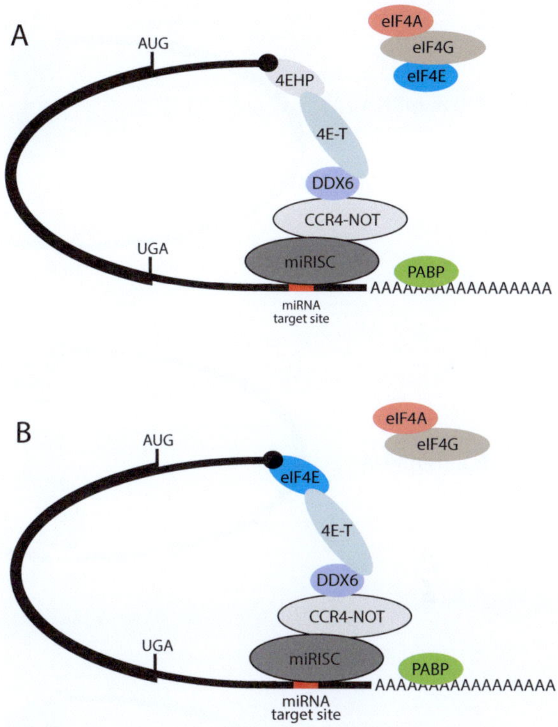


**Fig. 6.3** Schematic models for 4E-T involvement in microRNA-mediated translational repression. Of note, these models are not mutually exclusive. (**a**) A miRNA-targeted mRNA is bound by the miRISC, which in turn recruits the CCR4-NOT deadenylase complex. The CCR4-NOT/DDX6 complex recruits 4E-T and 4EHP, the latter functioning to displace eIF4E and associated translation factors (i.e. eIF4G and eIF4A) from the cap structure. (**b**) The CCR4-NOT/DDX6 complex recruits 4E-T, which binds to eIF4E, thereby displacing eIF4G. Figure modified from Duchaine and Fabian (2019) with permission from Cold Spring Harbor Laboratory Press © 2018

by tethering to their 3′ UTRs. CUP is recruited to oskar and nanos mRNAs via the RNA-binding proteins Bruno and Smaug (SMG), respectively, that bind to response elements in their 3′ UTRs (Wilhelm et al. 2003; Nakamura et al. 2004; Nelson et al. 2004). Thus, CUP represents a tethered translational repressor that simultaneously interacts with eIF4E and 3′ UTR-bound RBPs (Bruno or Smaug) in order to bridge the mRNA termini, displace eIF4G, and inhibit the translation of oskar and nanos mRNAs (Fig. 6.4a, b).

**Maskin** Maskin is an eIF4E-binding protein that plays an important role in regulating gene expression in *Xenopus* oocytes. Translational repression is pivotal during vertebrate oocyte maturation as immature oocytes are arrested at prophase of meiosis I (stage IV) where they synthesize large amounts of mRNA that are silenced and will serve to drive subsequent meiotic progression that takes place in the absence of transcription (Reyes and Ross 2016). Maskin is recruited to targeted mRNAs by directly interacting with the cytoplasmic polyadenylation element-binding (CPEB) protein, which binds to mRNAs containing cytoplasmic polyadenylation elements (CPEs) in their 3′ UTRs (Huang et al. 2006; Pique et al. 2008; Igea and Mendez 2010; Novoa et al. 2010). Specifically, Maskin is recruited by CPEB to maternal mRNAs with short poly(A) tails. There, Maskin simultaneously interacts with both CPEB and eIF4E, thereby preventing the association of eIF4E with eIF4G and consequently inhibiting mRNA translation (Stebbins-Boaz et al. 1999) (Fig. 6.4c).

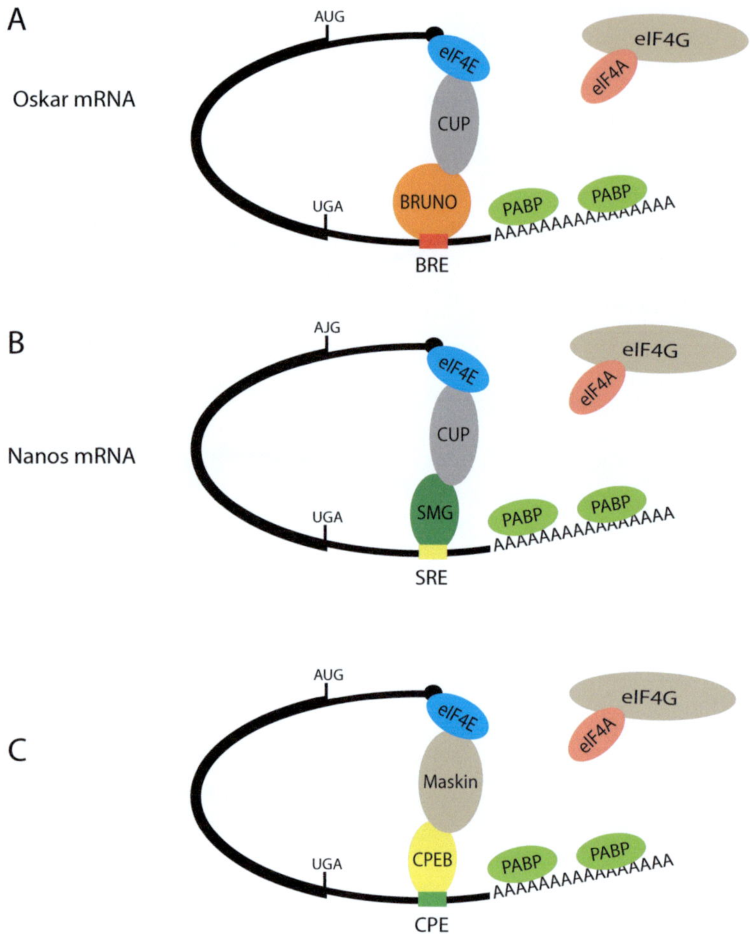


**Fig. 6.4** Tethered eIF4E-binding protein-mediated translational repression. (**a**) Oskar mRNA is translationally repressed by Bruno, which binds to the Bruno response element (BRE). There, Bruno interacts with the eIF4E-binding protein CUP, which binds to the 5′ cap-bound eIF4E in order to displace eIF4G and inhibit translation initiation. (**b**) Nanos mRNA is translationally repressed by Smaug (SMG), which binds to the Smaug response element (SRE). There, Smaug interacts with the eIF4E-binding protein CUP, which binds to the 5′ cap-bound eIF4E in order to displace eIF4G and inhibit translation initiation. (**c**) CPEB is recruited to CPE-containing mRNAs in late-stage Xenopus oocytes to repress their translation. CPEB binds to the eIF4E-binding protein Maskin, which binds to the 5′ cap-bound eIF4E in order to displace eIF4G and inhibit translation initiation

Maskin-mediated translational repression is then relieved in mature oocytes upon its phosphorylation on one or more of six major sites (T58, S152, S311, S343, S453, S638) by CDK1, which causes the release of eIF4E from Maskin (Barnard et al. 2005). Importantly, Minshall et al. showed that Maskin is only expressed after stage

IV and thus proposed another mode of silencing maternal transcripts in stages I–IV (Minshall et al. 2007). Instead, they found that CPEB interacted with a number of translational repressors, including DDX6 and 4E-T, as well as an eIF4E isoform (eIF4E1b). In contrast to eIF4E, eIF4E1b shows weak binding to both eIF4G and the cap structure. Based on these data, it was proposed that the CPEB/4E-T/eIF4E1b translational repression complex plays a role in early maternal silencing in growing oocytes.

**Neuroguidin** In addition to being expressed in Xenopus oocytes, CPEB is also expressed in neural tissues where it plays a role in regulating synaptic plasticity and memory formation (Darnell and Richter 2012; Rayman and Kandel 2017). Neuroguidin represents a neural-specific CPEB-interacting eIF4E-binding protein. Neuroguidin is not detected in Xenopus oocytes, however, ectopic expression of Neuroguidin in oocytes bound CPEB and led to the translational repression of CPE-containing mRNAs. In addition, knocking down Neuroguidin in the Xenopus embryo led to defects in neural crest migration and neural tube closure (Jung et al. 2006). Thus, Neuroguidin may act in a manner similar to Maskin to translationally repress CPEB-targeted mRNAs in neural tissues.

## 6.6 Conclusions and Future Perspectives

Overall, there is an abundance of biochemical and genetic evidence indicating that interactions between the mRNA 5′ and 3′ termini regulate mRNA translation. Notwithstanding these data, several key aspects of 5′–3′ mRNA communication remain to be elucidated. mRNA circularization via PABP-eIF4G enhances mRNA translation; yet, we do not know how stable these closed-loop structures are. Are they long-lasting interactions during protein synthesis, or transient structures that briefly form and then are disrupted upon translation initiation? Moreover, as much of the data behind this model have been generated using reporter mRNAs or in the context of select mRNA species, it remains to be determined whether all mRNAs require PABP–eIF4G contact or whether specific types of mRNAs are more dependent on mRNA circularization for their efficient translation (Archer et al. 2015; Thompson and Gilbert 2017).

While many observations favor the closed-loop model for promoting translation initiation, it is still a working model that is under investigation (Thompson and Gilbert 2017). Experiments in cell-free systems have shown that the translation of capped mRNAs lacking 3′ poly(A) tails can be stimulated upon the addition of free poly(A) RNA in trans (Borman et al. 2002). In addition, this stimulation was abolished upon the addition of a viral protein that disrupts the PABP–eIF4G interaction. Taken together, these data suggest that while the PABP–eIF4G interaction stimulates translation, it may not always act to circularize mRNAs. Recent investigations using cryo-electron tomography suggest that circular polysomes in cell-free systems can exist on mRNAs that lack both a 5′ cap structure and a 3′ poly

(A) tail (Afonina et al. 2014). Finally, mRNA closed-loop dynamics during translation have recently been investigated *in cellulo* using single-molecule resolution fluorescent in situ hybridization (smFISH) and super-resolution microscopy (Adivarahan 2018; Khong and Parker 2018). Both studies conclude that the 5′ and 3′ ends of actively translating mRNAs rarely co-localize, and that the distance between the mRNA termini increases as a function of ribosome occupancy. Thus, the mRNA closed-loop state may not be stable during translation, and the interaction between eIF4G and PABP may only occur during specific stages of the translation cycle and/or for a subset of mRNAs. In conclusion, while communication between the mRNA termini is a key aspect of translational control, it would be premature to close the book on the closed-loop model.

**Acknowledgments** We apologize for any directly related work that we have not cited in this review. This work was supported by a Canadian Institute of Health Research (CIHR) grant (MOP-130425) and a Natural Sciences and Engineering Research Council of Canada (NSERC) Discovery grant to M.R.F. (RGPIN-2015-03712), as well as Fonds de recherché du Québec-Santé (FRQS) Chercheur-Boursier Junior 1 salary award and a CIHR New Investigator award to M.R.F.

## References

Adivarahan S, Livingston N, Nicholson B, Rahman S, Wu B, Rissland OS, Zenklusen D (2018) Spatial organization of single mRNPs at different stages of the gene expression pathway. Mol Cell 72(4):727–738.e5

Afonina ZA, Myasnikov AG, Shirokov VA, Klaholz BP, Spirin AS (2014) Formation of circular polyribosomes on eukaryotic mRNA without cap-structure and poly(A)-tail: a cryo electron tomography study. Nucleic Acids Res 42:9461–9469

Amaya Ramirez CC, Hubbe P, Mandel N, Bethune J (2018) 4EHP-independent repression of endogenous mRNAs by the RNA-binding protein GIGYF2. Nucleic Acids Res 46:5792–5808

Amrani N, Ghosh S, Mangus DA, Jacobson A (2008) Translation factors promote the formation of two states of the closed-loop mRNP. Nature 453:1276–1280

Archer SK, Shirokikh NE, Hallwirth CV, Beilharz TH, Preiss T (2015) Probing the closed-loop model of mRNA translation in living cells. RNA Biol 12:248–254

Baglioni C, Vesco C, Jacobs-Lorena M (1969) The role of ribosomal subunits in mammalian cells. Cold Spring Harb Symp Quant Biol 34:555–565

Barnard DC, Cao QP, Richter JD (2005) Differential phosphorylation controls maskin association with eukaryotic translation initiation factor 4E and localization on the mitotic apparatus. Mol Cell Biol 25:7605–7615

Borman AM, Michel YM, Kean KM (2000) Biochemical characterisation of cap-poly(A) synergy in rabbit reticulocyte lysates: the eIF4G-PABP interaction increases the functional affinity of eIF4E for the capped mRNA 5′-end. Nucleic Acids Res 28:4068–4075

Borman AM, Michel YM, Malnou CE, Kean KM (2002) Free poly(A) stimulates capped mRNA translation in vitro through the eIF4G-poly(A)-binding protein interaction. J Biol Chem 277:36818–36824

Brandmann T, Fakim H, Padamsi Z, Youn JY, Gingras AC, Fabian MR, Jinek M (2018) Molecular architecture of LSM14 interactions involved in the assembly of mRNA silencing complexes. EMBO J 37

Chapat C, Jafarnejad SM, Matta-Camacho E, Hesketh GG, Gelbart IA, Attig J, Gkogkas CG, Alain T, Stern-Ginossar N, Fabian MR et al (2017) Cap-binding protein 4EHP effects translation silencing by microRNAs. Proc Natl Acad Sci U S A 114:5425–5430
Chen S, Gao G (2017) MicroRNAs recruit eIF4E2 to repress translation of target mRNAs. Protein Cell 8:750–761
Cho PF, Poulin F, Cho-Park YA, Cho-Park IB, Chicoine JD, Lasko P, Sonenberg N (2005) A new paradigm for translational control: inhibition via 5′-3' mRNA tethering by Bicoid and the eIF4E cognate 4EHP. Cell 121:411–423
Cho PF, Gamberi C, Cho-Park YA, Cho-Park IB, Lasko P, Sonenberg N (2006) Cap-dependent translational inhibition establishes two opposing morphogen gradients in Drosophila embryos. Curr Biol 16:2035–2041
Craig AW, Haghighat A, Yu AT, Sonenberg N (1998) Interaction of polyadenylate-binding protein with the eIF4G homologue PAIP enhances translation. Nature 392:520–523
Darnell JC, Richter JD (2012) Cytoplasmic RNA-binding proteins and the control of complex brain function. Cold Spring Harb Perspect Biol 4:a012344
Deo RC, Groft CM, Rajashankar KR, Burley SK (2002) Recognition of the rotavirus mRNA 3′ consensus by an asymmetric NSP3 homodimer. Cell 108:71–81
Dostie J, Ferraiuolo M, Pause A, Adam SA, Sonenberg N (2000) A novel shuttling protein, 4E-T, mediates the nuclear import of the mRNA 5′ cap-binding protein, eIF4E. EMBO J 19:3142–3156
Duchaine TF, Fabian MR (2019) Mechanistic insights into microRNA-mediated gene silencing. Cold Spring Harb Perspect Biol 11(3)
Ephrussi A, Dickinson LK, Lehmann R (1991) Oskar organizes the germ plasm and directs localization of the posterior determinant nanos. Cell 66:37–50
Fabian MR, White KA (2004) 5′-3' RNA-RNA interaction facilitates cap- and poly(A) tail-independent translation of tomato bushy stunt virus mrna: a potential common mechanism for tombusviridae. J Biol Chem 279:28862–28872
Fabian MR, White KA (2006) Analysis of a 3′-translation enhancer in a tombusvirus: a dynamic model for RNA-RNA interactions of mRNA termini. RNA 12:1304–1314
Ferraiuolo MA, Basak S, Dostie J, Murray EL, Schoenberg DR, Sonenberg N (2005) A role for the eIF4E-binding protein 4E-T in P-body formation and mRNA decay. J Cell Biol 170:913–924
Gallie DR (1991) The cap and poly(A) tail function synergistically to regulate mRNA translational efficiency. Genes Dev 5:2108–2116
Gao F, Kasprzak W, Stupina VA, Shapiro BA, Simon AE (2012) A ribosome-binding, 3′ translational enhancer has a T-shaped structure and engages in a long-distance RNA-RNA interaction. J Virol 86:9828–9842
Gazo BM, Murphy P, Gatchel JR, Browning KS (2004) A novel interaction of Cap-binding protein complexes eukaryotic initiation factor (eIF) 4F and eIF(iso)4F with a region in the 3-′-untranslated region of satellite tobacco necrosis virus. J Biol Chem 279:13584–13592
Gratia M, Sarot E, Vende P, Charpilienne A, Baron CH, Duarte M, Pyronnet S, Poncet D (2015) Rotavirus NSP3 is a translational surrogate of the poly(A) binding protein-poly(A) complex. J Virol 89:8773–8782
Gray NK, Coller JM, Dickson KS, Wickens M (2000) Multiple portions of poly(A)-binding protein stimulate translation in vivo. EMBO J 19:4723–4733
Groft CM, Burley SK (2002) Recognition of eIF4G by rotavirus NSP3 reveals a basis for mRNA circularization. Mol Cell 9:1273–1283
Guo L, Allen EM, Miller WA (2001) Base-pairing between untranslated regions facilitates translation of uncapped, nonpolyadenylated viral RNA. Mol Cell 7:1103–1109
Harb M, Becker MM, Vitour D, Baron CH, Vende P, Brown SC, Bolte S, Arold ST, Poncet D (2008) Nuclear localization of cytoplasmic poly(A)-binding protein upon rotavirus infection involves the interaction of NSP3 with eIF4G and RoXaN. J Virol 82:11283–11293
Huang Y-S, Kan M-C, Lin C-gL, Richter JD (2006) CPEB3 and CPEB4 in neurons: analysis of RNA-binding specificity and translational control of AMPA receptor GluR2 mRNA. EMBO J 25:4865–4876

Igea A, Mendez R (2010) Meiosis requires a translational positive loop where CPEB1 ensues its replacement by CPEB4. EMBO J 29:2182–2193
Igreja C, Peter D, Weiler C, Izaurralde E (2014) 4E-BPs require non-canonical 4E-binding motifs and a lateral surface of eIF4E to repress translation. Nat Commun 5:14
Iizuka N, Najita L, Franzusoff A, Sarnow P (1994) Cap-dependent and cap-independent translation by internal initiation of mRNAs in cell extracts prepared from Saccharomyces cerevisiae. Mol Cell Biol 14:7322–7330
Imataka H, Gradi A, Sonenberg N (1998) A newly identified N-terminal amino acid sequence of human eIF4G binds poly(A)-binding protein and functions in poly(A)-dependent translation. EMBO J 17:7480–7489
Jafarnejad SM, Chapat C, Matta-Camacho E, Gelbart IA, Hesketh GG, Arguello M, Garzia A, Kim SH, Attig J, Shapiro M et al (2018) Translational control of ERK signaling through miRNA/4EHP-directed silencing. Elife:7
Jonas S, Izaurralde E (2015) Towards a molecular understanding of microRNA-mediated gene silencing. Nat Rev Genet 16:421–433
Jung MY, Lorenz L, Richter JD (2006) Translational control by neuroguidin, a eukaryotic initiation factor 4E and CPEB binding protein. Mol Cell Biol 26:4277–4287
Kahvejian A, Svitkin YV, Sukarieh R, M'Boutchou MN, Sonenberg N (2005) Mammalian poly(A)-binding protein is a eukaryotic translation initiation factor, which acts via multiple mechanisms. Genes Dev 19:104–113
Kamenska A, Lu WT, Kubacka D, Broomhead H, Minshall N, Bushell M, Standart N (2014a) Human 4E-T represses translation of bound mRNAs and enhances microRNA-mediated silencing. Nucleic Acids Res 42:3298–3313
Kamenska A, Simpson C, Standart N (2014b) eIF4E-binding proteins: new factors, new locations, new roles. Biochem Soc Trans 42:1238–1245
Kamenska A, Simpson C, Vindry C, Broomhead H, Benard M, Ernoult-Lange M, Lee BP, Harries LW, Weil D, Standart N (2016) The DDX6-4E-T interaction mediates translational repression and P-body assembly. Nucleic Acids Res 44:6318–6334
Keyes LN, Spradling AC (1997) The Drosophila gene fs(2)cup interacts with otu to define a cytoplasmic pathway required for the structure and function of germ-line chromosomes. Development 124:1419–1431
Khong A, Parker R (2018) mRNP architecture in translating and stress conditions reveals an ordered pathway of mRNP compaction. J Cell Biol 217(12):4124–4140
Kimha J, Smith JL, Macdonald PM (1991) Oskar messenger-RNA is localized to the posterior pole of the drosophila oocyte. Cell 66:23–34
Kubacka D, Kamenska A, Broomhead H, Minshall N, Darzynkiewicz E, Standart N (2013) Investigating the consequences of eIF4E2 (4EHP) interaction with 4E-transporter on its cellular distribution in HeLa cells. PLoS One 8:e72761
Lasko P (2011) Posttranscriptional regulation in Drosophila oocytes and early embryos. Wiley Interdiscip Rev RNA 2:408–416
Le H, Tanguay RL, Balasta ML, Wei CC, Browning KS, Metz AM, Goss DJ, Gallie DR (1997) Translation initiation factors eIF-iso4G and eIF-4B interact with the poly(A)-binding protein and increase its RNA binding activity. J Biol Chem 272:16247–16255
Martineau Y, Derry MC, Wang X, Yanagiya A, Berlanga JJ, Shyu AB, Imataka H, Gehring K, Sonenberg N (2008) Poly(A)-binding protein-interacting protein 1 binds to eukaryotic translation initiation factor 3 to stimulate translation. Mol Cell Biol 28:6658–6667
Marzluff WF, Koreski KP (2017) Birth and death of histone mRNAs. Trends Genet 33:745–759
Marzluff WF, Wagner EJ, Duronio RJ (2008) Metabolism and regulation of canonical histone mRNAs: life without a poly(A) tail. Nat Rev Genet 9:843–854
Mathias AP, Williamson R, Huxley HE, Page S (1964) Occurrence and function of polysomes in rabbit reticulocytes. J Mol Biol 9:154–167
Merrick WC, Pavitt GD (2018) Protein synthesis initiation in eukaryotic cells. Cold Spring Harb Perspect Biol

Miller WA, White KA (2006) Long-distance RNA-RNA interactions in plant virus gene expression and replication. Annu Rev Phytopathol 44:447–467
Minshall N, Reiter MH, Weil D, Standart N (2007) CPEB interacts with an ovary-specific eIF4E and 4E-T in early Xenopus oocytes. J Biol Chem 282:37389–37401
Morita M, Ler LW, Fabian MR, Siddiqui N, Mullin M, Henderson VC, Alain T, Fonseca BD, Karashchuk G, Bennett CF et al (2012) A novel 4EHP-GIGYF2 translational repressor complex is essential for mammalian development. Mol Cell Biol 32:3585–3593
Nakamura A, Sato K, Hanyu-Nakamura K (2004) Drosophila cup is an eIF4E binding protein that associates with Bruno and regulates oskar mRNA translation in oogenesis. Dev Cell 6:69–78
Nelson MR, Leidal AM, Smibert CA (2004) Drosophila Cup is an eIF4E-binding protein that functions in Smaug-mediated translational repression. EMBO J 23:150–159
Neusiedler J, Mocquet V, Limousin T, Ohlmann T, Morris C, Jalinot P (2012) INT6 interacts with MIF4GD/SLIP1 and is necessary for efficient histone mRNA translation. RNA 18:1163–1177
Nicholson BL, White KA (2008) Context-influenced cap-independent translation of Tombusvirus mRNAs in vitro. Virology 380:203–212
Nicholson BL, White KA (2011) 3' Cap-independent translation enhancers of positive-strand RNA plant viruses. Curr Opin Virol 1:373–380
Nicholson BL, White KA (2014) Functional long-range RNA-RNA interactions in positive-strand RNA viruses. Nat Rev Microbiol 12:493–504
Nicholson BL, Wu B, Chevtchenko I, White KA (2010) Tombusvirus recruitment of host translational machinery via the 3' UTR. RNA 16:1402–1419
Nicholson BL, Zaslaver O, Mayberry LK, Browning KS, White KA (2013) Tombusvirus Y-shaped translational enhancer forms a complex with eIF4F and can be functionally replaced by heterologous translational enhancers. J Virol 87:1872–1883
Nishimura T, Padamsi Z, Fakim H, Milette S, Dunham WH, Gingras AC, Fabian MR (2015) The eIF4E-binding protein 4E-T is a component of the mRNA decay machinery that bridges the 5′ and 3' termini of target mRNAs. Cell Rep 11:1425–1436
Novoa I, Gallego J, Ferreira PG, Mendez R (2010) Mitotic cell-cycle progression is regulated by CPEB1 and CPEB4-dependent translational control. Nat Cell Biol 12:447–U482
Ozgur S, Basquin J, Kamenska A, Filipowicz W, Standart N, Conti E (2015) Structure of a human 4E-T/DDX6/CNOT1 complex reveals the different interplay of DDX6-binding proteins with the CCR4-NOT complex. Cell Rep 13:703–711
Peter D, Weber R, Sandmeir F, Wohlbold L, Helms S, Bawankar P, Valkov E, Igreja C, Izaurralde E (2017) GIGYF1/2 proteins use auxiliary sequences to selectively bind to 4EHP and repress target mRNA expression. Genes Dev 31:1147–1161
Philipps GR (1965) Haemoglobin synthesis and polysomes in intact reticulocytes. Nature 205:567–570
Pique M, Lopez JM, Foissac S, Guigo R, Mendez R (2008) A combinatorial code for CPE-mediated translational control. Cell 132:434–448
Preiss T, Hentze MW (1998) Dual function of the messenger RNA cap structure in poly(A)-tail-promoted translation in yeast. Nature 392:516–520
Ptushkina M, von der Haar T, Vasilescu S, Frank R, Birkenhager R, McCarthy JE (1998) Cooperative modulation by eIF4G of eIF4E-binding to the mRNA 5′ cap in yeast involves a site partially shared by p20. EMBO J 17:4798–4808
Rayman JB, Kandel ER (2017) Functional prions in the brain. Cold Spring Harb Perspect Biol 9
Reyes JM, Ross PJ (2016) Cytoplasmic polyadenylation in mammalian oocyte maturation. Wiley Interdisciplinary Reviews-RNA 7:71–89
Rom E, Kim HC, Gingras AC, Marcotrigiano J, Favre D, Olsen H, Burley SK, Sonenberg N (1998) Cloning and characterization of 4EHP, a novel mammalian eIF4E-related cap-binding protein. J Biol Chem 273:13104–13109
Safaee N, Kozlov G, Noronha AM, Xie J, Wilds CJ, Gehring K (2012) Interdomain allostery promotes assembly of the poly(A) mRNA complex with PABP and eIF4G. Mol Cell 48:375–386

Simon AE, Miller WA (2013) 3′ cap-independent translation enhancers of plant viruses. Annu Rev Microbiol 67:21–42

Singh RK, Liang D, Gajjalaiahvari UR, Kabbaj MHM, Paik J, Gunjan A (2010) Excess histone levels mediate cytotoxicity via multiple mechanisms. Cell Cycle 9:4236–4244

Stebbins-Boaz B, Cao QP, de Moor CH, Mendez R, Richter JD (1999) Maskin is a CPEB-associated factor that transiently interacts with eIF-4E. Mol Cell 4:1017–1027

Stupina VA, Meskauskas A, McCormack JC, Yingling YG, Shapiro BA, Dinman JD, Simon AE (2008) The 3′ proximal translational enhancer of Turnip crinkle virus binds to 60S ribosomal subunits. RNA 14:2379–2393

Talbert PB, Henikoff S (2017) Histone variants on the move: substrates for chromatin dynamics. Nat Rev Mol Cell Biol 18:115–126

Tan D, Marzluff WF, Dominski Z, Tong L (2013) Structure of histone mRNA stem-loop, human stem-loop binding protein, and 3 ' hExo ternary complex. Science 339:318–321

Tarun SZ Jr, Sachs AB (1995) A common function for mRNA 5′ and 3′ ends in translation initiation in yeast. Genes Dev 9:2997–3007

Tarun SZ Jr, Wells SE, Deardorff JA, Sachs AB (1997) Translation initiation factor eIF4G mediates in vitro poly(A) tail-dependent translation. Proc Natl Acad Sci U S A 94:9046–9051

Thompson MK, Gilbert WV (2017) mRNA length-sensing in eukaryotic translation: reconsidering the “closed loop” and its implications for translational control. Curr Genet 63:613–620

Treder K, Kneller EL, Allen EM, Wang Z, Browning KS, Miller WA (2008) The 3′ cap-independent translation element of Barley yellow dwarf virus binds eIF4F via the eIF4G subunit to initiate translation. RNA 14:134–147

Vende P, Piron M, Castagne N, Poncet D (2000) Efficient translation of rotavirus mRNA requires simultaneous interaction of NSP3 with the eukaryotic translation initiation factor eIF4G and the mRNA 3′ end. J Virol 74:7064–7071

von Der Haar T, Ball PD, McCarthy JE (2000) Stabilization of eukaryotic initiation factor 4E binding to the mRNA 5'-Cap by domains of eIF4G. J Biol Chem 275:30551–30555

von Moeller H, Lerner R, Ricciardi A, Basquin C, Marzluff WF, Conti E (2013) Structural and biochemical studies of SLIP1-SLBP identify DBP5 and eIF3g as SLIP1-binding proteins. Nucleic Acids Res 41:7960–7971

Wakiyama M, Imataka H, Sonenberg N (2000) Interaction of eIF4G with poly(A)-binding protein stimulates translation and is critical for Xenopus oocyte maturation. Curr Biol 10:1147–1150

Wang ZF, Whitfield ML, Ingledue TC, Dominski Z, Marzluff WF (1996) The protein that binds the 3′ end of histone mRNA: a novel RNA-binding protein required for histone pre-mRNA processing. Genes Dev 10:3028–3040

Wang Z, Treder K, Miller WA (2009) Structure of a viral cap-independent translation element that functions via high affinity binding to the eIF4E subunit of eIF4F. J Biol Chem 284:14189–14202

Wei CC, Balasta ML, Ren J, Goss DJ (1998) Wheat germ poly(A) binding protein enhances the binding affinity of eukaryotic initiation factor 4F and (iso)4F for cap analogues. Biochemistry 37:1910–1916

Wells SE, Hillner PE, Vale RD, Sachs AB (1998) Circularization of mRNA by eukaryotic translation initiation factors. Mol Cell 2:135–140

Wilhelm JE, Hilton M, Amos Q, Henzel WJ (2003) Cup is an eIF4E binding protein required for tooth the translational repression of oskar and the recruitment of Barentsz. J Cell Biol 163:1197–1204

# Chapter 7
# Bioinformatics Approaches to Gain Insights into cis-Regulatory Motifs Involved in mRNA Localization

**Louis Philip Benoit Bouvrette, Mathieu Blanchette, and Eric Lécuyer**

**Abstract** Messenger RNA (mRNA) is a fundamental intermediate in the expression of proteins. As an integral part of this important process, protein production can be localized by the targeting of mRNA to a specific subcellular compartment. The subcellular destination of mRNA is suggested to be governed by a region of its primary sequence or secondary structure, which consequently dictates the recruitment of *trans*-acting factors, such as RNA-binding proteins or regulatory RNAs, to form a messenger ribonucleoprotein particle. This molecular ensemble is requisite for precise and spatiotemporal control of gene expression. In the context of RNA localization, the description of the binding preferences of an RNA-binding protein defines a *motif*, and one, or more, instance of a given motif is defined as a *localization element* (zip code). In this chapter, we first discuss the *cis*-regulatory motifs previously identified as mRNA localization elements. We then describe motif representation in terms of entropy and information content and offer an overview of motif databases and search algorithms. Finally, we provide an outline of the motif topology of asymmetrically localized mRNA molecules.

**Keywords** RNA localization · RNA binding protein/RBP · cis-Regulatory motifs

---

L. P. Benoit Bouvrette
Institut de Recherches Cliniques de Montréal (IRCM), Montréal, QC, Canada

Département de Biochimie, Université de Montréal, Montréal, QC, Canada
e-mail: louis-philip.benoit-bouvrette@umontreal.ca

M. Blanchette
McGill School of Computer Science, McGill University, Montréal, QC, Canada
e-mail: mathieu.blanchette@mcgill.ca

E. Lécuyer (✉)
Institut de Recherches Cliniques de Montréal (IRCM), Montréal, QC, Canada

Département de Biochimie, Université de Montréal, Montréal, QC, Canada

Division of Experimental Medicine, McGill University, Montréal, QC, Canada
e-mail: Eric.Lecuyer@ircm.qc.ca



M. Oeffinger, D. Zenklusen (eds.), *The Biology of mRNA: Structure and Function*, Advances in Experimental Medicine and Biology 1203,
https://doi.org/10.1007/978-3-030-31434-7_7

## 7.1 General Introduction

In 1950, it was first hypothesized that RNA was synthesized in the nucleus and then transferred into the cytoplasm, where it was aggregating with other molecules (Jeener and Szafarz 1950). A better appreciation of the role of RNA was gained in 1961 when three publications revolutionized the way gene function was perceived by establishing messenger RNA (mRNA) as an information carrier in a transitional stage towards the synthesis of protein (Brenner et al. 1961; Gros et al. 1961; Jacob and Monod 1961). Following these breakthroughs, it was not immediately apparent whether mRNA could localize to specific subcellular sites. It was not until the mid-1980s that the first elements of the answer were identified when it was reported that the *actin* mRNA in ascidian oocytes and embryos was asymmetrically distributed (Jeffery et al. 1983). The discovery of additional localized RNAs implicated in processes such as embryonic patterning and cell migration led to the realization that regulated subcellular trafficking of mRNAs was biologically important (Cody et al. 2013; Hoek et al. 1998; Johnstone and Lasko 2001; Lecuyer et al. 2007; Martin and Ephrussi 2009; Paquin and Chartrand 2008). This work led to the model that mRNA transport is a multiple step process involving (1) the formation of a messenger ribonucleoprotein (mRNP) created by the association of an mRNA with RNA-binding proteins (RBPs), (2) the transport of this mRNP to a specific subcellular region, (3) the in situ anchoring of this mRNP and (4) the local translation of the mRNA to produce the required protein (Ainger et al. 1993; Wilhelm and Vale 1993). Since then, a broad diversity of mRNAs have been shown to be localized, through different mechanisms, in different cell types, organisms and developmental stages (Benoit Bouvrette et al. 2018; Crofts et al. 2004; Henry et al. 2010; Heym and Niessing 2012; Keiler 2011; Lasko 2012; Lecuyer et al. 2007; Medioni et al. 2012; Prodon et al. 2007; Serikawa et al. 2001; Wang et al. 2012; Zarnack and Feldbrugge 2010). With the advances of microscopy techniques, genomic approaches and, nowadays, bioinformatics modelling, it is now appreciated that a majority of mRNAs undergo regulated subcellular trafficking (Batish et al. 2012; Benoit Bouvrette et al. 2018; Blower et al. 2007; Eberwine et al. 2002; Farris et al. 2014; Hutten et al. 2014; Jambor et al. 2015; Lecuyer et al. 2007; Lefebvre et al. 2017; Mikl et al. 2011; Mili et al. 2008). This growing body of evidence has underlined the importance of RNA localization as a key aspect of post-transcriptional gene regulation while also emphasizing the potentially critical role played by *cis*-acting localization elements in this regulatory process.

This chapter is aimed at the informatics-enthusiast biologists with an interest in RNA localization and who are keen to gain insights in the processing and analysis of RNA biology data. While the methods described herein to study *cis*-regulatory motifs, and their instances, may be applied to many aspects of post-transcriptional gene regulation, the examples given are focused on the specifics of RNA localization analysis. Additionally, we do not aim to provide a complete picture of the diverse resources available, but we cover useful examples to help guide the reader.

## 7.2 Fundamental Aspects of RNA Localization

Gene expression is modulated by a wide array of regulatory events that can be mediated by compartment-specific ribonucleoprotein (RNP) complexes. These complexes are involved in all aspects of the mRNA life cycle, from synthesis, processing, editing, nuclear export, cytoplasmic localization, translation and degradation (Gerstberger et al. 2014). These events are interdependent and can occur in different locales of a cell, from precise intra-nuclear regions, where nascent transcripts are synthesized, to the targeting of mature transcripts to specific regions of the cytoplasm or extracellular milieu through secretion. An important facet of post-transcriptional gene regulation is the subcellular transit of mRNA, which may serve a variety of functions mechanistically. Firstly, when combined with localized translation, this process can serve to enrich protein products within a specific compartment of the cell in an efficient manner. Indeed, targeted translation has been proposed as a possible facilitator of the assembly of localized protein complexes (Batada et al. 2004; Kuriyan et al. 2007). Consistent with this notion, transcripts that encode functionally related proteins can have similar localization patterns, which, in turn, are often distinct among different functional classes (Jambor et al. 2015; Lecuyer et al. 2007; Wilk et al. 2016). Secondly, mRNA localization may also be important to avoid the aberrant targeting of protein products, which could have deleterious effects if they were to accumulate in certain regions of the cell. Interestingly, while RNA localization has been known to have a special relevance in polarized cells, especially neurons, it has also been described to be highly prevalent in a myriad of cell types and appears to be conserved evolutionarily (Benoit Bouvrette et al. 2018; Crofts et al. 2004; Henry et al. 2010; Heym and Niessing 2012; Keiler 2011; Lasko 2012; Lecuyer et al. 2007; Medioni et al. 2012; Prodon et al. 2007; Serikawa et al. 2001; Wang et al. 2012; Zarnack and Feldbrugge 2010).

At the molecular level, mRNA localization is coordinated by *cis*-regulatory motifs (CRMs), where one or more instances of these motifs, present within the RNA molecule itself, are referred to as localization elements or zip codes that mediate interaction with *trans*-acting factors (Bergalet and Lécuyer 2014; Jambhekar and DeRisi 2007) (Fig. 7.1). These CRMs are generally defined by their primary sequence and/or secondary/tertiary structure features (Van De Bor and Davis 2004). CRMs are thought to be recognized by RNA-binding proteins (RBPs) that seed the formation of mRNP complexes necessary for transit. RBPs form a prominent and deeply conserved family of regulatory proteins, which are classified based on their RNA-binding domains (RBDs) (Gerstberger et al. 2014). While RBDs often confer binding to single-stranded RNA sequences, some RBP subfamilies mediate binding to structured regions of the target RNA. Different mechanisms may exist in order to target mRNA molecules and to keep them in a translationally repressed state during transport (Dahm and Kiebler 2005). After nuclear export, an mRNP may acquire or discard a series of trans-regulatory factors (e.g. RBPs, miRNA) that will guide RNA fate by modulating its transport, translation and stability (Giorgi and Moore 2007). One of the major mechanisms

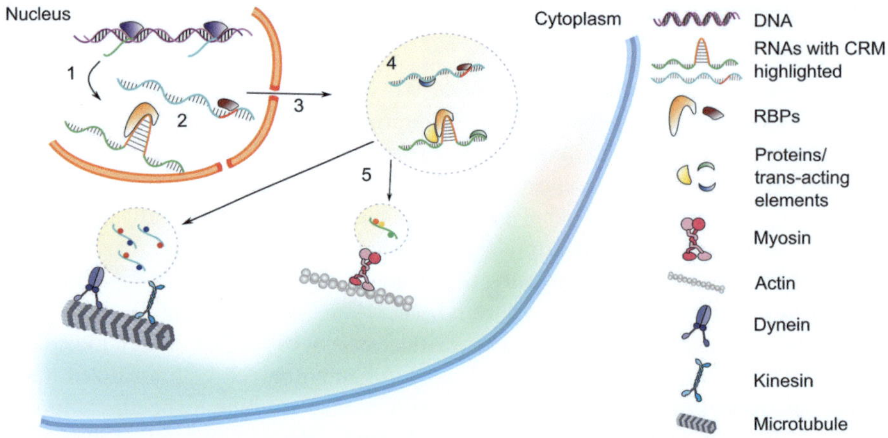


**Fig. 7.1** Distinct mRNA *cis*-regulatory motifs, acting as localization elements, guide the assembly with an RBP to form an mRNP that gets targeted to a specific subcellular region. Schema of RNA localization *cis*-regulatory motifs. Following transcription (**1**), mRNAs are bound in the nucleus by RBPs (**2**) that recognize CRMs formed by primary sequence (red) or secondary structure (orange) to form an mRNP. Following export into the cytoplasm via a nuclear pore (**3**), RBPs and *trans*-acting elements may be added, or removed, to remodel the mRNP and assemble it into RNA granules (**4**). These RNA granules associate with motor proteins and are transported by cytoskeletal elements towards their target subcellular location (**5**)

characterized to achieve subcellular targeting implies the direct trafficking of a localization-competent RNP by association with specific molecular motor proteins that direct transport along cytoskeletal networks in the cytoplasm (Bergalet and Lécuyer 2014; Tekotte and Davis 2002). Upon reaching its destination, the mRNP can be anchored and remodelled to enable translation to take place (Forrest and Gavis 2003).

In this section, we survey some of the better-documented CRMs implicated in the intracellular trafficking of RNA. For more comprehensive discussion of the functions and biological benefits of intracellular RNA trafficking or the molecular mechanisms involved, please refer to other recent reviews (Bergalet and Lécuyer 2014; Bovaird et al. 2018; Jambhekar and DeRisi 2007; Martin and Ephrussi 2009).

### *7.2.1* cis-*Regulatory Motifs Implicated in RNA Localization*

The characterization of CRMs involved in RNA localization is of great importance to gain insights into the mechanisms of this post-transcriptional regulatory process. CRMs are typically discrete intrinsic elements of information that can function independently from their host mRNA molecule, i.e. they can confer localization activity to a normally non-localized reporter RNA molecule (e.g. *gfp, lacz*). As such, CRMs can be identified via structure-function studies, by tracking the subcellular localization of fragments derived from an asymmetrically distributed mRNA, which is achieved by fusing such fragments to a reporter transcript. This chimeric transcript makes it possible to identify which region of an mRNA exhibits CRM activity and whether this component is sufficient for proper RNA targeting. For example, the vasopressin CRM was used to confer dendritic compartmentalization to *alpha-tubulin* mRNA, normally confined to the cell body (Prakash et al. 1997). This has allowed the delimitation of a number of CRMs from a wide array of localized mRNAs (Jambhekar and DeRisi 2007). In Table 7.1, we compile a summarized list of few CRMs known to be involved in mRNA localization.

Interestingly, while many localization CRMs have been mapped to the 3′ untranslated region (UTR) of mRNAs, some have also been characterized in the 5′ UTR or coding regions (Chartrand et al. 1999; Gavis and Lehmann 1992; Jambhekar and DeRisi 2007; Kowanda et al. 2016; Meer et al. 2012; Mowry and Melton 1992; Serano and Rubin 2003; Van De Bor and Davis 2004). In addition to their variability in distribution across the mRNA molecule, the CRMs can also exhibit heterogeneity both in their sequence length and structure. The relative length of a CRM can vary greatly between transcripts, with some being only a few nucleotides long while others running over kilobases of sequence (Bergalet and Lécuyer 2014; Jambhekar and DeRisi 2007). Moreover, as mentioned above, some CRMs are defined by simple primary sequence motifs or stem-loop elements, while others may be composed of more complex structural features, such as G-quadruplexes (Jambhekar and DeRisi 2007; Van De Bor and Davis 2004). For example, transcripts such as *beta-actin*, *nanos*, *MBP* or *vg1* have CRMs in the form of short primary sequence elements (Afroz et al. 2014; Ainger et al. 1993; Allen et al. 2003; Bergsten et al. 2001; Betley et al. 2004; Chao et al. 2010; Czaplinski et al. 2005; Deshler et al. 1997, 1998; Farina et al. 2003; Forrest and Gavis 2003; Gautreau et al. 1997; Gavis et al. 1996a, b; Gavis and Lehmann 1992; Gu et al. 2002; Hoek et al. 1998; Huttelmaier et al. 2005; Kislauskis and Singer 1992; Kress et al. 2004; Lerit and Gavis 2011; Munro et al. 1999; Oleynikov and Singer 2003; Pan et al. 2009; Patel et al. 2012; Ross et al. 1997; Shestakova et al. 2001; Zhang et al. 2001). In most cases, the CRMs of these mRNAs are composed of multiple regions that may act sequentially or in concert to direct localization. In particular, *Drosophila nanos* mRNA bears four CRMs spanning a 280-nucleotide region of its 3′ UTR, which govern localization in a combinatorial way and ultimately function in the patterning of the anterior-posterior body axis (Afroz et al. 2014; Bergsten et al. 2001; Forrest and Gavis 2003; Gavis et al. 1996a, b; Gavis and Lehmann 1992; Lerit and Gavis 2011). By contrast, transcripts such as *Anxa2*, *ASH1*, *bicoid*, *CamKIIa* and *Gurken* have structural CRMs (Afroz et al. 2014; Bertrand et al. 1998; Blichenberg et al. 2001; Bullock and Ish-Horowicz 2001; Chartrand et al. 1999, 2002; Ferrandon et al. 1997; Gonzalez et al. 1999; Heym and Niessing 2012; Kugler and Lasko 2009; Long et al. 1997, 2000; Macdonald and Kerr 1997, 1998; Macdonald et al. 1993; Macdonald and Struhl 1988; Mayford et al. 1996; Mori et al. 2000; Rihan et al. 2017; Saunders and Cohen 1999; Snee et al. 2005; Takizawa et al. 1997; Van De Bor et al. 2005; Weil et al. 2006). In particular, in *Drosophila* oogenesis, the localization of *bicoid* mRNA is driven by a 650-nucleotide segment of its 3′ UTR, for which five domains

**Table 7.1** Summary of cis-regulatory motifs involved in localized transcript

| mRNA (gene name) | *cis*-Regulatory motifs (localization element names) | Type of motifs | Position in mRNA (5′ UTR, CDS, 3′ UTR) | Recognized by RNA binding protein | Organisms | References |
|---|---|---|---|---|---|---|
| Arc | DTE | Primary sequence | 3′ UTR | Multiple | Mammals | Dynes and Steward (2007, 2012) |
| ASH1 | E1, E2A, E2B, E3 (43 nt stem-loop) | Secondary structure | 3′ UTR, CDS, 5′ UTR | She2p | Yeast | Bertrand et al. (1998), Chartrand et al. (2002, 1999), Gonzalez et al. (1999), Jambhekar and DeRisi (2007), Jambhekar et al. (2005), Kruse et al. (2002), Long et al. (1997, 2000), Olivier et al. (2005), Takizawa et al. (1997) |
| Anxa2 | G-quadruplex (18 nt region) | Secondary structure | 3′ UTR | SMN | Mouse | Rihan et al. (2017) |
| **β**-actin | 54 nt region | Primary sequence | 3′ UTR | ZBP1, ZBP2 | Chicken, human | Chao et al. (2010), Farina et al. (2003), Gu et al. (2002), Katz et al. (2012) Kislauskis et al. (1994), Pan et al. (2009), Patel et al. (2012), Ross et al. (1997), Shestakova et al. (2001), Welshhans and Bassell (2011), Yamagishi et al. (2009a, b), Zhang et al. (2001) |
| Bicoid | Domain III; stem-loop IV–V; BLE1 | Primary sequence; secondary structure | 3′ UTR | Staufen | *Drosophila* | Ferrandon et al. (1997), Kugler and Lasko (2009), Macdonald and Kerr (1997, 1998), Macdonald et al. (1993), Macdonald and Struhl (1988), Snee et al. (2005), Weil et al. (2006) |
| Bitesize | BLR | Primary sequence | CDS | | *Drosophila* | Serano and Rubin (2003) |

| | | | | | | |
|---|---|---|---|---|---|---|
| Bsg25D | 2 sequence elements | Primary sequence | CDS, 3′ UTR | | *Drosophila* | Kowanda et al. (2016) |
| CamKIIa | G-quadruplex (30 nt region) | Secondary structure | 3′ UTR | Staufen, hnRNP U, PSF, FMRP | Mammals | Blichenberg et al. (2001), Dictenberg et al. (2008), Huang et al. (2003), Knowles et al. (1996), Mayford et al. (1996), Mori et al. (2000) |
| Fatvg | FVLE (25 nt region) | Primary sequence | 3′ UTR | | *Xenopus* | Chan et al. (1999, 2001) |
| GIRK2 | YCAY element | Primary sequence | Introns and 3′ UTR | Nova | Mammals | Racca et al. (2010) |
| Gurken | GLE1 (35 nt), GLS (64 nt stem-loop) | Primary sequence; secondary structure | 5′ end of ORF | Egalitarian | *Drosophila* | Bullock et al. (2010), Dienstbier et al. (2009), [Kugler 2009 #164], Saunders and Cohen (1999), Van De Bor et al. (2005) |
| hairy | Structural element | Secondary structure | | Egalitarian | *Drosophila* | Bullock et al. (2010), Dienstbier et al. (2009) |
| k10 | TLS | Primary sequence; secondary structure | 3′ UTR | Egalitarian | *Drosophila* | Bullock et al. (2010), Cheung et al. (1992), Dienstbier et al. (2009), Serano and Cohen (1995) |
| MAP2 | DTE (640 nt region) | Secondary structure | 3′ UTR | | Rat | Blichenberg et al. (1999, 2001) |
| MBP | A2RE (11 nt region) | Primary sequence | 3′ UTR | hnRNP A2 | Rat | Ainger et al. (1993, 1997), Hoek et al. (1998), Munro et al. (1999) |
| Nanos | 4 redundant regions | Primary sequence | 3′ UTR | Multiples | *Drosophila* | Forrest and Gavis (2003) [Kugler 2009 #164], Bergsten et al. (2001), Gavis et al. (1996a, b, Gavis and Lehmann (1992) |
| Neurogranin | CPE (170 nt region) | Primary sequence | 3′ UTR | | Mammals | Mori et al. (2000) |
| orb | 280 nt region | Primary sequence | 3′ UTR | | *Drosophila* | Lantz and Schedl (1994) |

(continued)

**Table 7.1** (continued)

| mRNA (gene name) | *cis*-Regulatory motifs (localization element names) | Type of motifs | Position in mRNA (5′ UTR, CDS, 3′ UTR) | Recognized by RNA binding protein | Organisms | References |
|---|---|---|---|---|---|---|
| Oskar | EJC, 150 nt region, SOLE | Primary sequence; secondary structure | 5′ and 3′ UTR | Y14 Staufen, hnRNP A/B | *Drosophila* | Brendza et al. (2000), Hachet and Ephrussi (2001), Hachet and Ephrussi (2004), Huynh et al. (2004), Kim et al. (2014), Kim-Ha et al. (1991, 1993), Kugler and Lasko (2009), Ryu et al. (2017), St Johnston et al. (1991), Yano et al. (2004), Zimyanin et al. (2008) |
| sensorin | 66 nt stem-loop region | Secondary structure | 5′ UTR | | | Gomes et al. (2014), Meer et al. (2012) |
| tau | 91 nt region | Primary sequence | 3′ UTR | | Rat | Aranda-Abreu et al. (1999), Aronov et al. (2001), Behar et al. (1995) |
| Vasopressin | ORF fragment | Primary sequence | 3′ UTR | | Rat | Mohr et al. (1995), Prakash et al. (1997) |
| Vg1 | Vm1, E1–E4 (>300 nt region) | Primary sequence | 3′ UTR | 40KoVe, hnRP U, Vg1RBP/Vera, Kinesin-1, Kinesin-2 | *Xenopus* | Betley et al. (2004), Czaplinski et al. (2005), Deshler et al. (1997, 1998), Gautreau et al. (1997), Kress et al. (2004), Messitt et al. (2008), Mowry and Melton (1992) |
| Xcat2 | MCLE (250 nt region), GGLE (164 nt region) | Primary sequence | 3′ UTR | | *Xenopus* | Kloc et al. (2000), Zhou and King (1996a, b) |
| XNIF | 300 nt region | Primary sequence | 5′ UTR | | *Xenopus* | Claußen et al. (2004) |
| Xlsirt | 71–81 nt repeats | Primary sequence | 3′ UTR | | *Xenopus* | Allen et al. (2003) |
| Xvelo | 75 nt stem-loop | Primary sequence; secondary structure | 5′ and 3′ end | | *Xenopus* | Claussen and Pieler (2004) |

of secondary structure have been shown to cooperate at the various steps of the transport process (Ferrandon et al. 1997; Kugler and Lasko 2009; Macdonald and Kerr 1997, 1998; Macdonald et al. 1993; Macdonald and Struhl 1988; Snee et al. 2005; Weil et al. 2006). Lastly, it is common to observe that multiple elements of different motifs cooperate in a combinatorial fashion and act at distinct steps of the localization process (Gautreau et al. 1997; Mori et al. 2000). On the other hand, recurring copies of a single motif can act synergistically to promote individual steps (Ainger et al. 1997; Deshler et al. 1997). While these examples convey the diversity of CRM topological organization within localized mRNA molecules, it has been difficult to glean consensus sequence or structure features within families of mRNAs that share similar localization properties. It is important to note that the variability in CRM features might be in part due to the experimental complexity inherent to their study, often requiring painstaking structure-function mapping via sequence trimming and mutagenesis. As such, in many cases, the characterization of minimal regions that define specific CRMs may have been imprecise.

Evidence supports the notion that RNAs have similar localization phenotypes in different cell types and species, suggesting that some CRMs might be evolutionarily conserved and operating via similar pathways (Benoit Bouvrette et al. 2018; Bullock and Ish-Horowicz 2001). For example, strong correlations in the distribution profiles of ~2500 mRNA orthologs between human and *Drosophila* were recently characterized, with shared general similarities with respect to their UTR and coding sequence lengths (Benoit Bouvrette et al. 2018). With the development of new experimental approaches to characterize subcellular transcriptomes, such as CeFra-seq or APEX-RIP (Benoit Bouvrette et al. 2018; Kaewsapsak et al. 2017), and the datasets generated, this establishes the basis for the implementation of bioinformatics approaches to map putative sequence motifs that may drive RNA localization.

## 7.3 Representation and Information Content of Sequence Motifs

RNA sequence motifs, regardless of their biological functions, can be viewed, from a more mathematical point of view, as blocks of regulatory information. This notion that information can be quantitatively measured is important as it allows for the modelling and discovery of additional instances of a given sequence motif. Here, we define a sequence motif as a specific pattern that is common to a set of DNA, RNA or protein molecules, which are presumed to share particular biological properties or regulatory logic. In the case of RNA localization regulation, the sequence motifs can be the states and patterns that modulate the interaction of a transcript with specific RPBs that direct its targeting to a given subcellular destination. Below, we discuss the various ways by which RNA regulatory motifs can be represented and provide an overview of the different approaches used to map putative regulatory motifs.

There are numerous ways to describe sequence motifs within biological molecules in order to accurately annotate the binding preferences of a given RBP

(Fig. 7.2). For instance, one of the first biological motifs identified was the TATA box, which was identified by aligning gene promoter elements and transcription start sites and observing an over-representation of that short DNA substring. Therefore, the simplest representation of a motif is stating it as a short sequence. Similarly, if we were interested in the A2RE motif found in RNA targets of the HNRNP A2 protein, we could align multiple sequences containing the motifs and search for a cognate subsequence (Fig. 7.2a). The consensus, or canonical, sequence is obtained by selecting the most frequent nucleotide (or amino acid in the case of proteins) observed at each position (Fig. 7.2b). While this is an adequate way of modelling a motif, it is insufficient to fully capture its essence or identify other naturally occurring motifs, because RBPs tend to have flexible binding preferences. A motif is usually described as exact (precise), or degenerate (weak), according to the

**A**

```
>Human_alphaCaMKII_A2RE
GCCAGGAGCCA
>Human_Putative_A2RE-like_sequence
GAGAAGGAGGG
>Human_Neurogranin_A2RE
CUGUUGAGGGC
>Human_ARC_A2RE
GCGGACGAGGA
```

**B**

GCGAAGANGGA

**C**

SHSDDSRRSSV

**D**

| | | | | | | | | |
|---|---|---|---|---|---|---|---|---|
| A | 0.00 | 0.25 | 0.00 | 0.50 | 0.50 | 0.00 | ... | 0.50 |
| U | 0.00 | 0.25 | 0.00 | 0.25 | 0.25 | 0.00 | ... | 0.00 |
| G | 0.75 | 0.00 | 0.75 | 0.25 | 0.25 | 0.75 | ... | 0.25 |
| C | 0.25 | 0.50 | 0.25 | 0.00 | 0.00 | 0.25 | ... | 0.25 |

**E**

**Fig. 7.2** Various formats can be used to describe the A2RE motif. (**a**) Aligned fasta sequences of the A2RE localization element in four different human mRNA. (**b**) Consensus motif of (**a**) showing the most represented nucleotide at each position. Ambiguous nucleotides, where all bases are equally represented, are noted as "N." (**c**) IUPAC representation of (**a**). (**d**) Truncated position weight matrix (PWM) showing the percentage of each base observed at each position of (**a**). (**e**) Sequence logo, assuming a uniform background nucleotide probability

amount of deviations observed between its different instances. For example, the motif bound by HNRNP K is the fixed subsequence GCCGAC, which is considered an exact motif (Dominguez et al. 2018). On the other hand, HNRNP A2 mediates trafficking of RNAs containing the A2RE motif, which display greater diversity and is therefore more degenerate (Fig. 7.2a). One way to capture variations among instances is by way of a regular expression. For example, a *cis*-regulatory sequence motif might be formulated as [A][G][U][U or G][A][G], which can be abbreviated by the International Union of Pure and Applied Chemistry (IUPAC) nomenclature as AGU**K**AG, where K is the shorthand for either appearing nucleotide U or G (Fig. 7.2c). Most scripting languages handle the search for regular expression (regex) well. Here the search for AGUKAG could simply be encoded as AGU [UG]AG.

Many alternative ways exist to describe a motif, of which the most popular is the position weight matrix (PWM), which is further described here (Fig. 7.2d) (Liefooghe et al. 2006; Sandve and Drablos 2006; Staden 1984). This is a matrix with four rows (one for each base A, U, G, C) and width $k$ equal to the number of bases in the motif. A PWM assumes that each position has its own probability distribution over nucleotides and that the choices of nucleotide at different positions are independent. This means that the columns of a PWM can be thought of as a set of independent multinomial distributions. This allows for the easy calculation of the probability of a subsequence given a PWM, done by simply multiplying each relevant probability. For example, the probability of the sequence $S =$ CUG would be calculated by multiplying the probability of having a "C" in position 1, a "U" in position 2 and a "G" in position 3. Taking the three first positions of the PWM of Fig. 7.2d, this would be $0.25 \times 0.25 \times 0.75 = 0.0468$.

The level of specificity (or, inversely, flexibility) of a PWM is an important property that is captured in terms of the information theoretic notions of *information content* and *entropy*. Consider a given column of a PWM, with nucleotide probabilities $P_n$ ($n =$ A, C, G, U). The *Shannon entropy* of a probability distribution is defined as $H(P) = -\sum_{n=1}^{4} P_n \log_2 P_n$. This will yield a non-negative value, measured in bits. A bit represents the amount of information necessary to select between two equiprobable options (Machta 1999; Schneider 2010). For DNA and RNA, which are each made of 4 bases, this value will be between 0 and $\log_2 4 = 2$, whereas for protein motifs it can reach $\log_2 20 \cong 4.32$. Since entropy is a measure of uncertainty, when $P_n$ assigns a probability of 1 to a particular nucleotide, the entropy of $P_n$ will be 0 bits, as there is no uncertainty. On the contrary, when all four bases are equiprobable, the entropy will be 2 bits. It requires 2 bits of information to determine which of the four bases occurs at that position. The first 1 bit of decision divides the set by half (e.g. purine vs. pyrimidine), leaving only 2 choices, A/G or C/T. A related notion is that of the *information content* $H$ of a distribution $P$ (e.g. a column of a PWM) against a certain background distribution $B$ (e.g. the genome-wide nucleotide frequencies), defined as $H(P) = \sum_{n=1}^{4} P_n \log_2 \frac{P_n}{B_n}$. The information content of $P$ against $B$, also known as the Kullback-Leibler divergence (Gupta et al. 2007) between the two distributions, is a measure of how different the two distributions are. Note that when $B$ is a uniform distribution ($B_n = 0.25$ for $n =$ A, C, G, U), $H(P, B) = 2 - I(P)$.

An elegant way to visually represent a PWM while conveying its information content is called the graphical sequence logo (Fig. 7.2e) (Schneider and Stephens 1990; Shaner et al. 1993). In a sequence logo, each position of the motif is represented as a stack of nucleotides, whose total height corresponds to the information content at that position. The height of each nucleotide is proportional to its probability at that position. Therefore, the sequence logo provides a rapid visual portrayal of the conservation and composition of each position in a motif (Crooks et al. 2004).

Knowing the information content of a motif is useful when searching for additional instances as a motif with $n$ bits of information will occur about once in every $2^n$ bases of random sequence. For example, the six-mer GCCCAC motif of HNRNP K has an information content of 12 bits (6 base motifs with 2 bits of information each); it is expected that a putative motif instance for this RBP will be observed in an RNA sequence every $2^{12} = 4096$ bases (assume a uniform background), close to what has been described before (Paziewska et al. 2004). By contrast the information content of the more degenerate HNRNP K motif [GC]CCCAC is $\log_2(2) + 5 \times \log_2(4) = 11$ and would be expected to occur twice as frequently as $2^{11} = 2048$. It is easy to see that GCCCAC or CCCCAC can occur two times more often than GCCCAC alone. However, this frequency of putative motif instance estimation is different than the frequency of actual RBP binding sites, as the former could include identifications of motif instances as false-positive binding sites and therefore be much larger than the latter.

## 7.4 Algorithms and Tools for Finding Motifs

### 7.4.1 Fundamentals of Major Motif Discovery Algorithms

One important question in bioinformatics applied to the study of RNA is: How can one extract known and unknown regulatory motifs from an ensemble of given sequences? This question comes in two flavours. *Motif scanning* aims to predict new instances of one or more known motifs in a given sequence. For example, one may use this approach to identify, in a given mRNA sequence, candidate binding sites for an RBP with a known PWM. *De novo motif discovery*, on the contrary, aims to determine, from a set of sequences thought to be co-regulated (e.g. identified through a CLIP-seq experiment on a given RBP) or co-localized, the motif(s) that may best capture the binding preferences of the RBPs involved.

Motif scanning is simple and fast. When searching for matches given a PWM in a given sequence longer than $k$, the score of the $k$-mer starting at every possible position in the sequence is evaluated as shown above, and high-scoring sites are reported (Beckstette et al. 2006; Kel et al. 2003; Matys et al. 2003; Sandelin et al. 2004). The main issue is to decide on a score threshold above which sites should be reported. Various strategies have been proposed, aiming to maximize the sensitivity of the scan while maintaining an acceptable level of false positives (Giudice et al. 2016; Kel et al. 2003; Liu et al. 2017; Paz et al. 2014). One such approach is illustrated in the next section.

De novo motif discovery typically falls within one of three types: enumerative algorithms, probabilistic optimization and deterministic optimization.

The first, and perhaps simplest, de novo motif discovery approach is designated as an *enumerative*, or dictionary, approach. In its basic form, it aims at discovering motifs represented as strict consensus sequences. For every possible consensus sequence $w$ of length $k$ (user-defined), these algorithms contrast the number of occurrences of $w$ in a set of positive sequences (e.g. isolated RNA from a subcellular compartment), compared to a control set (unlocalized or random sequences). Enrichment within the positive set is then quantified statistically, to obtain an enrichment $p$-value. While effective, this approach is based on exact occurrence of specific strings of characters and is often too restrictive for a sensible application in biology where proteins generally bind RNA via degenerate motifs. As such, it is possible that none of the motifs would occur often enough to be observed in a statistically significant fashion. Fortunately, it is possible to generalize the method by being more flexible on the definition of the motifs to search. This alternative approach to the enumeration algorithm can be achieved by either using regular expression or allowing an explicit number of mismatches (Carlson et al. 2007; Fauteux et al. 2008; Korn et al. 1977; MacIsaac and Fraenkel 2006; Pavesi et al. 2004; Queen et al. 1982; Sandve and Drablos 2006; Sinha and Tompa 2003).

A second approach for finding motifs de novo is the *probabilistic optimization* strategy, which aims at inferring a PWM from a set of co-regulated sequences. It is perhaps best exemplified by the Gibbs sampling algorithm, one of the earlier motif detection methods (Lawrence et al. 1993; MacIsaac and Fraenkel 2006). It works by first selecting a random position in each sequence and building a PWM from them. It further selects a sequence at random to scan and score all possible sites in this sequence using this predetermined PWM. It can then select a new motif site and update the motif instances and the weight matrix accordingly. Finally, the algorithm iterates over the last steps until a convergence is reached. This algorithm works well to find de novo motifs since a real motif is expected to be overrepresented and therefore should be encountered more often when searching at random, which will bias the original weight matrix. Updating the matrix will further lean it towards finding more motifs, until convergence. Since there is a random element involved, one caveat is that while it will always find a motif, there is no certainty that it will always converge towards the same motif.

A third strategy for finding de novo motifs, similar to Gibbs sampling, makes use of a *deterministic optimization* of the PWM for describing a motif and the binding probabilities for its associated sites and is referred to as the expectation maximization (EM) strategy (Bailey and Elkan 1994; Bailey and Elkan 1995; Lawrence and Reilly 1990; MacIsaac and Fraenkel 2006). EM class algorithms are often used for learning probabilistic models in problems that involve hidden states. In a motif-finding tool, this can be defined as the position(s) where the motif occurs in each sequence.

Sequences can have 0, 1 or multiple occurrences of a given motif. This approach has the advantage of simultaneously identifying the position and characteristics of a motif. Briefly, this is achieved by initializing a weight matrix with a single *k*-mer and a subset of the background frequencies. Then, by scanning the possible space of motifs for each *k*-mer in the sequence set, it calculates the probability that this *k*-mer was generated by the motifs from the matrix, rather than by the background distribution. The matrix then gets updated based on these probabilities. A new and refined motif is therefore produced by alternating the calculation of the probability of each site based on the current matrix and calculating the new matrix based on these probabilities. By performing multiple iterations, this algorithm converges towards a maximum value for the motifs' matrix.

The algorithms described above are aimed at identifying de novo motifs. It is essential to consider that there is an understated yet important difference between searching for known and de novo motifs. While searching for known motifs in a set of sequences can be of great interest, the ultimate result will solely reveal which of these motifs are present and their position in the sequences. Conversely, a de novo motif search is done by querying the sequences to identify which motifs are most enriched. This should be taken into consideration as it influences the interpretation of the results. For example, performing a de novo search on a set of sequences could result in the proper identification of the GAGAAGGAGGG in the human putative A2RE-like sequence (similar to Fig. 7.2a). On the other hand, if an unrelated known motif search was performed on these same sequences using a database of genome-wide annotations of transcription factors like JASPAR, hits like the myeloid zinc finger 1 (MZF1), whose canonical motif is GAGGGG, would be identified, perhaps erroneously, despite having a low *p*-value (Khan et al. 2018). While biologically counterintuitive, this example shows the limits of motif searches. This demonstrates that motif search can be reduced to local multiple string alignments where context is easily lost at the algorithm level but should be kept in consideration when performing such analyses.

While the two approaches aim to do different things, as one seeks to annotate sequences with known motifs and the other seeks to discover new motifs, they are often complementary. One decisive advantage of known motif searches is when the ensemble of sequences is limited as the accuracy of de novo searches can be reduced in such cases. For example, a de novo search is impossible on a single sequence. Otherwise, de novo searches are often thought to be less limiting. One common way to palliate this dilemma is to first perform a robust de novo motif search and then complete a detailed comparison of these hits to a database of known motifs. Tools to achieve this, like HOMER or the MEME suite, methodologies and examples are detailed in the next sections.

To add to the complexity of robust identification of CRMs involved in localization, RNA often possesses additional *cis*-regulatory elements found scattered throughout its sequence, which may be needed for other aspects of post-transcriptional regulation, such as splicing and stability regulation. This can make it challenging to assign a specific localization function to a given signature motif. Furthermore, certain RBPs might bind only very short motifs that are quite prevalent in biological sequences (e.g. there might be cases where a CRM necessary and sufficient for localization is only 3 nucleotides long). One major challenge will be to distinguish these real but small motifs, from a background of specious motifs, for example, stemming from common repeat elements bearing little information content. In other words, the challenge rapidly becomes to distinguish the true positive among the large number of false positives created by these short motifs that can be found throughout the sequence space.

### 7.4.2 *Overview of Existing Computational Tools to Search for CRMs*

Most bioinformatics tools available nowadays tend to be developed through open collaborations and are offered with open source licences, thus allowing the source code to be used, modified or shared under defined terms and conditions, often free of charge, especially for academic uses. They are mostly available only on Linux or Mac OS operating systems and available on platforms such as web-based version control repository hosting services (e.g. GitHub, Bitbucket). Furthermore, as there is often little use for an elaborate graphical user interface, they are predominantly offered as command-line tools (e.g. using Terminal, iterm). This provides the most flexibility and allows for a wide range of customizable options. The running time and memory requirements of these algorithms can be quite high; therefore, it is often advisable to rely on high-performance computers (HPCs) allowing the use of parallel processing, which are generally accessible through major universities or private vendors (e.g. AWS). To be more accessible, many tools are offered as online databases and web servers, where analyses can be run without any local installation. However, web servers often come with strict limitations regarding the size of the inputs and local installation becomes necessary for larger-scale analyses.

In Table 7.2, we compile a non-exhaustive list of motif scanning and de novo motif discovery tools available to the community. These tools can be used, for example, to identify motifs that are likely to be candidates for potential regulatory roles in modulating different features of the RNA life cycle, including localization control. Dissecting the exact functions of a particular motif therefore requires the implementation of biological assays to assess the impact of the motif on RNA processing or activity (e.g. the use of reporter assays and site-specific mutagenesis to disrupt candidate motifs).

As there are an ever-increasing number of biologically validated motifs identified, databases are a valuable first place to search. The RNA-Binding Protein DataBase (RBPDB) is a large, manually curated, database grouping published observations of experimentally defined motifs (Cook et al. 2010a). This database has the advantage of allowing one to search by RNA-binding domain (RBD) or by species or to use it as a web server to scan an RNA sequence for putative RBP binding sites. Along the same line, the Catalog of Inferred Sequence Binding Proteins of RNA (CISBP-

**Table 7.2** A selection of motif databases, web servers and search algorithms

| Tools | Type | URL | Motif types | References |
|---|---|---|---|---|
| ATtRAC | Database, web server | https://attract.cnic.es | Primary sequence | Giudice et al. (2016) |
| CISBP-RNA | Database, web server | http://cisbp-rna.ccbr.utoronto.ca | Primary sequence | Ray et al. (2013) |
| DeepBind | Database; stand-alone | http://tools.genes.toronto.edu/deepbind/ | Primary sequence | Alipanahi et al. (2015) |
| Gibbs sampling | Algorithm | | Primary sequence | Lawrence et al. (1993) |
| GRAPHprot | Stand-alone | http://www.bioinf.uni-freiburg.de/Software/GraphProt/ | Primary sequence; secondary structure | Maticzka et al. (2014) |
| Homer | Stand-alone | http://homer.ucsd.edu/homer/index.html | Primary sequence | Heinz et al. (2010) |
| LESMoN | Stand-alone | http://cs.mcgill.ca/~blanchem/LESMoN/ | Primary sequence | Lavallée-Adam et al. (2017) |
| MatrixREDUCE | Stand-alone | https://systemsbiology.columbia.edu/matrixreduce | Primary sequence | Foat et al. (2005, 2006), Ward and Bussemaker (2008) |
| MEME | Stand-alone | http://meme-suite.org | Primary sequence | Bailey et al. (2009), Bailey and Elkan (1994) |
| MEMERIS | Stand-alone | http://www.bioinf.uni-freiburg.de/~hiller/MEMERIS/ | Primary sequence | Hiller et al. (2006) |
| MotifMap-RNA[a] | Database, web server | http://motifmap-rna.ics.uci.edu | Primary sequence | Liu et al. (2017) |
| RBPDB | Database, web server | http://rbpdb.ccbr.utoronto.ca | Primary sequence | Cook et al. (2010b) |
| RBPmap | Web server | http://rbpmap.technion.ac.il | Primary sequence | Paz et al. (2014) |
| RCK | Stand-alone | http://cb.csail.mit.edu/cb/rck/ | Primary sequence | Orenstein et al. (2016) |
| RNAcontext | Web server; stand-alone | http://www.cs.toronto.edu/~hilal/rnacontext/ | Primary sequence; secondary structure | Kazan et al. (2010) |
| ssHMM | Stand-alone | https://github.molgen.mpg.de/heller/ssHMM | Primary sequence; secondary structure | Heller et al. (2017) |

[a]oRNAment Database http://rnabiology.ircm.qc.ca/oRNAment/Primary sequence; Benoit Bouvrette et al. (2019)

RNA) is a database of RBP motifs and specificities derived from the impressive work compiling the results of systematic RNAcompete experiments. RNAcompete is a method through which the consensus binding motifs of ~300 RBPs were characterized through an in vitro selection assay in which purified RBPs were incubated with a random RNA pool, followed by the profiling of the RNA molecules selectively bound by the RBP (Ray et al. 2013).

A separate database that extends RBPDB and CISBP-RNA, and which has rapidly established itself as a gold standard, is the "A daTabase of experimentally validated RNA-binding proteins and AssoCiated moTifs" (ATtRACT) resource (Giudice et al. 2016). This database currently compiles information on 370 RBPs and 1583 manually curated consensus RBP binding motifs, in addition to having integrated updates and information about protein-RNA complexes as described in the Protein Data Bank (PDB) database (Gene Ontology 2015). As with other databases, ATtRACT also provides the capacity to search for motifs in target sequences. Finally, MotifMap-RNA is another database and web server that expands on RBPDB/CISBP-RNA and allows for genome-wide motif searches (Liu et al. 2017). While most databases described also offer web server capabilities to scan sequences and search for potential motifs, these tend to be limited. RBPmap is a web server that improves upon the scanning of sequences. Building on motifs compiled in all the previously mentioned databases and with the possibility to input additional user-defined motifs, this algorithm can be quite efficient in predicting and mapping binding sites (Paz et al. 2014).

In order to gain more insights into CRMs, de novo motif search tools are a great complement to established motif databases. These algorithms typically use only the sequence, and do not consider structure, when calling a motif. A first suite of tools for de novo motif discovery is the Hypergeometric Optimization of Motif EnRichment (HOMER) (Heinz et al. 2010). HOMER is a powerful tool that identifies motifs by looking for subsequences with differential enrichment between two sets of sequences. While it is advised to use a background of meaningful sequences (e.g. localized vs. non-localized), the background set can be simply random sequences. Interestingly, HOMER will also make some attempts to compare the motifs observed to a database of known motifs and will identify similarities. When only one group of sequences is available, Multiple EM for Motif Elicitation (MEME) is perhaps best suited. It is a suite of tools that implement multiple motif-finding algorithms, each with their own specificities for sequence search and motif discovery, analysis and comparison. It builds upon the EM algorithm described in Sect. 7.4.1 (Bailey et al. 2009; Bailey and Elkan 1994). Alternatively, MatrixREDUCE is a motif discovery algorithm that was originally designed to infer the binding specificity of transcription factors from microarray data (Foat et al. 2005, 2006; Ward and Bussemaker 2008), but can also be applied to the study of RNA sequence motifs. Local Enrichment of Sequence Motifs in biological Networks (LESMoN) takes a different approach by being an enumerative motif discovery algorithm that integrates gene set enrichment and biological network analysis (Lavallée-Adam et al. 2017).

While primary sequence is a critical component of *cis* elements, RNA secondary and tertiary structures can also be key features that can influence the binding to *trans*-regulatory machineries. Indeed, depending on the type of RNA-binding domain (RBD) they contain, RBPs can bind RNA based on primary sequence or structural motifs, although the most abundant classes of RBPs tend to bind specific primary sequence motifs (Gerstberger et al. 2014; Ray et al. 2013). As such, some regulatory motif prediction algorithms are taking structural prediction information into account. For example, the MEMERIS algorithm is built on the same principle as MEME but searches for RNA motifs enriched in any type of single-stranded regions (e.g. the loop of a hairpin). This has been shown to improve RNA-binding site predictions (Hiller et al. 2006). Expanding on the idea that approaches making use of RNA sequence and structure can be used for better motif predictions, the RNAcontext tool integrates predictions on whether a nucleotide is paired, in a hairpin loop or unstructured region, to help define putative regulatory elements (Kazan et al. 2010).

Machine-learning frameworks are proving to be quite efficient for identifying RBP binding preferences. In that category, GRAPHprot is able to detect motifs by taking into consideration both sequence and structure (Maticzka et al. 2014). Alternatively, DeepBind, a state of the art in sequence models, only considers sequence and not structure but has been shown to perform better than GRAPHprot (Alipanahi et al. 2015). RCK is an elegant machine-learning algorithm that takes into account both sequence and structure and has established itself as an efficient and scalable tool for robust motif discovery (Orenstein et al. 2016). Another tool named sequence-structure hidden Markov model (ssHMM) searches for motif based on a statistical model named hidden Markov model (HMM) and Gibbs sampling, which it performs while integrating the sequence and structure preference of an RBP (Heller et al. 2017).

Some algorithms have also been developed specifically to provide answers on localization. DeepLncRNA is a machine-learning algorithm that predicts the subcellular localization of lncRNA considering only its sequences (Gudenas and Wang 2018). Finally, RNATracker is a novel algorithm that takes advantage of deep neural network using both sequence and structural information to infer subcellular distribution of transcripts (Yan et al. 2019).

Individually, the results obtained from these databases, web servers and standalone algorithms must be analysed with great caution, as they are likely to produce a very large number of false-positive predictions. This is unavoidable, given the low information content of certain motifs. Cross-validation of results from multiple tools, detailed literature consideration and experimental validation via mutational analysis or reporter assays is therefore of the utmost importance.

## 7.5 Examples of Motif Discovery Applications

In order to exemplify the most important concepts addressed in this chapter, we performed different known and de novo motif searches on the complete human coding transcriptome (i.e. all portions of an mRNA) and between two sets of sequences that were observed to be localized to either the nucleus or the cytoplasm of human HepG2 cells (Benoit Bouvrette et al. 2018; Lefebvre et al. 2017; Benoit Bouvrette et al. 2019).

We first sought to assess the general distribution of motifs for 70 RBPs (listed in Fig. 7.3a) for which PWMs were obtained by RNAcompete (Ray et al. 2013). Sites

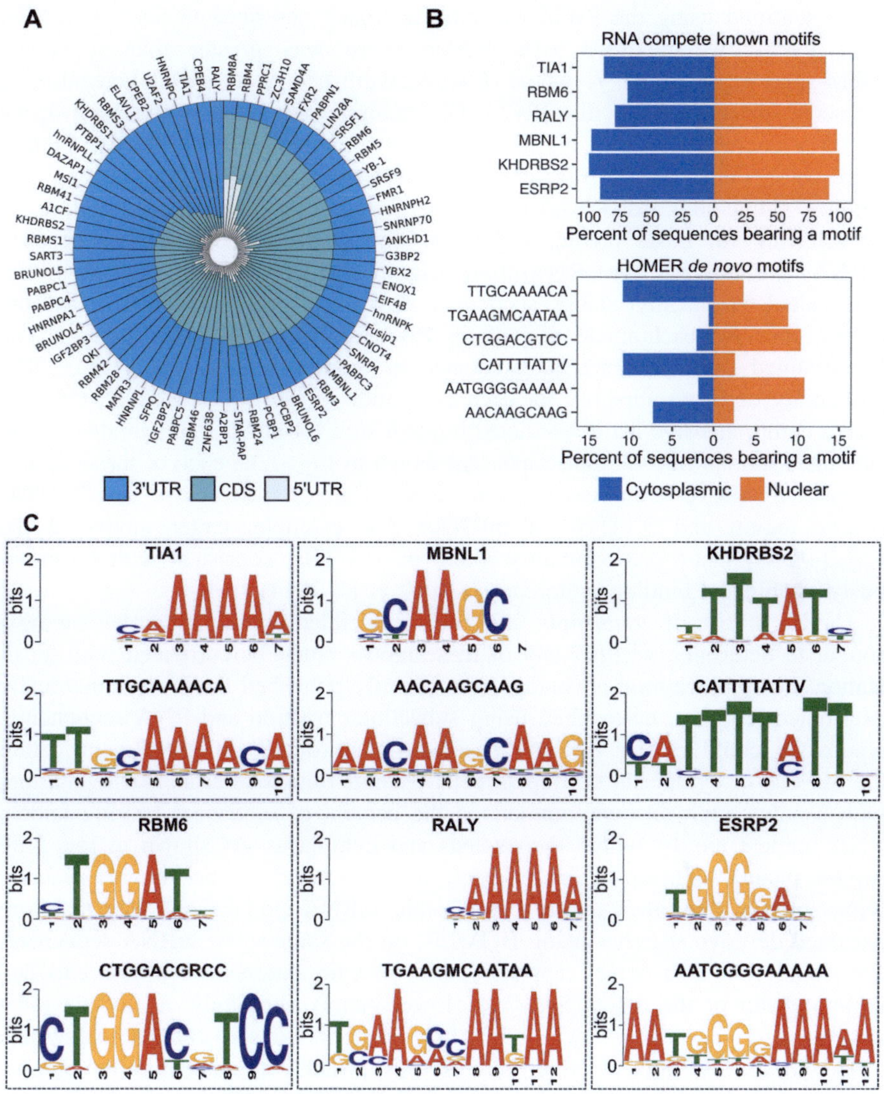


**Fig. 7.3** Global overview of known and *de novo* motifs and their putative role in RNA localization. (**a**) Circos plot showing the relative regionalization towards the 5′ UTR, coding sequence (CDS) and 3′ UTR of 70 known motifs from RNAcompete. (**b**) Histogram showing the percent of localized sequences, either enriched in the cytoplasm or the nucleus, harbouring a known motif from an RNAcompete experiment (upper panel) and from a de novo motif search using HOMER (lower panel) in their 3′ UTR. (**c**) Sequence logos comparing the known motif from an RNAcompete experiment to the de novo motif identified by HOMER in the 3′ UTR of localized sequence, in the normal strand or its reverse complement (**b**)

were identified using the PWM scanning approach described in Sect. 7.4.1. For each PWM, we recorded sites whose score was greater than a certain PWM-specific threshold *T*, where *T* was established as the 99th percentile of the score distribution for that PWM. For example, for a PWM of length 4, we would calculate the score of all 256 possible 4-mers and kept only the two highest scores as a threshold. As RNAcompete motifs were designed for preferentially binding single-stranded RNA, we further reduce the list of putative motifs by selecting for those predicted to lie within single-stranded regions of each mRNA. For this we used RNAplfold, a gold standard RNA folding algorithm that calculates locally stable secondary structures and outputs base pairing probabilities for each nucleotide of an RNA of interest (Lorenz et al. 2011). We retained only predicted sites located in regions with a higher than 90% probability of being unpaired for each nucleotide of the k-mer. This provided us with a comprehensive list of predicted binding sites for all 70 RBPs in all 179,236 annotated human mRNA transcripts. As shown in Fig. 7.3a, each of these 70 sets of putative CRMs exhibited variable distribution profiles across the 5′ UTR, coding region and 3′ UTR of mRNAs. For example, target motifs of the CPEB4 protein are predominantly found in 3′ UTRs, consistent with its previously established binding preferences (Afroz et al. 2014).

Having a list of transcripts and their embedded motif instances, we next sought to determine whether any of these motifs could be correlated with localization. For this we took advantage of a recently published list of asymmetrically distributed mRNAs, determined using subcellular fraction and RNA sequencing, where we could cluster mRNAs based on their degree of enrichment within the nucleus or cytoplasm of HepG2 cells (Benoit Bouvrette et al. 2018). Starting with a naïve approach, we enumerated the percent of sequences bearing known RNAcompete motifs, within the nucleus and cytoplasm. As shown in Fig. 7.3b (upper panel), the top 6 interrogated motifs tended to be roughly equally represented within nuclear and cytoplasmic mRNA populations. We therefore executed de novo searches using HOMER, on the same set of mRNA sequences. By doing so, it becomes apparent that specific subsequences are enriched in one group or the other (Fig. 7.3b, lower panel). Strikingly, all the de novo motifs identified are longer than the ones previously defined using the RNAcompete in vitro pipeline. Interestingly, when we compared the known motifs with those found de novo using TOMTOM, a motif comparison tool available in the MEME suite, we observed significant similarities between the two sets of results (Gupta et al. 2007). Indeed, these 6 motifs of length 7 derived from RNAcompete data can be embedded in the longer motifs identified by HOMER (Fig. 7.3c). We can conclude from this that short motifs may not contain enough information to differentiate sets of RNA with distinctive biological features or behaviours. However, supplementing such analyses with de novo motif prediction strategies offers a promising avenue to identify biologically relevant CRM involved in localization.

## 7.6 Conclusion

As outlined in this review, mRNA localization has been shown to be a key layer of post-transcriptional gene regulation that impacts a wide array of biological processes. The targeting of a transcript to a precise subcellular location involves a complex coaction between a variety of CRMs, RBPs and additional factors to form an mRNP. Nevertheless, there is much to be discovered regarding the necessary and sufficient region of each mRNA dictating their subcellular distribution. Mathematical tools, such as information content and entropy, have been adapted to address the representation of biological motifs, like PWMs and sequence logos. This has laid the groundwork for the implementation of computational procedures, such as motif enumeration, that may help in deciphering and classifying individual CRMs. Already, a variety of programs exist that use these tools and procedures with the aim of filtering true motifs within a given subset of sequences. We demonstrated that it was possible to identify putative motifs involved in localization through the execution of these programs on sets of asymmetrically distributed transcripts. By combining the resulting motif inferences with classical molecular biology experiments, such as reporter assays, it is but a question of time before we have a more comprehensive knowledge of the regulatory code driving mRNA subcellular localization.

## References


Afroz T, Skrisovska L, Belloc E, Guillen-Boixet J, Mendez R, Allain FH (2014) A fly trap mechanism provides sequence-specific RNA recognition by CPEB proteins. Genes Dev 28:1498–1514

Ainger K, Avossa D, Morgan F, Hill SJ, Barry C, Barbarese E, Carson JH (1993) Transport and localization of exogenous myelin basic protein mRNA microinjected into oligodendrocytes. J Cell Biol 123:431–441

Ainger K, Avossa D, Diana AS, Barry C, Barbarese E, Carson JH (1997) Transport and localization elements in myelin basic protein mRNA. J Cell Biol 138:1077–1087

Alipanahi B, Delong A, Weirauch MT, Frey BJ (2015) Predicting the sequence specificities of DNA-and RNA-binding proteins by deep learning. Nat Biotechnol 33:831–839

Allen L, Kloc M, Etkin LD (2003) Identification and characterization of the Xlsirt cis-acting RNA localization element. Differentiation 71:311–321

Aranda-Abreu GE, Behar L, Chung S, Furmeaux H, Ginzburg I (1999) Embryonic lethal abnormal vision-like RNA-binding proteins regulate neurite outgrowth and tau expression in PC12 cells. J Neurosci 19:6907–6917

Aronov S, Aranda G, Ginzburg I (2001) Axonal tau mRNA localization coincides with tau protein in living neuronal cells and depends on axonal targeting signal. J Neurosci 21:6577–6587

Bailey TL, Elkan C (1994) Fitting a mixture model by expectation maximization to discover motifs in bipolymers. Proceedings of the second international conference on intelligent systems for molecular biology, pp 28–36

Bailey TL, Elkan C (1995) Unsupervised learning of multiple motifs in biopolymers using expectation maximization. Mach Learn 21:51–80

Bailey TL, Boden M, Buske FA, Frith M, Grant CE, Clementi L, Ren J, Li WW, Noble WS (2009) MEME Suite: tools for motif discovery and searching. Nucleic Acids Res 37:W202–W208
Batada NN, Shepp LA, Siegmund DO (2004) Stochastic model of protein–protein interaction: why signaling proteins need to be colocalized. Proc Natl Acad Sci U S A 101:6445–6449
Batish M, van den Bogaard P, Kramer FR, Tyagi S (2012) Neuronal mRNAs travel singly into dendrites. Proc Natl Acad Sci U S A 109:4645–4650
Beckstette M, Homann R, Giegerich R, Kurtz S (2006) Fast index based algorithms and software for matching position specific scoring matrices. BMC Bioinformatics 7:389
Behar L, Marx R, Sadot E, Barg J, Ginzburg I (1995) cis-acting signals and trans-acting proteins are involved in tau mRNA targeting into neurites of differentiating neuronal cells. Int J Dev Neurosci 13:113–127
Benoit Bouvrette LP, Cody NAL, Bergalet J, Lefebvre FA, Diot C, Wang X, Blanchette M, Lecuyer E (2018) CeFra-seq reveals broad asymmetric mRNA and noncoding RNA distribution profiles in Drosophila and human cells. RNA 24:98–113
Benoit Bouvrette LP, Bovaird S, Blanchette M, Lecuyer E (2019) oRNAment: a database of putative RNA binding protein target sites in the transcriptomes of model species. Nucleic Acids Res (in press)
Bergalet J, Lécuyer E (2014) The functions and regulatory principles of mRNA intracellular trafficking. In: Yeo G (ed) Systems biology of RNA binding proteins. Advances in experimental medicine and biology, vol 825. Springer, New York
Bergsten SE, Huang T, Chatterjee S, Gavis ER (2001) Recognition and long-range interactions of a minimal nanos RNA localization signal element. Development 128:427–435
Bertrand E, Chartrand P, Schaefer M, Shenoy SM, Singer RH, Long RM (1998) Localization of ASH1 mRNA particles in living yeast. Mol Cell 2:437–445
Betley JN, Heinrich B, Vernos I, Sardet C, Prodon F, Deshler JO (2004) Kinesin II mediates Vg1 mRNA transport in Xenopus oocytes. Curr Biol 14:219–224
Blichenberg A, Schwanke B, Rehbein M, Garner CC, Richter D, Kindler S (1999) Identification of a cis-acting dendritic targeting element in MAP 2 mRNAs. J Neurosci 19:8818–8829
Blichenberg A, Rehbein M, Müller R, Garner CC, Richter D, Kindler S (2001) Identification of a cis-acting dendritic targeting element in the mRNA encoding the alpha subunit of Ca2+/calmodulin-dependent protein kinase II. Eur J Neurosci 13:1881–1888
Blower MD, Feric E, Weis K, Heald RJ (2007) Genome-wide analysis demonstrates conserved localization of messenger RNAs to mitotic microtubules. J Cell Biol 179:1365–1373
Bovaird S, Patel D, Padilla JA, Lecuyer E (2018) Biological functions, regulatory mechanisms, and disease relevance of RNA localization pathways. FEBS Lett 592:2948–2972
Brendza RP, Serbus LR, Duffy JB, Saxton WM (2000) A function for kinesin I in the posterior transport of oskar mRNA and Staufen protein. Science 289:2120–2122
Brenner S, Jacob F, Meselson M (1961) An unstable intermediate carrying information from genes to ribosomes for protein synthesis. Nature 190:576–581
Bullock SL, Ish-Horowicz D (2001) Conserved signals and machinery for RNA transport in Drosophila oogenesis and embryogenesis. Nature 414:611–616
Bullock SL, Ringel I, Ish-Horowicz D, Lukavsky PJ (2010) A'-form RNA helices are required for cytoplasmic mRNA transport in Drosophila. Nat Struct Mol Biol 17:703–709
Carlson JM, Chakravarty A, DeZiel CE, Gross RH (2007) SCOPE: a web server for practical de novo motif discovery. Nucleic Acids Res 35:W259–W264
Chan AP, Kloc M, Etkin LD (1999) fatvg encodes a new localized RNA that uses a 25-nucleotide element (FVLE1) to localize to the vegetal cortex of Xenopus oocytes. Development 126:4943–4953
Chan AP, Kloc M, Bilinski S, Etkin LD (2001) The vegetally localized mRNA fatvg is associated with the germ plasm in the early embryo and is later expressed in the fat body. Mech Dev 100:137–140
Chao JA, Patskovsky Y, Patel V, Levy M, Almo SC, Singer R (2010) ZBP1 recognition of β-actin zipcode induces RNA looping. Genes Dev 24:148–158
Chartrand P, Meng XH, Singer RH, Long RM (1999) Structural elements required for the localization of ASH1 mRNA and of a green fluorescent protein reporter particle in vivo. Curr Biol 9:333–336

Chartrand P, Meng XH, Huttelmaier S, Donato D, Singer RH (2002) Asymmetric sorting of ash1p in yeast results from inhibition of translation by localization elements in the mRNA. Mol Cell 10:1319–1330

Cheung HK, Serano TL, Cohen RS (1992) Evidence for a highly selective RNA transport system and its role in establishing the dorsoventral axis of the Drosophila egg. Development 114:653–661

Claussen M, Pieler T (2004) Xvelo1 uses a novel 75-nucleotide signal sequence that drives vegetal localization along the late pathway in Xenopus oocytes. Dev Biol 266:270–284

Claußen M, Horvay K, Pieler T (2004) Evidence for overlapping, but not identical, protein machineries operating in vegetal RNA localization along early and late pathways in Xenopus oocytes. Development 131:4263–4273

Cody NA, Iampietro C, Lecuyer E (2013) The many functions of mRNA localization during normal development and disease: from pillar to post. Wiley Interdiscip Rev Dev Biol 2:781–796

Cook KB, Kazan H, Zuberi K, Morris Q (2010a) RBPDB: a database of RNA-binding specificities. Nucleic Acids Res 39:D301–D308

Cook KB, Kazan H, Zuberi K, Morris Q, Hughes TR (2010b) RBPDB: a database of RNA-binding specificities. Nucleic Acids Res 39:D301–D308

Crofts AJ, Washida H, Okita TW, Ogawa M, Kumamaru T, Satoh H (2004) Targeting of proteins to endoplasmic reticulum-derived compartments in plants. The importance of RNA localization. Plant Physiol 136:3414–3419

Crooks GE, Hon G, Chandonia JM, Brenner SE (2004) WebLogo: a sequence logo generator. Genome Res 14:1188–1190

Czaplinski K, Köcher T, Schelder M, Segref A, Wilm M, Mattaj IW (2005) Identification of 40LoVe, a Xenopus hnRNP D family protein involved in localizing a TGF-beta-related mRNA during oogenesis. Dev Cell 8:505–515

Dahm R, Kiebler M (2005) Cell biology: silenced RNA on the move. Nature 438:432–435

Deshler JO, Highett MI, Schnapp BJ (1997) Localization of Xenopus Vg1 mRNA by Vera protein and the endoplasmic reticulum. Science 276:1128–1131

Deshler JO, Highett MI, Abramson T, Schnapp BJ (1998) A highly conserved RNA-binding protein for cytoplasmic mRNA localization in vertebrates. Curr Biol 8:489–496

Dictenberg JB, Swanger SA, Antar LN, Singer RH, Bassell GJ (2008) A direct role for FMRP in activity-dependent dendritic mRNA transport links filopodial-spine morphogenesis to fragile X syndrome. Dev Cell 14:926–939

Dienstbier M, Boehl F, Li X, Bullock SL (2009) Egalitarian is a selective RNA-binding protein linking mRNA localization signals to the dynein motor. Genes Dev 23:1546–1558

Dominguez D, Freese P, Alexis MS, Su A, Hochman M, Palden T, Bazile C, Lambert NJ, Van Nostrand EL, Pratt GA et al (2018) Sequence, structure, and context preferences of human RNA binding proteins. Mol Cell 70(854–867):e859

Dynes JL, Steward O (2007) Dynamics of bidirectional transport of Arc mRNA in neuronal dendrites. J Comp Neurol 500:433–447

Dynes JL, Steward O (2012) Arc mRNA docks precisely at the base of individual dendritic spines indicating the existence of a specialized microdomain for synapse-specific mRNA translation. J Comp Neurol 520:3105–3119

Eberwine J, Belt B, Kacharmina JE, Miyashiro K (2002) Analysis of subcellularly localized mRNAs using in situ hybridization, mRNA amplification, and expression profiling. Neurochem Res 27:1065–1077

Farina KL, Hüttelmaier S, Musunuru K, Darnell R, Singer R (2003) Two ZBP1 KH domains facilitate β-actin mRNA localization, granule formation, and cytoskeletal attachment. The Journal of cell *160*:77–87

Farris S, Lewandowski G, Cox CD, Steward O (2014) Selective localization of arc mRNA in dendrites involves activity- and translation-dependent mRNA degradation. J Neurosci 34:4481–4493

Fauteux F, Blanchette M, Stromvik MV (2008) Seeder: discriminative seeding DNA motif discovery. Bioinformatics 24:2303–2307

Ferrandon D, Koch I, Westhof E, Nusslein-Volhar C (1997) RNA–RNA interaction is required for the formation of specific bicoid mRNA 3' UTR–STAUFEN ribonucleoprotein particles. EMBO J 16:1751–1997

Foat BC, Houshmandi SS, Olivas WM, Bussemaker HJ (2005) Profiling condition-specific, genome-wide regulation of mRNA stability in yeast. Proc Natl Acad Sci U S A 102:17675–17680

Foat BC, Morozov AV, Bussemaker HJ (2006) Statistical mechanical modeling of genome-wide transcription factor occupancy data by MatrixREDUCE. Bioinformatics 22:e141–e149

Forrest KM, Gavis ER (2003) Live imaging of endogenous RNA reveals a diffusion and entrapment mechanism for nanos mRNA localization in Drosophila. Curr Biol 13:1159–1168

Gautreau D, Cote CA, Mowry KL (1997) Two copies of a subelement from the Vg1 RNA localization sequence are sufficient to direct vegetal localization in Xenopus oocytes. Development 124:5013–5020

Gavis ER, Lehmann R (1992) Localization of nanos RNA controls embryonic polarity. Cell 71:301–313

Gavis ER, Curtis D, Lehmann R (1996a) Identification of cis-acting sequences that control nanos RNA localization. Dev Biol 176:36–50

Gavis ER, Lunsford L, Bergsten SE, Lehmann R (1996b) A conserved 90 nucleotide element mediates translational repression of nanos RNA. Development 122:2791–2800

Gene Ontology C (2015) Gene Ontology Consortium: going forward. Nucleic Acids Res 43: D1049–D1056

Gerstberger S, Hafner M, Tuschl T (2014) A census of human RNA-binding proteins. Nat Rev Genet 15:829–845

Giorgi C, Moore MJ (2007) The nuclear nurture and cytoplasmic nature of localized mRNPs. Semin Cell Dev Biol 18:186–193

Giudice G, Sánchez-Cabo F, Torroja C, Lara-Pezzi E (2016) ATtRACT—a database of RNA-binding proteins and associated motifs. Database

Gomes C, Merianda TT, Lee S, Yoo S, Twiss JL (2014) Molecular determinants of the axonal mRNA transcriptome. Dev Neurobiol 74:218–232

Gonzalez I, Buonomo SB, Nasmyth K, von Ahsen U (1999) ASH1 mRNA localization in yeast involves multiple secondary structural elements and Ash1 protein translation. Curr Biol 9:337–340

Gros F, Hiatt H, Gilbert W, Kurland CG, Risebrough RW, Watson JD (1961) Unstable ribonucleic acid revealed by pulse labelling of Escherichia coli. Nature 190:581–585

Gu W, Pan F, Zhang H, Bassell GJ, Singer RHJ (2002) A predominantly nuclear protein affecting cytoplasmic localization of β-actin mRNA in fibroblasts and neurons. J Cell Biol 156:41–51

Gudenas BL, Wang L (2018) Prediction of LncRNA subcellular localization with deep learning from sequence features. Sci Rep 8:16385

Gupta S, Stamatoyannopoulos JA, Bailey TL, Noble WS (2007) Quantifying similarity between motifs. Genome Biol 8:R24

Hachet O, Ephrussi A (2001) Drosophila Y14 shuttles to the posterior of the oocyte and is required for oskar mRNA transport. Curr Biol 11:1666–1674

Hachet O, Ephrussi A (2004) Splicing of oskar RNA in the nucleus is coupled to its cytoplasmic localization. Nature 428:959–963

Heinz S, Benner C, Spann N, Bertolino E, Lin YC, Lasslo P, Cheng JX, Murre C, Singh H, Glass KK (2010) Simple combinations of lineage-determining transcription factors prime cis-regulatory elements required for macrophage and B cell identities. Mol Cell 38:576–589

Heller D, Krestel R, Ohler U, Vingron M, Marsico A (2017) ssHMM: extracting intuitive sequence-structure motifs from high-throughput RNA-binding protein data. Nucleic Acids Res 45:11004–11018

Henry JJ, Perry KJ, Fukui L, Alvi N (2010) Differential localization of mRNAs during early development in the mollusc, Crepidula fornicata. Integr Comp Biol 50:720–733

Heym RG, Niessing D (2012) Principles of mRNA transport in yeast. Cell Mol Life Sci 69:1843–1853

Hiller M, Pudimat R, Busch A, Backlofen R (2006) Using RNA secondary structures to guide sequence motif finding towards single-stranded regions. Nucl Acids Res 34(17):e117

Hoek KS, Kidd GJ, Carson JH, Smith R (1998) hnRNP A2 selectively binds the cytoplasmic transport sequence of myelin basic protein mRNA. Biochemistry 37:7021–7029

Huang YS, Carson JH, Barbarese E, Richter D (2003) Facilitation of dendritic mRNA transport by CPEB. Genes Dev 17:638–653

Huttelmaier S, Zenklusen D, Lederer M, Dictenberg J, Lorenz M, Meng X, Bassell GJ, Condeelis J, Singer RH (2005) Spatial regulation of beta-actin translation by Src-dependent phosphorylation of ZBP1. Nature 438:512–515

Hutten S, Sharangdhar T, Kiebler M, Hutten S, Sharangdhar T, Kiebler M (2014) Unmasking the messenger. RNA Biol 11:992–997

Huynh J-R, Munro TP, Smith-Litière K, Lepesant J-A, St Johnston D (2004) The Drosophila hnRNPA/B homolog, Hrp48, is specifically required for a distinct step in osk mRNA localization. Dev Cell 6:625–635

Jacob F, Monod J (1961) Genetic regulatory mechanisms in the synthesis of proteins. J Mol Biol 3:318–356

Jambhekar A, DeRisi JL (2007) Cis-acting determinants of asymmetric, cytoplasmic RNA transport. RNA 12:625–642

Jambhekar A, McDermott K, Sorber S, Shepard KA, Vale RD, Takizava P, DeRisi JL (2005) Unbiased selection of localization elements reveals cis-acting determinants of mRNA bud localization in Saccharomyces cerevisiae. Proc Natl Acad Sci U S A 102:18005–18010

Jambor H, Surendranath V, Kalinka AT, Mejstrik P, Saalfeld S, Tomancak P (2015) Systematic imaging reveals features and changing localization of mRNAs in Drosophila development. Elife 4

Jeener R, Szafarz D (1950) Relation between the rate of renewal and the intracellular localization of ribonucleic acid. Arch Biochem Biophys 26:54–67

Jeffery WR, Tomlinson CR, Brodeur RD (1983) Localization of actin messenger RNA during early ascidian development. Dev Biol 99:408–417

Johnstone O, Lasko P (2001) Translational regulation and RNA localization in Drosophila oocytes and embryos. Annu Rev Genet 35:365–406

Kaewsapsak P, Shechner DM, Mallard W, Rinn JL, Ting AY (2017) Live-cell mapping of organelle-associated RNAs via proximity biotinylation combined with protein-RNA crosslinking. Elife 6

Katz ZB, Wells AL, Park HY, Wu B, Shenoy SM, Singer R (2012) β-Actin mRNA compartmentalization enhances focal adhesion stability and directs cell migration. Genes Dev 26:1885–1890

Kazan H, Ray D, Chan ET, Hughes TR (2010) RNAcontext: a new method for learning the sequence and structure binding preferences of RNA-binding proteins. PLoS Comput Biol 6: e1000832

Keiler KC (2011) RNA localization in bacteria. Curr Opin Microbiol 14:155–159

Kel AE, Gossling E, Reuter I, Cheremushkin E, Kel-Margoulis OV, Wingender E (2003) MATCH: a tool for searching transcription factor binding sites in DNA sequences. Nucleic Acids Res 31:3576–3579

Khan A, Fornes O, Stigliani A, Gheorghe M, Castro-Mondragon JA, van der Lee R, Bessy A, Cheneby J, Kulkarni SR, Tan G et al (2018) JASPAR 2018: update of the open-access database of transcription factor binding profiles and its web framework. Nucleic Acids Res 46:D1284

Kim J, Lee J, Lee S, Lee B, Kim-Ha J (2014) Phylogenetic comparison of oskar mRNA localization signals. Biochem Biophys Res Commun 444:98–103
Kim-Ha J, Smith JL, Macdonald PM (1991) oskar mRNA is localized to the posterior pole of the Drosophila oocyte. Cell *66*:23–35
Kim-Ha J, Webster PJ, Smith JL, Macdonald PM (1993) Multiple RNA regulatory elements mediate distinct steps in localization of oskar mRNA. Development 119:169–178
Kislauskis EH, Singer RH (1992) Determinants of mRNA localization. Curr Opin Cell Biol 4:975–978
Kislauskis EH, Zhu X, Singer RH (1994) Sequences responsible for intracellular localization of beta-actin messenger RNA also affect cell phenotype. J Cell Biol 127:441–451
Kloc M, Bilinski S, Pui-Yee Chan A, Etkin LD (2000) The targeting of Xcat2 mRNA to the germinal granules depends on a cis-acting germinal granule localization element within the 3'UTR. Dev Biol 217:221–229
Knowles RB, Sabry JH, Martone ME, Deerink T, Ellisman M, Bassell GJ, Kosik K (1996) Translocation of RNA granules in living neurons. J Neurosci 16:7812–7820
Korn LJ, Queen CL, Wegman MN (1977) Computer analysis of nucleic acid regulatory sequences. Proc Natl Acad Sci U S A 74:4401–4405
Kowanda M, Bergalet J, Wieczorek M, Brouhard G, Lécuyer E, Lasko P (2016) Loss of function of the Drosophila Ninein-related centrosomal protein Bsg25D causes mitotic defects and impairs embryonic development. Biol Open 5:1040–1051
Kress TL, Yoon YJ, Mowry KL (2004) Nuclear RNP complex assembly initiates cytoplasmic RNA localization. J Cell Biol 165:203–211
Kruse C, Jaedicke A, Beaudouin J, Böhl F, Ferring D, Güttler T, Ellenberg J, Jansen R-P (2002) Ribonucleoprotein-dependent localization of the yeast class V myosin Myo4p. J Cell Biol 159:971–982
Kugler JM, Lasko P (2009) Localization, anchoring and translational control of oskar, gurken, bicoid and nanos mRNA during Drosophila oogenesis. Fly 3:15–28
Kuriyan J, Eisenberg D, Kuriyan J, Eisenberg D (2007) The origin of protein interactions and allostery in colocalization. Nature 450:983–990
Lantz V, Schedl P (1994) Multiple cis-acting targeting sequences are required for orb mRNA localization during Drosophila oogenesis. Mol Cell Biol 14:2235–2342
Lasko P (2012) mRNA localization and translational control in Drosophila oogenesis. Cold Spring Harb Perspect Biol 4:1–15
Lavallée-Adam M, Cloutier P, Coulombe B (2017) Functional 5′ UTR motif discovery with LESMoN: local enrichment of sequence motifs in biological networks. Nucleic Acids Res 45:10415–10427
Lawrence CE, Reilly AA (1990) An expectation maximization (EM) algorithm for the identification and characterization of common sites in unaligned biopolymer sequences. Proteins Struct Funct Genet 7:41–51
Lawrence CE, Altschul SF, Boguski MS, Liu JS, Neuwald AF, Wootton JC (1993) Detecting subtle sequence signals: a Gibbs sampling strategy for multiple alignment. Science *262*:208–214
Lecuyer E, Yoshida H, Parthasarathy N, Alm C, Babak T, Cerovina T, Hughes TR, Tomancak P, Krause HM (2007) Global analysis of mRNA localization reveals a prominent role in organizing cellular architecture and function. Cell 131:174–187
Lefebvre FA, Cody NAL, Benoit-Bouvrette LP, Bergalet J, Wang X, Lecuyer E (2017) CeFra-seq: Systematic mapping of RNA subcellular distribution properties through cell fractionation coupled to deep-sequencing. Methods 126:138–148
Lerit DA, Gavis ER (2011) Transport of germ plasm on astral microtubules directs germ cell development in Drosophila. Curr Biol 21:439–448
Liefooghe A, Touzet H, Varré JS (2006) Large scale matching for position weight matrices. Annual symposium on combinatorial pattern matching
Liu Y, Sun S, Bredy T, Wood M, Spitale RC, Baldi P (2017) MotifMap-RNA: a genome-wide map of RBP binding sites. Bioinformatics 33:2029–2031

Long RM, Singer RH, Meng X, Gonzalez I, Nasmyth K, Jansen R-P (1997) Mating type switching in yeast controlled by asymmetric localization of ASH1 mRNA. Science 277:383–387

Long RM, Gu W, Lorimer E, Singer RH, Chartrand P (2000) She2p is a novel RNA-binding protein that recruits the Myo4p-She3p complex to ASH1 mRNA. EMBO J 19:6592–6601

Lorenz R, Bernhart SH, Honer Zu Siederdissen C, Tafer H, Flamm C, Stadler PF, Hofacker IL (2011) ViennaRNA Package 2.0. Algorithms. Mol Biol 6:26

Macdonald PM, Kerr K (1997) Redundant RNA recognition events in bicoid mRNA localization. RNA 3:1413–1420

Macdonald PM, Kerr K (1998) Mutational analysis of an RNA recognition element that mediates localization of bicoid mRNA. Mol Cell Biol 18:3788–3795

Macdonald PM, Struhl G (1988) cis-acting sequences responsible for anterior localization of bicoid mRNA in Drosophila embryos. Nature 336:595–598

Macdonald PM, Kerr K, Smith JL, Leask A (1993) RNA regulatory element BLE1 directs the early steps of bicoid mRNA localization. Development 118:1233–1243

Machta J (1999) Entropy, information, and computation. Am J Phys 62:1074–1077

MacIsaac KD, Fraenkel E (2006) Practical strategies for discovering regulatory DNA sequence motifs. PLoS Comput Biol 2:e36

Martin KC, Ephrussi A (2009) mRNA localization: gene expression in the spatial dimension. Cell 136:719–730

Maticzka D, Lange SJ, Costa F, Backlofen R (2014) GraphProt: modeling binding preferences of RNA-binding proteins. Genome Biol 15(1):R17

Matys V, Fricke E, Geffers R, Gossling E, Haubrock M, Hehl R, Hornischer K, Karas D, Kel AE, Kel-Margoulis OV et al (2003) TRANSFAC: transcriptional regulation, from patterns to profiles. Nucleic Acids Res 31:374–378

Mayford M, Baranes D, Podsypania K, Kandel E (1996) The 3'-untranslated region of CaMKIIα is a cis-acting signal for the localization and translation of mRNA in dendrites. Proc Natl Acad Sci U S A 93:13250–13255

Medioni C, Mowry K, Besse F (2012) Principles and roles of mRNA localization in animal development. Development 139:3263–3276

Meer EJ, Wang D, Kim S, Barr I, Guo F, Martin KC, Meer EJ, Wang D, Kim S, Barr I et al (2012) Identification of a cis-acting element that localizes mRNA to synapses. Proc Natl Acad Sci U S A 109:4639–4644

Messitt TJ, Gagnon JA, Kreiling JA, Pratt CA, Yoon YJ, Mowry KL (2008) Multiple kinesin motors coordinate cytoplasmic RNA transport on a subpopulation of microtubules in Xenopus oocytes. Dev Cell 15:426–436

Mikl M, Vendra G, Kiebler MA (2011) Independent localization of MAP 2, CaMKIIalpha and beta-actin RNAs in low copy numbers. EMBO Rep 12:1077–1084

Mili S, Moissoglu K, Macara IG (2008) Genome-wide screen reveals APC-associated RNAs enriched in cell protrusions. Nature 453:115–121

Mohr E, Morris JF, Richter D (1995) Differential subcellular mRNA targeting: deletion of a single nucleotide prevents the transport to axons but not to dendrites of rat hypothalamic magnocellular neurons. Proc Natl Acad Sci U S A 92:4377–4381

Mori Y, Imaizumi K, Katayama T, Yoneda T, Tohyama M (2000) Two cis-acting elements in the 3' untranslated region of alpha-CaMKII regulate its dendritic targeting. Nat Neurosci 3:1079–1084

Mowry KL, Melton DA (1992) Vegetal messenger RNA localization directed by a 340-nt RNA sequence element in Xenopus oocytes. Science 255:991–994

Munro TP, Magee RJ, Kidd GJ, Carson JH, Barbarese E, Smith LM, Smith R (1999) Mutational analysis of a heterogeneous nuclear ribonucleoprotein A2 response element for RNA trafficking. J Biol Chem 274:34389–34395

Oleynikov Y, Singer RH (2003) Real-time visualization of ZBP1 association with beta-actin mRNA during transcription and localization. Curr Biol 13:199–207

Olivier C, Poirier G, Gendron P, Boisgontier A, Major M, Chartrand P (2005) Identification of a conserved RNA motif essential for She2p recognition and mRNA localization to the yeast bud. Mol Cell Biol 25:4752–4766

Orenstein Y, Wang Y, Berger B (2016) RCK: accurate and efficient inference of sequence-and structure-based protein–RNA binding models from RNAcompete data. Bioinformatics 32:i351–i359

Pan F, Hüttelmaier S, Singer RH, Gu W (2009) ZBP2 Facilitates Binding of ZBP1 to β-Actin mRNA during Transcription. Mol Cell Biol 27:8340–8351

Paquin N, Chartrand P (2008) Local regulation of mRNA translation: new insights from the bud. Trends Cell Biol 18:105–111

Patel VL, Mitra S, Harris R, Buxbaum AR, Lionnet T, Brenowitz M, Girvin M, Levy M, Almo SC, Singer RH (2012) Spatial arrangement of an RNA zipcode identifies mRNAs under post-transcriptional control. Genes Dev 26:43–53

Pavesi G, Mereghetti P, Mauri G, Pesole G (2004) Weeder Web: discovery of transcription factor binding sites in a set of sequences from co-regulated genes. Nucleic Acids Res 32:W199–W203

Paz I, Kosti I, Ares M Jr, Cline M (2014) RBPmap: a web server for mapping binding sites of RNA-binding proteins. Nucleic Acids Res 42:W361–W367

Paziewska A, Wyrwicz LS, Bujnicki JM, Bomsztyk K, Ostrowski J (2004) Cooperative binding of the hnRNP K three KH domains to mRNA targets. FEBS Lett 577:134–140

Prakash N, Fehr S, Mohr E, Richter D (1997) Dendritic localization of rat vasopressin mRNA: ultrastructural analysis and mapping of targeting elements. Eur J Neurosci 9:523–532

Prodon F, Yamada L, Shirae-Kurabayashi M, Nakamura Y, Sasakura Y (2007) Postplasmic/PEM RNAs: a class of localized maternal mRNAs with multiple roles in cell polarity and development in ascidian embryos. Developmental dynamic 236:1698–1715

Queen C, Wegman MN, Korn LJ (1982) Improvements to a program for DNA analysis: a procedure to find homologies among many sequences. Nucleic Acids Res 10:449–456

Racca C, Gardiol A, Eom T, Ule J, Triller A, Darnell RB (2010) The neuronal splicing factor nova co-localizes with target RNAs in the dendrite. Front Neural Circuits 4:5

Ray D, Kazan H, Cook KB, Weirauch MT, Najafabadi HS, Li X, Gueroussov S, Albu M, Zheng H, Yang A et al (2013) A compendium of RNA-binding motifs for decoding gene regulation. Nature 499:172–177

Rihan K, Antoine E, Maurin T, Bardoni B, Bordonné R, Soret J, Rage F (2017) A new cis-acting motif is required for the axonal SMN-dependent Anxa2 mRNA localization. RNA 23:899

Ross AF, Oleynikov Y, Kislauskis EH, Taneja KL, Singer R (1997) Characterization of a beta-actin mRNA zipcode-binding protein. Mol Cell Biol 17:2158–2165

Ryu YH, Kenny A, Gim Y, Snee M, Macdonald PMJ (2017) Multiple cis-acting signals, some weak by necessity, collectively direct robust transport of oskar mRNA to the oocyte. J Cell Sci 130 (18):3060–3071

Sandelin A, Alkema W, Engstrom P, Wasserman WW, Lenhard B (2004) JASPAR: an open-access database for eukaryotic transcription factor binding profiles. Nucleic Acids Res 32:D91–D94

Sandve GK, Drablos F (2006) A survey of motif discovery methods in an integrated framework. Biol Direct 1:11

Saunders C, Cohen RS (1999) The role of oocyte transcription, the 5'UTR, and translation repression and derepression in Drosophila gurken mRNA and protein localization. Mol Cell 3:43–54

Schneider TD (2010) A brief review of molecular information theory. Nano Commun Netw 1:173–180

Schneider TD, Stephens RM (1990) Sequence logos: a new way to display consensus sequences. Nucleic Acids Res 18:6097–6100

Serano TL, Cohen RS (1995) A small predicted stem-loop structure mediates oocyte localization of Drosophila K10 mRNA. Development 121:3809–3818

Serano J, Rubin GM (2003) The Drosophila synaptotagmin-like protein bitesize is required for growth and has mRNA localization sequences within its open reading frame. Proc Natl Acad Sci 100:13368–13373

Serikawa KA, Porterfield DM, Mandoli DF (2001) Asymmetric subcellular mRNA distribution correlates with carbonic anhydrase activity in Acetabularia acetabulum. Plant Physiol 125:900–911

Shaner MC, Blair IM, Schneider TD (1993) Sequence logos: A powerful, yet simple, tool. IEEE 1:813–821

Shestakova EA, Singer RH, Condells J (2001) The physiological significance of β-actin mRNA localization in determining cell polarity and directional motility. Proc Natl Acad Sci 98:7045–7050

Sinha S, Tompa M (2003) YMF: A program for discovery of novel transcription factor binding sites by statistical overrepresentation. Nucleic Acids Res 31:3586–3588

Snee MJ, Arn EA, Bullock SL, Macdonald PM (2005) Recognition of the bcd mRNA localization signal in Drosophila embryos and ovaries. Mol Cell Biol 25:1501–1510

St Johnston D, Beuchle D, Nüsslein-Volhard C (1991) Staufen, a gene required to localize maternal RNAs in the Drosophila egg. Cell 66:51–63

Staden R (1984) Computer methods to locate signals in nucleic acid sequences. Nucleic Acids Res 12:505–519

Takizawa PA, Sil A, Swedlow JR, Herskowitz I, Vale RD (1997) Actin-dependent localization of an RNA encoding a cell-fate determinant in yeast. Nature 389:90–93

Tekotte H, Davis I (2002) Intracellular mRNA localization: motors move messages. Trends Genet 18:636–642

Van De Bor V, Davis I (2004) mRNA localisation gets more complex. Curr Opin Cell Biol 16:300–307

Van De Bor V, Hartswood E, Jones C, Finnegan D, Davis I (2005) gurken and the I factor retrotransposon RNAs share common localization signals and machinery. Dev Cell 9:51–62

Wang ET, Cody NA, Jog S, Biancolella M, Wang TT, Treacy DJ, Luo S, Schroth GP, Housman DE, Reddy S et al (2012) Transcriptome-wide regulation of pre-mRNA splicing and mRNA localization by muscleblind proteins. Cell 150:710–724

Ward LD, Bussemaker HJ (2008) Predicting functional transcription factor binding through alignment-free and affinity-based analysis of orthologous promoter sequences. Bioinformatics 24:i165–i171

Weil TT, Forrest KM, Gavis ER (2006) Localization of bicoid mRNA in late oocytes is maintained by continual active transport. Dev Cell 11:251–262

Welshhans K, Bassell GJ (2011) Netrin-1-induced local β-actin synthesis and growth cone guidance requires zipcode binding protein 1. J Neurosci 31:9800–9813

Wilhelm JE, Vale RD (1993) RNA on the move: the mRNA localization pathway. J Cell Biol 123:269–274

Wilk R, Hu J, Blotsky D, Krause HM (2016) Diverse and pervasive subcellular distributions for both coding and long noncoding RNAs. Genes Dev 30:594–609

Yamagishi M, Ishihama Y, Shirasaki Y, Kurama H, Funatsu T (2009a) Single-molecule imaging of beta-actin mRNAs in the cytoplasm of a living cell. Exp Cell Res 315:1142–1147

Yamagishi M, Shirasaki Y, Funatsu T (2009b) Size-dependent accumulation of mRNA at the leading edge of chicken embryo fibroblasts. Biochem Biophys Res Commun 390:750–754

Yan Z, Lecuyer E, Blanchette M (2019) Prediction of mRNA subcellular localization using deep recurrent neural networks. Proceedings of ISMB 2019

Yano T, López de Quinto S, Matsui Y, Shevchenko A, Shevchenko A, Ephrussi A (2004) Hrp48, a Drosophila hnRNPA/B homolog, binds and regulates translation of oskar mRNA. Dev Cell 6:637–648

Zarnack K, Feldbrugge M (2010) Microtubule-dependent mRNA transport in fungi. Eukaryot Cell 9:982–990

Zhang HL, Eom T, Oleynikov Y, Shenoy SM, Liebelt DA, Dictenberg JB, Singer RH, Bassell GJ (2001) Neurotrophin-induced transport of a beta-actin mRNP complex increases beta-actin levels and stimulates growth cone motility. Neuron 31:261–275

Zhou Y, King ML (1996a) Localization of Xcat-2 RNA, a putative germ plasm component, to the mitochondrial cloud in Xenopus stage I oocytes. Development 122:2947–2953

Zhou Y, King ML (1996b) RNA transport to the vegetal cortex of Xenopus oocytes. Dev Biol 179:173–183

Zimyanin VL, Belaya K, Pecreaux J, Gilchrist MJ, Clark A, Davis I, St Johnston D (2008) In vivo imaging of oskar mRNA transport reveals the mechanism of posterior localization. Cell 134:843–853

# Chapter 8
# RNA Granules and Their Role in Neurodegenerative Diseases

**Hadjara Sidibé and Christine Vande Velde**

**Abstract** In recent years, cytoplasmic RNA granules, which are micron-sized membrane-less entities formed by phase separation, have progressively gained recognition as essential constituents of neuronal RNA metabolism. Stress granules form under adverse growth conditions in order to protect nontranslating mRNA, shift translation toward the production of prosurvival factors, as well as potentially serve as hubs for intracellular signaling. In contrast, processing bodies play a role in RNA degradation in both stressed and homeostatic conditions. Lastly, transport granules permit, as their name indicates, the transport of mRNA within neurons. All of these granule subtypes are required for proper neuronal function; thus, impairments in their regulation and/or composition are expected to be deleterious. Here, we review these cytoplasmic RNA granule subtypes and discuss how they have been implicated in some neurodegenerative diseases.

**Keywords** Neurodegeneration · RNA metabolism · Stress granules · Transport granules · Processing bodies · Amyotrophic lateral sclerosis · Alzheimer's disease · Tauopathy

## 8.1 Introduction

Intracellular compartmentalization serves to organize and regulate biochemical processes. While lipid-bound organelles such as mitochondria and lysosomes have long been described, the study of membrane-less organelles, such as RNA granules, is relatively recent. The discovery of these membrane-less organelles introduced a new level of cellular regulation. Indeed, these granules are essential for a variety of

H. Sidibé · C. Vande Velde (✉)
Department of Neurosciences, Université de Montréal, Montréal, QC, Canada

Centre Hospitalier de l'Université de Montréal (CHUM) Research Centre, Montréal, QC, Canada
e-mail: hadjara.sidibe@umontreal.ca; c.vande.velde@umontreal.ca



M. Oeffinger, D. Zenklusen (eds.), *The Biology of mRNA: Structure and Function*, Advances in Experimental Medicine and Biology 1203,
https://doi.org/10.1007/978-3-030-31434-7_8

cellular processes including oogenesis, neuronal plasticity, and RNA metabolism (Anderson and Kedersha 2006). RNA granules are microscopic foci composed of both protein and RNA and are implicated in a wide range of post-transcriptional processes including RNA stabilization, repression, and transport (Anderson and Kedersha 2009). RNA granules can be classified as nuclear or cytoplasmic. Nuclear RNA granules that include splicing speckles, Cajal bodies, and paraspeckles play an important role in the precise regulation of gene expression in the nucleus. The present chapter will focus on how cytoplasmic RNA granules participate in the fine-tuning of gene expression. Cytoplasmic RNA granule subtypes include stress granules (SGs), processing bodies (PBs), and transport granules. Each of these granules holds specific functions linked to their characteristic composition (Anderson and Kedersha 2006; Sephton et al. 2011; Alami et al. 2014).

In recent years, cytoplasmic RNA granules have been increasingly recognized as essential to the regulation of neuronal gene expression. Dysfunction of these granules and/or their key components is associated with several neurodegenerative diseases including amyotrophic lateral sclerosis (ALS), frontotemporal dementia (FTD), and Alzheimer's disease (AD) (Ash et al. 2014; Wolozin and Apicco 2015; Maziuk et al. 2017). Here, we will briefly describe the three main cytoplasmic RNA granule subtypes and expand on how they may be relevant to ALS, FTD, and AD.

### 8.1.1 RNA-Binding Proteins

RNA-binding proteins (RBPs) are a class of proteins intrinsically related to RNA functions. mRNAs are not synthesized as lone molecules. At every step of their life, transcripts are associated with RBPs and other factors in messenger ribonucleoprotein (mRNP) complexes. This association is crucial for the control of gene expression. While some RBPs remain bound to an RNA until its degradation, others bind transiently to influence specific processes (Dreyfuss et al. 2002). RBPs regulate and/or participate in a wide range of processes, from splicing and alternative polyadenylation to mRNA localization, storage, and degradation (Glisovic et al. 2008). Ultimately, RBPs are major players in translational control (Glisovic et al. 2008). The actions exerted by RBPs have profound impacts on normal cellular physiology as well as on several pathologies, including neurodegenerative disease (Cestra et al. 2017). For the longest time, RBPs have been considered as ubiquitously expressed. However, it has recently been revealed that their expression can vary remarkably in an age- and tissue-dependent manner (McKee et al. 2005; Masuda et al. 2009). Interestingly, the activities of some RBPs are specific to the brain (McKee et al. 2005). Thus, changes in the levels of RBPs or defects in these proteins are suggested to have a singular effect on the brain and play a major role in the development of some neurodegenerative diseases.

Many RBPs feature a conserved structure which includes specialized RNA recognition motifs (RRMs) that dictate RNA binding specificity (Lunde et al.

2007) and low complexity domains (LCDs) that are highly enriched in select amino acids (Radó-Trilla and Albà 2012; Harrison and Shorter 2017). While the mechanisms underlying RNA binding specificity are as yet poorly understood but seem to be influenced by electrostatic interactions and steric inhindrance, recent studies have shed light on the importance of the LCD (Lukavsky et al. 2013; Lin et al. 2015; Van Treeck and Parker 2018). These domains are most often composed of glycine-rich regions and regions enriched in polar uncharged residues such as asparagine, glutamine, tyrosine, and serine (Sun et al. 2011; Maniecka and Polymenidou 2015). The repetition of these amino acids allows hydrophobic interactions between LCDs, leading to protein-protein interactions. These weak interactions are highly affected by molecular crowding, temperature, and aliphatic molecules; thus, tight regulation of these interactions is expected (Lin et al. 2015; Molliex et al. 2015; Nott et al. 2015; Patel et al. 2015).

RBPs associated with their target mRNAs can assemble into macrostructures referred to as RNA granules via protein-protein, RNA-protein, and RNA-RNA interactions mediated by the polar amino acids and the nucleic acids (Lin et al. 2015; Molliex et al. 2015; Van Treeck and Parker 2018). The cytoplasmic foci thereby formed are, thus, membrane-less and contain regulatory components and translationally repressed mRNAs. They play a major role in mRNA metabolism; thus, any malfunction can be detrimental to the cell and the entire organism. While each RNA granule contains key specific components that are linked to the purported function of the granule, it is appreciated that there is some overlap of components between RNA granule subtypes (Anderson and Kedersha 2006). This is likely due to interactions between granules which facilitate the sharing and exchange of components (Anderson and Kedersha 2008). Thus, although each granule has specific functions, they are all intricately linked, leading to the hypothesis that common mRNA regulatory pathways operate in diverse mRNA granule subtypes (Anderson and Kedersha 2006).

### *8.1.2 RBP Aggregation*

Most proteins, including RBPs, need to fold into well-defined three-dimensional structures in order to properly perform their functions (Kim et al. 2013b). Misfolded proteins often form insoluble aggregates that are presumably cytotoxic and are suggested to contribute to various neurodegenerative diseases (Ciryam et al. 2015). In order to properly fold, a protein must overcome numerous challenges. First, it must find the binding partners that facilitate its folding in a highly crowded environment prone to promiscuous interactions. Relevant here, many RBPs are present in the cell at supersaturated concentrations, rendering them prone to aggregation (Ciryam et al. 2015). Mutations can further increase the misfolding of certain RBPs and thus increase their propensity to aggregate (Labbadia and Morimoto 2015; Conicella et al. 2016). In normal physiology, multimeric protein assemblies are often required to achieve functionality; thus controlled protein aggregation is essential.

With this in mind, we will explore the controlled aggregation of RNPs that is critical for RNA granules and then scrutinize where loss of this regulation may contribute to the pathogenesis of select neurodegenerative diseases.

## 8.2 Stress Granules

Cells are continuously exposed to numerous stress-inducing stimuli. Changes in pH, viral or bacterial exposure, lack of metabolites, and increased oxidative stress are just a few of the many adverse processes which cells need to overcome in order to survive. To cope with these challenges, cells are equipped with mechanisms to conserve energy and protect macromolecules. One of these involves the temporary shutdown of global translation and the concomitant prioritized production of enzymes and chaperones which function to counteract the stressful stimulus. This is facilitated by the formation of stress granules (SGs). These micron-sized, highly dynamic structures provide protection to translationally arrested polyadenylated mRNAs and some essential components of the translational machinery (Anderson and Kedersha 2008). SGs are suggested to be a triage site, such that transcripts to be preserved until the stress abates are sorted from the ones that are to be degraded during stress. It is presumed that this sequestration protects the molecules from degradation until the stress subsides (Anderson and Kedersha 2008). The prevailing viewpoint is that the storage of mRNAs, rather than degradation, permits the rapid reinitiation of translation following dissolution of the SGs without the costly process of resynthesis. Given that energy stores might be compromised post-stress exposure, from an energetic point of view, the storage and release of SG components represent a more attractive solution than protein degradation and de novo synthesis (Anderson and Kedersha 2002a).

In mammalian cells exposed to a dramatic environmental change (stress), the formation of canonical SGs is triggered by the phosphorylation of eIF2α by one of four kinases and the consequent specific translational arrest of non-stress-related transcripts (Anderson and Kedersha 2002b). Each kinase is activated by a specific stimulus that is linked to a specific survival process. eIF2α phosphorylation leads to polysome disassembly and the release of translation initiation factors, ribosomal subunits, and mRNPs which can assemble into SGs (Wolozin 2012). Certain RBPs, such as Ras GTPase-activating protein-binding protein 1 (G3BP1) and T-cell-restricted intracellular antigen-1 (TIA-1), nucleate SGs as driven by their LCDs. Note that a subset of SGs can be generated in an eIF2α-independent manner and are referred to as non-canonical SGs. In this case, translation is also specifically inhibited but the phosphorylation of eIF2α is not necessary (Dang et al. 2006; Moujaber et al. 2017). The mechanism for polysome disassembly and promotion of non-canonical SG formation has yet to be uncovered.

As previously mentioned, one of the main features of RBPs is the presence of LCDs which facilitate the protein-protein and RNA-protein interactions that are implicated in SG formation. SGs assemble due to increased cytoplasmic concentration of RBPs which enhances self-assembly mediated by hydrophobic interactions (Kim et al. 2013a). This promotes the phase separation of RNA-protein complexes within the cytoplasm, resulting in the formation of micron-sized dynamic inclusions with droplet-like properties (Molliex et al. 2015; Courchaine et al. 2016). The generation of these dynamic foci requires the presence of both RNA and highly hydrophobic LCDs. Thus, the concentration and intrinsic properties of RBPs and RNAs are highly regulated in order to tightly control SG formation (Kroschwald et al. 2015; Smith et al. 2016). It is important to also mention the importance of the cytoskeleton in the formation of SGs and the other cytoplasmic RNA granules. Indeed, while protein-protein, RNA-protein, and RNA-RNA interactions allow the phase separation of molecules at short distances, motor proteins and microtubules are important for the gathering/fusing of mRNPs over longer distances (Perez-Pepe et al. 2018). Thus, RBPs serve as the link between mRNPs and motor proteins.

SGs have recently been proposed to be composed of two compartments: a less concentrated shell, made of loosely interacting molecules, and an internal, more stable/less dynamic core structure (Jain et al. 2016). According to this, two assembly models can be expected (Wheeler et al. 2016). The first model suggests that SGs initially assemble into individual mRNPs which will form droplet-like structures. The supersaturation of RBPs in the internal part of the droplet will drive the transition into less dynamic structures over time and the formation of inner cores (Jain et al. 2016). However, SG core size does not change over time, suggesting that this model may not be valid (Wheeler et al. 2016). The second model proposes that nontranslating mRNPs first coalesce into stable cores so as to create a nucleating platform for the growth of a more dynamic, less dense shell around these cores. Then, the coalescence of individual core/shell assemblies into larger ones gives rise to SGs (Wheeler et al. 2016). The first step of this model would coincide with what has been previously referred to as primary aggregation which is driven by SG nucleating proteins such as G3BP1 and TIA-1. By extension, the shell would correspond with the previously described secondary aggregation which represents coalescence of smaller SGs into larger assemblies and the recruitment of various other RBPs and/or SG accessory components (Kedersha et al. 2005; McDonald et al. 2011; Aulas et al. 2012; Aulas and Vande Velde 2015). This biphasic architecture of SGs provides certain advantages for cells. For example, one might expect that the lower-density shell compartment allows for an easier and more dynamic exchange of RNAs between the cytoplasm, ribosome, and other RNA granule subtypes (Wheeler et al. 2016). In contrast, those RNAs located within the cores may be sequestered and not exchanged. This interesting idea remains to be tested.

Due to their liquid-like properties, the isolation of SGs and study of their composition are difficult. Microscopic analyses have historically been the only reliable technique to determine SG protein composition. These studies have demonstrated that foci contain not only RBPs and polyadenylated mRNAs but also some components of the translational machinery. RBPs such as TDP-43, FUS, SMN, Staufen, PABP, TIA-1, G3BP1 and G3BP2 are localized to SGs (Kedersha et al. 1999; Anderson and Kedersha 2002a, b; Tourriere et al. 2003; Hua and Zhou 2004;

McDonald et al. 2011). Most of these RBPs are essential for RNA metabolism such that their dysfunction is deleterious for cells (and whole organisms). Small ribosomal subunits are also selectively recruited to SGs as well as the translation factors eIF2, eIF3 and eIF4E (Anderson and Kedersha 2002a, b; Kedersha and Anderson 2002; Kedersha et al. 2002).

Recently, a biochemical approach to isolate and define SG composition has been described (Khong et al. 2017a; Wheeler et al. 2017). As SGs are closely linked to translational control, studies aimed at describing RNA composition have been primarily focused on the polyadenylated mRNAs that localize to SGs (Kedersha et al. 2002). A recent study from Khong et al. (2017c) estimates that SG cores are composed of 42,000 RNAs, of which 80% were mRNAs, representing 10% of the total cellular pool. Surprisingly, no single RNA represents more than 1% of the total RNA isolated from these substructures. If one recalls the paradigm (that SGs are to store translationally arrested mRNAs), then this result is not expected. It has long been thought that SGs sequestered specific mRNAs in order to regulate specific cellular pathways. However, these new findings challenge that view, instead suggesting that no specific pathway is selectively preserved in SGs.

It is well known that SG protein composition varies according to the stress condition (Kedersha and Anderson 2002; Kedersha et al. 2002). It seems that SG mRNA composition also follows the same trend. For example, transcripts encoding the glycolysis-related enzyme glyceraldehyde 3-phosphate dehydrogenase (GAPDH) are excluded from SGs induced by heat, osmotic, and ER stress, while mRNA encoding the transferrin receptor TFRC is targeted only to heat shock-induced SGs (Khong et al. 2017c). That SG transcriptomes and proteomes change as a function of stress suggests that cells modulate the molecular triage depending on the insult. It can be further speculated that each stress condition triggers the selective SG recruitment of particular RBPs with specific bound transcripts in need of preservation (Khong et al. 2017c). Given that each stress condition is expected to generate specific types of damage, it makes sense that different transcripts and RBPs would require protection in order to facilitate the return to a normal physiological state as quickly as possible after the stress clears. However, this hypothesis also suggests that there is an active mechanism for the selection and targeting of certain RBPs and their bound mRNAs to SGs. The mechanism underlying this selection remains to be discovered. One hypothesis is that RNA post-transcriptional modifications influence mRNA localization to specific SGs. This hypothesis stems from the observation that G3BP1 and FMR1, two proteins that are considered as SG markers, are differentially engaged with transcripts harboring $N^6$-methyladeosine modifications ($m^6A$). While G3BP1 is repelled by this modification, FMR1 preferably binds these methylated RNAs, implying that FMR1-positive granules may be enriched with $m^6A$, while G3BP1 granules are not (Edupuganti et al. 2017). However, both proteins are suggested to, in some cases, form a complex rendering the interpretation of these results more complicated (Wu et al. 2016). Very interestingly, $m^6A$ modifications are predominantly observed in the brain and their levels increase with adulthood and neuronal maturation, suggesting that age influences the transcriptome of G3BP1 and FMR1 granules; this influence could play a role in neurodegenerative disorders (Meyer et al. 2012; Edupuganti et al. 2017).

Khong et al. further uncovered a number of characteristics of SG core-associated transcripts. In general, they found that transcripts enriched in SG cores have a short half-life, have lower GC content, and are of low translational efficiency (Khong et al. 2017c). Indeed, these mRNA are generally less abundant in the cell and less stable (Radhakrishnan and Green 2016). In addition, SG core-enriched mRNAs averaged 7.1 kb in length, which is larger than non-enriched transcripts. These longer RNAs may facilitate phase separation and RNP condensation since longer sequences may impart increased opportunities for RBP or RNA-RNA interactions (Khong et al. 2017c; Van Treeck and Parker 2018).

While the data generated in this study are intriguing and yielded important information on SG biology, it is important to remember that the isolation method is based on complexes involving G3BP1, which have been interpreted to be SG cores. Given that super-resolution microscopy indicates that cores measure 0.0066 $\mu m^3$, but that SGs are estimated to be 10 $\mu m^3$ (Khong et al. 2017b), it seems that a major proportion of SGs remains to be sampled. New approaches to isolate and characterize SGs are required to obtain a more comprehensive appraisal of SG composition.

SGs are systematically disassembled within a few hours after their initial appearance (Alberti et al. 2017). The disassembly of SGs allows for the recycling of SG components such as signaling molecules, mRNAs, and 40S ribosomal subunits which otherwise would need to be resynthesized de novo. The exact mechanism underlying SG clearance has not yet been established. However, the current literature suggests two possible mechanisms. First, SGs are proposed to disassemble in a two-step process: the dissolution of the less rigid shell followed by the disassembly and/or clearance of the cores by autophagy. This mechanism implies many cellular advantages. Indeed, as the shell would presumably allow for an exchange between molecular components of the shell and the cytosol, the dissolution of the shell could permit the rapid reintroduction of mRNAs and translation initiation factors to the translational machinery. Once the shell is disbanded, the degradation machinery would be able to access and degrade the cores. Another possibility is that the SG shell is enriched in degradation factors that can directly dissolve the core when the stress is resolved. SG clearance requires factors involved in the proteasomal system and autophagic pathway, especially VCP/p97, HDAC6 and SYK, all of which are themselves localized to SGs (Kwon et al. 2007; Buchan et al. 2013; Krisenko et al. 2015). Interestingly, both HDAC6 and VCP/p97 are key components of the aggresome, which functions to sequester misfolded/aggregated proteins. Thus, it is has been demonstrated that SGs can be cleared via autophagy (Kawaguchi et al. 2003; Kitami et al. 2006; Boyault et al. 2007; Kwon et al. 2007; Buchan et al. 2013; Ling et al. 2013; Meyer and Weihl 2014; Ganassi et al. 2016; Mateju et al. 2017; Chitiprolu et al. 2018). Indeed, the recruitment of the pro-inflammatory tyrosine kinase SYK into SGs also promotes autophagy-mediated SG clearance via the phosphorylation of SG proteins (Krisenko et al. 2015). It has also been reported that siRNA-mediated depletion of VCP/p97 in mammalian cells, as well as VCP chemical inhibition, impairs SG clearance and short-term inhibition of the ubiquitin proteasome system (UPS) triggers the accumulation of ubiquitinated SGs positive for HDAC6 (Kwon et al. 2007; Buchan et al. 2013).

Alternatively, SG cores may be dissolved via chaperone activity. The yeast ortholog of HSP70 and the small HSPs HSPB1/HSP27 and HSPB8/HSP22 are all localized to SGs (Collier and Schlesinger 1986; Scharf et al. 1998; Kedersha et al. 1999; Walters et al. 2015; Ganassi et al. 2016). Furthermore, while cells overexpressing HSP70 are incapable of forming SGs, cells depleted of this chaperone fail to disassemble their SGs and have impaired translational recovery post-stress (Mazroui et al. 2007). Moreover, HSP70 detangles protein aggregates in order to facilitate refolding by HSP100 (Liberek et al. 2008). Thus, it is possible that these chaperones facilitate disaggregation of SG core proteins after stress. The exact mechanism governing SG disassembly is not yet fully understood and may entail one or more of these pathways.

## 8.3 Processing Bodies

Processing bodies (PBs) are cytoplasmic aggregates of 100–300 nm in diameter and are present in both basal and stress conditions (Eulalio et al. 2007b). PBs were identified following the discovery that Dcp1, Dcp2, and the miRNA pathway component GW182 colocalize into cytoplasmic foci distinct from SGs and also containing translationally arrested mRNAs (Bashkirov et al. 1997; Ingelfinger et al. 2002; van Dijk et al. 2002; Eystathioy et al. 2003). As PBs are observed even in physiological conditions, it has been suggested that they contribute to the coordination of basal cellular functions by sequestering transcripts and preventing translation (Decker and Parker 2012). A few models have been proposed regarding PB functions. First, it is hypothesized that PBs allow the condensation of mRNPs and their association with the decapping and mRNA decay machinery. This is premised on the observation that PBs are enriched in components of the mRNA-decapping machinery (Eulalio et al. 2007c; Parker and Sheth 2007), the AU-rich element (ARE)-mRNA decay, and nonsense-mediated decay (NMD) pathways (Kedersha et al. 2005; Franks and Lykke-Andersen 2008). Thus, it is supposed that PBs are a site of mRNA degradation. Second, it has been suggested that PBs are an mRNA storage site. This is based on data demonstrating that some mRNPs are targeted to PBs in order to silence specific pathways according to cellular requirements. Indeed, this model is supported by recent data demonstrating that some PB-enriched mRNAs are protected from 5′-truncation (Hubstenberger et al. 2017), thus inhibiting their degradation. Also, the coalescence of RBPs and their associated mRNA targets yields an increase in the catalytic activity of condensing enzymes such as DDX6, a central component of PBs (Ayache et al. 2015). Finally, it has been demonstrated that DDX6 along with its partner for translational repression 4E-T are essential for PB formation, but not DDX6 partners which are linked to RNA decay machineries (Ayache et al. 2015). Moreover, the dissolution of PBs is not associated with any impairment in mRNA decay or RNA-mediated gene silencing (Eulalio et al. 2007b). It is suggested that the decay and silencing processes initiate in the cytoplasm in soluble complexes that aggregate to form PBs (Eulalio et al. 2007b ). In any case, PBs seem to not be primarily responsible for the translational repression of mRNAs but, instead, are the result of translational arrest. Indeed, yeast and in vitro studies have demonstrated that mRNAs released from polysomes are not sufficient to trigger PB formation, while interactions with the decapping factors Dcp2, Pdc1, and Edc3 promote PB formation via phase separation (Eulalio et al. 2007b; Fromm et al. 2014).

PBs are dynamic structures formed by the assembly of translationally repressed mRNPs, suggesting that their formation is directly proportional to the cytoplasmic concentration of translationally repressed transcripts (Franks and Lykke-Andersen 2008). Indeed, PB assembly is RNA-dependent (Decker et al. 2007). In addition, pharmacological arrest of translation in human, yeast, and *Drosophila* cells enhances PB formation (Cougot et al. 2004; Teixeira et al. 2005; Eulalio et al. 2007a, b, c). GW182 and Ge-1, a component of the miRNA pathway and a decapping cofactor, respectively, are suggested to act as scaffolds in PB assembly, as their depletion leads to the dissolution of PBs, while their overexpression leads to the formation of enlarged PBs (Eulalio et al. 2007a). As well, the proteins Edc3 and Lsm4 promote physical interactions between mRNPs and their coalescence into PBs in yeast and *Drosophila* (Decker et al. 2007; Ling et al. 2008; Reijns et al. 2008). Interestingly, deletion of the LCDs of either Edc3 or Lsm4 completely abrogates PB assembly, demonstrating the importance of these domains in the biogenesis of RNA granules (Decker et al. 2007; Reijns et al. 2008). Edc3 and Lsm4 are components of two separate complexes, the decapping complex and the Lsm-Pat1 complex (which drives mRNA decay), respectively, both involved in RNA degradation (Decker et al. 2007). Thus, it is suggested that PBs are formed via the assembly of non-translationally active mRNPs and driven by proteins involved in mRNA degradation. As alluded to earlier, it remains unknown how mRNAs are selected for recruitment to PBs. Recent work indicates that PBs contain mRNAs relevant to select pathways such that PB formation is perhaps intended to repress mRNA regulons (Hubstenberger et al. 2017; Wang et al. 2018). This leads to several other questions, including how are RNAs selected for PB targeting? Does a single PB contain only a single type of mRNA or various members of the same regulon? And lastly, do basal PBs and stress-induced PBs serve the same cellular function? Future work will undoubtedly address many of these concepts.

A comparison between PB and SG core proteomes reveals that 75% of PB proteins and 91% of SG proteins are granule-specific (Jain et al. 2016; Hubstenberger et al. 2017). Thus, even if SGs and PBs have several common proteins, their primary composition is unique, which may indicate specific functions and a rare redundancy between these granule subtypes. PBs are three times denser than SGs and are two-fold enriched in RBPs compared to SGs (Hubstenberger et al. 2017). It is hypothesized that this may be due to the inherent promiscuity of PB proteins such that they can simultaneously bind numerous mRNAs, which may further bridge additional proteins. For example, DDX6 interacts with half of all known PB proteins, which is consistent with its central role in PBs (Hubstenberger et al. 2017). Of interest, the proteins in common between PBs and SGs are mainly related to translational arrest. Moreover, PBs are enriched in decapping and mRNA degradation proteins (Hubstenberger et al. 2017). Indeed, as previously mentioned the decapping-associated protein Edc3 and the degradation-associated protein Lsm4 are localized to PBs. To this, we can add the decapping agents Dhh1p, Pat1p, DDX6 and the nucleases Xrn1 and Ccr4p (Teixeira and Parker 2007). Other RBPs, such as HuR, TIA-1 and Staufen, are also found in these entities (as well as SGs) (Parker and Sheth 2007). The exact functions of most PB proteins are as yet unknown; however, some have been suggested to play a role in translational arrest, while others are proposed to have a structural or scaffolding role. Finally, the human Argonaute proteins and GW182, key components of miRNA-mediated gene silencing, are also localized to PBs suggesting a role for PBs in mRNA degradation via miRNA silencing. This is reinforced by the observation that Ago1 and Ago2 coimmunoprecipitate with Dcp1a and Dcp2, respectively, and that depletion of GW182 disrupts PBs and impairs miRNA-mediated gene silencing (Jakymiw et al. 2005; Liu et al. 2005a; Sen and Blau 2005).

Protein-encoding RNAs are enriched in PBs, more so than non-coding RNAs (Hubstenberger et al. 2017). In addition, mRNAs encoding housekeeping genes are excluded from PBs, while protein-encoding transcripts that require regulation are enriched in PBs. This is in accord with the suggestion that PBs are associated with translation control (Hubstenberger et al. 2017). mRNAs involved in miRNA processing (Ago1, 2, and 3 and MOV) are also enriched in PBs. In addition, it is now known that PB-enriched mRNAs are typically bound by RBPs in their 3′UTRs. In contrast, RNAs bound by RBPs in their coding regions are depleted from PBs. Also, as polyA tail length is typically correlated with mRNA stabilization, it has been suggested that PBs are composed of mRNA lacking their polyA tail. Indeed, the absence of the polyA tail has been considered a main difference between SGs and PBs. However, it has been recently determined that mRNAs in PBs may contain heterogeneous polyA tail lengths (Hubstenberger et al. 2017). Furthermore, it seems that a third of cellular mRNAs are enriched in PBs, suggesting that specific mRNAs are targeted to PBs (Hubstenberger et al. 2017). In the future, it will be interesting to understand the mechanism which selectively targets mRNAs to PBs.

The exact mechanisms underlying PB disassembly are poorly understood. Evidence suggests that PBs can undergo one of two fates: either they can disassemble into smaller mRNP units that re-enter the translational pool or they can be degraded via autophagy (Sheth and Parker 2003; Parker and Sheth 2007; Franks and Lykke-Andersen 2008). The first step of the disassembly involves the release of mRNPs. As PBs are supposed to assemble via the linkage of mRNPs by proteins involved in mRNA degradation, it is possible that mRNP release is linked to these proteins. Indeed, loss of the adhesive function of key degradative components is expected to be associated with PB disassembly. The idea that mRNAs can be released from PBs for translation stems from observations in yeast where mRNAs assembled in PBs upon glucose starvation can later re-enter the translational pool once glucose levels are restored (Brengues et al. 2005). Moreover, when translation is increased such as during neuronal stimulation, the number of PBs is reportedly decreased (Zeitelhofer et al. 2008). Interestingly, autophagy is also associated with PB disassembly. The E3 ubiquitin ligase TRIM21, which is involved in the ubiquitin-dependent protein degradation pathway, is localized to PBs. The deubiquitinating enzyme UPSP4, while not localized to PBs, strongly interacts with TRIM21 and modulates PB number (Zheng et al. 2011). UPSP4 also interacts with the PB proteins DCP1a, Edc3 and Lsm4, all of which are essential for PB formation and activities. Thus, these interactions suggest that ubiquitination and protein degradation is linked to PB disassembly. Also, ATG2, which is essential for autophagosome formation, interacts with the PB proteins DDX6 and MOV10 (Zheng et al. 2011). The functional relevancy of these interactions is not yet determined. However, it is plausible that autophagosomes play a role in PB clearance. This is supported by results showing that ATG2 function is required for the targeting of liquid droplets to autophagosomes (Velikkakath et al. 2012).

Microscopic analyses indicate that PBs and SGs are often found closely juxtapositioned, referred to as docked (Kedersha et al. 2005; Aulas et al. 2015). The exact reason for this interaction is not yet totally understood but is hypothesized to permit the transfer of select mRNAs from SGs to PBs for degradation. This interaction between SGs and PBs leads to some interesting questions. First, what maintains them as distinct granules despite juxtapositioning? It is possible that the specific proteins contained within SGs and PBs confer differences in surface tension of the droplets, thus rendering the granules non-miscible. Then, it is possible that an internal reorganization of both foci could lead to interactions between common components and thus mediate mRNA exchange. The docking between SGs and PBs also raises another question: if SGs and PBs do exchange components, how is it that certain mRNPs/RBPs are selected for this? Moreover, if the recent new view that PBs function to repress mRNA regulons is accepted, how is it that transcripts are translationally repressed in two different locations and what governs this decision? Much work remains to fully understand the function and mechanisms of granule interactions.

## 8.4 Transport Granules

Neurons are highly polarized cells with elaborate processes extending long distances. The distance between the cell body and the distal synaptic bouton, as well as the functional and morphological differences between dendrites and the axon, impose numerous challenges that neurons need to overcome (Hirokawa 2006). The optimal physiology of neurons highly depends on the trafficking of mRNAs, in a translationally dormant state, from the soma to synapses. This is accomplished via packaging into transport RNA granules (also referred to as transport or transfer RNPs, tRNPs or neuronal granules) (Knowles et al. 1996). These transport granules facilitate local protein translation which is critical for neuronal activity thus permitting neurons to rapidly respond to their ever-changing environment (Tada and Sheng

2006; Alami et al. 2014). Indeed, the local modulation of the proteome allows a direct modification of the synapse and efficient storage of information in the brain and is essential for axon guidance and nerve regeneration (Klann and Dever 2004; Willis et al. 2005). Precise synapse development, formation, and maintenance are important for accurate neuronal network activity and normal brain function and critically rely on these specialized granules, which further provide for synaptic independence of somal transcriptome/proteome changes.

Transcripts synthesized in the nucleus are packaged with regulatory RBPs and referred to as mRNPs. Once in the cytoplasm, depending on their fate, mRNPs can either be directly translated in the soma, stored, degraded or assembled into transport granules and delivered to neuronal processes (Steward and Levy 1982; Singh et al. 2015). Mechanistically, mRNAs are transported in transport granules via the following distinct steps (Doyle and Kiebler 2011). First, mRNAs are bound by RBPs via specific sequences in their 3′UTR, referred as zipcodes or *cis*-acting localization elements. These elements are very heterogeneous as they can be either a sequence of 5–6 nucleotides or complex secondary structures. In addition, an individual mRNA may contain a variety of zipcodes (Doyle and Kiebler 2011). These sequences act as molecular targeting signals that, when recognized by specific *trans*-acting RBPs, prepare the RNA for delivery to a specific subcellular compartment. Following this binding, RBPs self-assemble to form microscopically visible granules. In addition to serving as a scaffold for transport granule formation, the implicated RBPs protect the client mRNAs during transport and act as intermediates between the mRNAs and the cytoskeleton to facilitate active transport. Indeed, transport granules targeted to their final destination are bound to microtubules and remain anchored until translational activation (Doyle and Kiebler 2011).

Transport granules contain translationally arrested mRNAs associated with regulatory RBPs and noncoding RNAs (Doyle and Kiebler 2011). The exact composition of these granules is unknown. However, evidence suggests that their composition is context-dependent (Krichevsky and Kosik 2001; Kiebler and Bassell 2006). It is thought that distinct stimuli activate the transport of specific families of mRNAs, via activation of RBPs that recognize particular zipcodes. The molecular mechanism underlying this specific activation is yet to be understood, but it is speculated that groups of mRNAs which comprise part of the same pathway and/or work in cooperation harbor the same zipcode so as to permit their collective transport and activation (Farina et al. 2003; Doyle and Kiebler 2012). This is supported by data demonstrating that each particle is composed of a particular subgroup of mRNAs associated with specific RBPs (Doyle and Kiebler 2011, 2012). Many of the proteins that have been reported in these granules are either related to transport, RNA regulation or protein synthesis. The most studied core transport granule proteins include Staufen 1, Staufen 2, Fragile X mental retardation protein (FMRP), and Zipcode binding protein 1 (ZBP1) (Kiebler and Bassell 2006; Doyle and Kiebler 2011).

Staufen proteins are RBPs implicated in mRNA localization, silencing, and decay (Vessey et al. 2008; Gong and Maquat 2011). While Staufen 1 is ubiquitously expressed, Staufen 2 is preferentially expressed in the brain (Mallardo et al. 2003;

Monshausen et al. 2004). Both proteins are directly involved in synaptic plasticity. Indeed, Staufen 1 is required for the late phase of long-term potentiation (LTP), while Staufen 2 is involved in metabotropic glutamate receptor (mGluR)-dependent long-term depression (Lebeau et al. 2008, 2011a, b). Both are mechanisms which require synaptic remodelling. Moreover, Staufen-positive granules move bidirectionally between the soma and dendrites (Köhrmann et al. 1999). Thus, Staufen depletion impairs mRNA transport to synapses and negatively impacts synaptic plasticity (Chatel-Chaix et al. 2004; Lebeau et al. 2008, 2011a, b; Vessey et al. 2008).

FMRP is mainly expressed in the brain and is implicated in a number of mRNA regulation steps (Zalfa et al. 2006; Bassell and Warren 2008). As an RBP, it is involved in activity-dependent mRNA transport to dendrites and the local translation of transported transcripts (Bassell and Warren 2008; Dictenberg et al. 2008). Indeed, FMRP associates with the molecular motor kinesin following mGluR activation (Dictenberg et al. 2008). In addition, loss-of-function studies demonstrate that FMRP is required for synaptic protein translation and optimal dendritic spine morphology and synaptic function (Bassell and Warren 2008; Dictenberg et al. 2008). Taken together, these studies indicate that FMRP serves dual functions in active mRNA transport and local translation (Bassell and Warren 2008).

Finally, ZBP1 is also one of the most studied transport granule *trans*-acting factors. ZBP family members were initially reported as interferon-inducible tumor-associated proteins essential for pro-inflammatory response and post-transcriptional gene regulation of the oncogenes MYC, CD44, and βTrCP1 (Noubissi et al. 2006; Stöhr et al. 2006; Vikesaa et al. 2006; Kuriakose and Kanneganti 2017). However, in neurons, ZBP1 plays a pivotal role in mRNA transport and local translation. For example, the protein mediates the transport of β-actin mRNA to axonal growth cones and neurites via the recognition of a loop within the 3′UTR (Kiebler and Bassell 2006; Kim et al. 2015). Disruption of the secondary structure or impairment of the binding via antisense oligonucleotides disrupts β-actin, but not CamKII, mRNA localization (Zhang et al. 2001; Eom et al. 2003; Kim et al. 2015). This differential regulation implies that ZBP1 can distinguish different structures and/or sequences within its targets. Once granules arrive at their final destination, ZBP1 is phosphorylated by the kinase Src, triggering the simultaneous release of β-actin mRNA and its subsequent translation to further drive growth of the cone (Huttelmaier et al. 2005; Lin and Holt 2007). Finally, upon synaptic stimulation, ZBP1 targets β-actin mRNA into dendritic spines, further demonstrating the importance of ZBP1 for β-actin mRNA trafficking (Tiruchinapalli et al. 2003; Welshhans and Bassell 2011).

## 8.5 Neurodegenerative Disease

Neurodegenerative disease is a term used to describe a variety of incurable and debilitating conditions leading to the progressive loss of neurons (Buratti and Baralle 2009; Wolozin and Apicco 2015). The consequences of neuronal loss are irreversible and devastate cognitive and physical function. To date, the etiologies of many cases of prevalent neurodegenerative diseases are still unknown. While some of these disorders are considered rare and having genetic origins and thus are typically familially inherited, a significant proportion lack any specific etiology and are referred to as sporadic. Aging and age-related loss of neuronal functions are the key descriptors of many neurodegenerative diseases (Buratti and Baralle 2009). Impairment in RNA metabolism, including RNA granule dysfunctions, has been suggested as one of the leading causes of neurodegeneration (Shukla and Parker 2016; Alberti et al. 2017; Cestra et al. 2017; Frankel et al. 2017; Harrison and Shorter 2017; Lechler and David 2017).

Links between RBP dysfunctions and impaired granule dynamics exist in the context of several neurodegenerative diseases. For example, the protein Huntingtin (HTT), whose gene arbors a pathological number of CAG expansions in Huntington's disease, is associated with mRNA transport in neurons via transport granules and coimmunoprecipitates with the PB resident protein Ago2, and its pathological deposits colocalize with the SG protein TIA-1 in transformed cell lines (Savas et al. 2008; Savas et al. 2010; Bentmann et al. 2013). Similarly, pathogenic CAG expansions in *ATXN2*, which are causative for spinocerebellar ataxia type 2, yield a mutant form of ATXN2 that impairs SG and PB assembly and is expected to affect transport granule physiology (Nonhoff et al. 2007; Paul et al. 2018). Here, we will specifically describe the contribution of RNA granule dysfunctions to the neurodegenerative diseases ALS, FTD, and AD (Tables 8.1, 8.2 and 8.3).

ALS is a fatal neurodegenerative disease characterized by the progressive degeneration of upper and lower motor neurons (Al-Chalabi and Hardiman 2013). FTD is a form of dementia characterized by the selective atrophy of the frontal and anterior temporal lobes of the brain (Neary et al. 2005). The comorbidity between these two diseases and the shared genetic risk factors suggests that ALS and FTD are part of the same continuum of neurodegenerative disorders (Guerreiro et al. 2015). Intracellular protein aggregates positive for RBPs and/or RNA granule markers are a common pathological hallmark of both diseases. Emerging evidence suggests that impaired RNA processing and disrupted protein homeostasis are two major pathogenic pathways for these diseases (De Conti et al. 2017).

Alzheimer's disease (AD) is the most common neurodegenerative disease (Hebert et al. 2003). It is characterized by the presence of intracellular neurofibrillary tangles (NFTs) composed of the hyperphosphorylated protein tau and extracellular plaques containing amyloid beta (Aβ) peptide and is associated with brain atrophy (Wenk 2003; Tiraboschi et al. 2004). AD belongs to a class of diseases referred as tauopathies. These disorders are characterized by the abnormal accumulation of hyperphosphorylated tau into NFTs. Interestingly, some cases of FTD, including frontotemporal dementia with Parkinsonism linked to chromosome-17 (FTDP-17), are classified as tauopathies. The following paragraphs will discuss the role of the RNA granule dysfunction in the context of ALS/FTD and Alzheimer's disease.

**Table 8.1** Summary of SG dysfunction caused by neurodegenerative disease-related proteins

| Protein | Expression | SG-related dysfunctions | Models | References |
|---|---|---|---|---|
| APP (Aβ plaques) | Basal expression | • Impaired microglial function due to sequestration of microglial factors into SGs<br>• Formation of large and long-lasting SGs positive for G3BP1and TIA-1 | N9<br>BV-2<br>Primary microglia<br>AD patient cortex | Ghosh and Geahlen (2015) |
| C9orf72 | Protein overexpression | • Spontaneous SG formation | N2a<br>Cortical neurons | Maharjan et al. (2017) |
| | Genetic deletion | • Increased cell death<br>• Impaired SG formation<br>• Reduced levels of TIA-1, HuR, and G3BP1 | C9orf72 KO N2a | Maharjan et al. (2017) |
| | RNA foci overexpression | • Increased cell death<br>• Sequestration of RBPs | SH-SY5Y<br>N2a | Lee et al. (2013)<br>Maharjan et al. (2017) |
| | DPR GR/PR overexpression | • Formation of insoluble inclusions<br>• Formation of spontaneous SGs with reduced dynamism<br>• Sequestration of SG-related RBPs | *Drosophila*<br>U2-OS<br>HeLa | Lee et al. (2016)<br>Lin et al. (2016)<br>Lopez-Gonzalez et al. (2016) |
| FUS | Mutant overexpression (R521G, R495X, and G515X) | • Cytoplasmic localization<br>• Enhanced and irreversible aggregation<br>• Increased recruitment to SGs | HEK-293<br>Zebrafish embryos | Bosco et al. (2010) |
| | Mutant overexpression (R518K, R521G, and R521H) | • Cytoplasmic localization<br>• Increased recruitment to SGs<br>• Protein inclusions colabel with SG markers | N2a<br>Primary motor neurons | Gal et al. (2010) |
| | Mutant overexpression (P525L) | • Nuclear import disruption<br>• Protein aggregation | HeLa<br>Primary neurons | Dormann et al. (2010) |
| | Mutant basal expression (R521C) | • Inclusions containing SG proteins | FTLD-FUS patients Spinal cord | Dormann et al. (2010) |
| | Recombinant protein (R522G, P525L, G495X, and FUS501) | • Impaired liquid droplet phase transition | In vitro | Murakami et al. (2015) |
| | Mutant overexpression (G156E, S96del) | • Sequestration of SMN and STAU-1<br>• Reduced levels of SMN and STAU-1 RNPs | *C. elegans*<br>In vitro | Murakami et al. (2015) |
| | Recombinant protein (G156E) | • Acceleration of droplet to fibril conversion | In vitro | Patel et al. (2015) |

(continued)

**Table 8.1** (continued)

| Protein | Expression | SG-related dysfunctions | Models | References |
|---|---|---|---|---|
| hnRNP A1 | Mutant basal expression (D262V and D262N) | • Cytoplasmic inclusions<br>• Cytoplasmic fibers<br>• Loss of nuclear localization<br>• VCP cytoplasmic accumulation<br>• Partial colocalization with TDP-43 and hnRNP A2/B1 inclusions<br>• Ubiquitin and p62 immunoreactive | MSP patients muscles | Kim et al. (2013a, b) |
| | Recombinant proteins (D262V and D262N) | • Acceleration of the fibrillization | In vitro | |
| | Mutant overexpression (D262V and D262N) | • Cytoplasmic inclusions<br>• Enhanced recruitment to SGs | HeLa<br>*Drosophila* | |
| hnRNP A2/B1 | Mutant basal expression (D290V) | • Cytoplasmic inclusions<br>• Cytoplasmic fibers<br>• Loss of nuclear localization<br>• VCP cytoplasmic accumulation<br>• Partial colocalization with TDP-43 and hnRNP A1 inclusions | MSP patient muscles | |
| | Recombinant proteins (D290V) | • Acceleration of the fibrillation process | In vitro | |
| | Mutant overexpression (D290V) | • Cytoplasmic inclusions<br>• Enhanced recruitment to SG | HeLa<br>*Drosophila* | |
| | Mutant overexpression (P301L) | • Formation of G3BP-positive SGs in cells without phospho-tau immunoreactivity<br>• Impaired TIA-1 interaction network<br>• Increased SG size and number with disease course<br>• Increased levels of TIA-1, G3BP1, TTP, and TDP-43 with disease course<br>• Loss of G3BP1 and TIA-1 colocalization<br>• TIA-1-positive SG colocalization with hyperphosphorylated tau inclusions | rTg4510 tau JNPL3 mice | |

(continued)

**Table 8.1** (continued)

| Protein | Expression | SG-related dysfunctions | Models | References |
|---|---|---|---|---|
| | | • TDP-43 and FUS cytoplasmic inclusions in brain of animals with moderate pathology | | |
| | Basal expression | • Loss of G3BP and TIA-1 colocalization<br>• TIA-1 colocalization with tau pathology in frontal cortex of AD and FTLD-17 patients<br>• TIA-1-positive inclusions in neurons, microglia, and astrocytes of AD patients | Frontal cortex AD patients | |
| | WT and mutant overexpression (P301L) | • Increased number of SGs<br>• Increased size of SGs | HT22 | Vanderweyde et al. (2016) |
| | WT and mutant overexpression (TauE14 and P301L) | • Internalized tau recruited to SGs in a TIA-1-dependent manner<br>• Alteration of SG dynamics by internalized tau<br>• Lack of TTP in TAU-induced SGs | HEK293T<br>N2a | Brunello et al. (2016) |
| | Recombinant protein (repeat region of 4R-Tau) | • Phase separation, potential transition into fibrils | In vitro | Ambadipudi et al. (2017) |
| TDP-43 | Mutant RRM-GFP overexpression (M337 V) | • Formation of condensed droplet with fewer internal bubbles than WT<br>• Impairment of the liquid properties of the droplets | Hek293 | Schmidt and Rohatgi (2016) |
| | Recombinant C-terminal domain (A321G, Q331K A326P, and A321V) | • Abrogation of phase separation (A321G and Q331K)<br>• Decreased phase separation (M337V)<br>• Increased phase separation (A326P, A321V) | In vitro | Conicella et al. (2016) |
| | Mutant overexpression (G294A, A315T, Q331K, and Q343R) | Following oxidative stress:<br>• Increased stress-induced aggregation<br>• Increased insoluble protein aggregates | BE-M17<br>HEK 293 | Liu-Yesucevitz et al. (2010) |

(continued)

**Table 8.1** (continued)

| Protein | Expression | SG-related dysfunctions | Models | References |
|---|---|---|---|---|
| | Mutant overexpression (G294A. A315T, G348C, and N3920S) | Following osmotic stress:<br>• Formation of larger SGs | HEK 293 | Dewey et al. (2011) |
| | Basal expression | • Presence of TIA-1 and eIF3 in pathological TDP-43 inclusions | Spinal cord ALS patients<br>Frontal cortex FTLD-U patients | Liu-Yesucevitz et al. (2010) |
| | Depletion | • Impairment SG assembly and disassembly<br>• SG number and size reduction<br>• Impaired SG-PB docking<br>• Altered expression TIA-1 and G3BP1 | HeLa<br>SK-N-SH | McDonald et al. (2011)<br>Aulas et al. (2012) |
| | Mutant basal expression (R361S) | Following oxidative stress:<br>• Decreased SG number<br>• Reduced G3BP levels | Patients lymphoblasts | McDonald et al. (2011) |
| TIA-1 | Recombinant proteins (P362L, A381T, and E384K) | • Impaired mobility | In vitro | MacKenzie et al. (2017) |
| | Mutant overexpression (P362L, A381T, and E384K) | Following heat shock:<br>• Delayed SG disassembly<br>• Increased TDP-43-SG insolubility | HeLa | |
| | WT and mutant overexpression (P362L) | • Increased tau insolubility, misfolding, and incorporation into SGs | Tau−/− hippocampal WT | Vanderweyde et al. (2016) |
| VCP | Mutant basal expression (R155H) | • Ubiquitinated TDP-43 and VCP inclusions | Neurons, frontal lobe of FTLD patients | Gitcho et al. (2009) |
| | Mutant overexpression (R155H) | • TDP-43 cytoplasmic localization | U2-OS | Ju et al. (2009) |
| | Mutant overexpression (R155H) | • ER stress<br>• Increased ubiquitination<br>• Increased cell death<br>• Impaired proteasomal activity<br>• TDP-43 cytoplasmic relocalization | SH-SY5Y | Gitcho et al. (2009) |
| | Mutant overexpression (R155H and A232E) | • TDP-43 cytoplasmic localization | *Drosophila*<br>HEK293<br>Mouse cortical neurons | Ritson et al. (2010) |

(continued)

**Table 8.1** (continued)

| Protein | Expression | SG-related dysfunctions | Models | References |
|---|---|---|---|---|
| | Mutant overexpression (A232E) | • Accumulation of high molecular weight TDP-43 isoform that interacts with VCP<br>• TDP-43 cytoplasmic accumulation to TIA-1 positive SGs | Transgenic mouse model | Rodriguez-Ortiz et al. (2013) |
| | Mutant overexpression (A232E) | • TDP-43 cytoplasmic inclusions | SH-SY5Y | |
| | Depletion | • Synaptic and locomotive defects | *Drosophila* | Azuma et al. (2014) |
| TDP-43 | Mutant RRM-GFP overexpression (M337V) | • Formation of condensed droplet with fewer internal bubbles than WT<br>• Impairment of the liquid properties of the droplets | Hek293 | Schmidt and Rohatgi (2016) |
| | Recombinant C-terminal domain (A321G, Q331K A326P, and A321V) | • Abrogation of phase separation (A321G and Q331K)<br>• Decreased phase separation (M337V)<br>• Increased phase separation (A326P, A321V) | In vitro | Conicella et al. (2016) |
| | Mutant overexpression (G294A, A315T, Q331K, and Q343R) | Following oxidative stress:<br>• Increased stress-induced aggregation<br>• Increased insoluble protein aggregates | BE-M17<br>HEK 293 | Liu-Yesucevitz et al. (2010) |
| | Mutant overexpression (G294A. A315T, G348C, and N3920S) | Following osmotic stress:<br>• Formation of larger SGs | HEK 293 | Dewey et al. (2011) |
| | Basal expression | • Presence of TIA-1 and eIF3 in pathological TDP-43 inclusions | Spinal cord ALS patients<br>Frontal cortex FTLD-U patients | Liu-Yesucevitz et al. (2010) |
| | Depletion | • Impairment SG assembly and disassembly<br>• SG number and size reduction<br>• Impaired SG-PB docking<br>• Altered expression TIA-1 and G3BP1 | HeLa<br>SK-N-SH | McDonald et al. (2011)<br>Aulas et al. (2012) |

(continued)

**Table 8.1** (continued)

| Protein | Expression | SG-related dysfunctions | Models | References |
|---|---|---|---|---|
| | Mutant basal expression (R361S) | Following oxidative stress:<br>• Decreased SG number<br>• Reduced G3BP levels | Patients lymphoblasts | McDonald et al. (2011) |
| TIA-1 | Recombinant proteins (P362L, A381T, and E384K) | • Impaired mobility | In vitro | Mackenzie et al. (2017) |
| | Mutant overexpression (P362L, A381T, and E384K) | Following heat shock:<br>• Delayed SG disassembly<br>• Increased TDP-43-SG insolubility | HeLa | |
| | WT and mutant overexpression (P362L) | • Increased tau insolubility, misfolding, and incorporation into SGs | Tau−/− hippocampal WT | Vanderweyde et al. (2016) |
| VCP | Mutant basal expression (R155H) | • Ubiquitinated TDP-43 and VCP inclusions | Neurons, frontal lobe of FTLD patients | Gitcho et al. (2009) |
| | Mutant overexpression (R155H) | • TDP-43 cytoplasmic localization | U2-OS | Ju et al. (2009) |
| | Mutant overexpression (R155H) | • ER stress<br>• Increased ubiquitination<br>• Increased cell death<br>• Impaired proteasomal activity<br>• TDP-43 cytoplasmic relocalization | SH-SY5Y | Gitcho et al. (2009) |
| | Mutant overexpression (R155H and A232E) | • TDP-43 cytoplasmic localization | *Drosophila*<br>HEK293<br>Mouse cortical neurons | Ritson et al. (2010) |
| | Mutant overexpression (A232E) | • Accumulation of high molecular weight TDP-43 isoform that interacts with VCP<br>• TDP-43 cytoplasmic accumulation to TIA-1 positive SGs | Transgenic mouse model | Rodriguez-Ortiz et al. (2013) |
| | Mutant overexpression (A232E) | • TDP-43 cytoplasmic inclusions | SH-SY5Y | |
| | Depletion | • Synaptic and locomotive defects | *Drosophila* | Azuma et al. (2014) |
| Ubiquilin-2 | Mutant overexpression (P497S and P506T) | • Age-dependent aggregation in brain and spinal cord<br>• TDP-43 cytoplasmic inclusions<br>• Slowed protein degradation | Transgenic mouse model | Le et al. (2016) |

**Table 8.2** Summary of PB dysfunction potentially involved in neurodegenerative diseases

| Protein | Function | Potential PB dysfunctions | Evidence | References |
|---|---|---|---|---|
| 4E-T | Translational repressor | • Disrupt PB formation/maintenance<br>• Impair translational inhibition<br>• Impair RNA silencing and degradation | • Depletion leads to PB dissolution | Ayache et al. (2015) |
| ATG2 | Autophagy-related protein: autophagosome formation | • Impair PB dissolution | • Interact with PB proteins | Zheng et al. (2011)<br>Velikkakath et al. (2012) |
| Ago1/ Ago2 | miRNA/siRNA silencing | • Impair miRNA-mediated silencing | • Lack of siRNA binding domain leads to impaired Ago2 localization to PBs | Liu et al. (2005a, b) |
| C9orf72 | – | • Disrupt PB formation | • Localizes to PB | Maharjan et al. (2017) |
| Dcp1/ Dcp2 | Decapping enzyme/catalytic subunit of decapping enzyme | • Disrupt PB formation<br>• Impaired RNA silencing and degradation | • Dcp2 KO induces smaller PBs | Lykke-Andersen (2002)<br>Ingelfinger et al. (2002), van Dijk et al. (2002) |
| DDX6 | Decapping complex<br>CPEB1 translation-repression complex | • Disrupt PB formation/maintenance<br>• Impair RNA silencing and degradation<br>• Impair translational inhibition impairment<br>• Impair miRNA-mediated silencing impairment | • Deletion impairs translational repression<br>• Repression stimulates cellular translation<br>• Participates in miRNA-mediated silencing<br>• Participates in deadenylation<br>• Depletion leads to PB dissolution | Chu and Rana (2006)<br>Mathys et al. (2014)<br>Holmes et al. (2004)<br>Ayache et al. (2015) |
| Dhh1p | Decapping activator | • Disrupt PB formation<br>• Impair RNA degradation | • Modulate PB composition and assembly | Teixeira and Parker (2007) |
| Edc3 | Decapping activator | • Disrupt PB formation<br>• Impair RNA degradation | • Deletion of LCDs impairs PB assembly | Decker et al. (2007)<br>Reijns et al. (2008)<br>Reijns et al. (2008) |
| G3BP1 | RBP | • Impair docking with SGs | • Depletion reduces docking and increases PB number | Aulas et al. (2015) |

(continued)

**Table 8.2** (continued)

| Protein | Function | Potential PB dysfunctions | Evidence | References |
|---|---|---|---|---|
| Ge-1 | Decapping cofactor | • Impair RNA silencing and degradation<br>• Impair translational inhibition<br>• Disrupt PB formation | • Overexpression causes accumulation of deadenylated mRNA<br>• Enhances decapping in vitro<br>• Impaired DCP1/2 localization to PBs | Yu et al. (2005)<br>Fenger-Gron et al. (2005) |
| GW182 | miRNA pathway | • Disrupt PB formation<br>• Impair miRNA-mediated silencing | • Depletion leads to PB dissolution and impaired miRNA-mediated silencing | Eystathioy et al. (2003) |
| Lsm4 | Lsm-Pat1 complex | • Disrupt PB formation<br>• Impair RNA degradation | • Deletion of LCDs impairs PB assembly | Decker et al. (2007)<br>Reijns et al. (2008) |
| MOV | miRNA/siRNA silencing | • Impair miRNA-mediated silencing | • Participates in the cleavage of miRNA precursors<br>• Participates in the incorporation of miRNAs into Ago2 complexes, and cleavage of complementary target RNA | Meister et al. (2005) |
| Pat1p | Decapping activator and translational repressor | • Disrupt PB formation<br>• Impair RNA degradation<br>• Translational inhibition | • Modulate PB composition and assembly | Teixeira and Parker (2007) |
| Staufen | RBP | • Defect in the RNA localization to PB<br>• Impair PB formation/maintenance<br>• Impair RNA silencing and degradation | • Localizes to PBs | Anderson and Kedersha (2006) |
| TIA-1 | RBP | • Defect in the RNA localization to PBs<br>• Impair PB formation/maintenance<br>• Impair RNA silencing and degradation | • Localizes to PBs | David Gerecht et al. (2010) |
| TDP-43 | RBP | • Defect in the RNA localization to PBs<br>• Impair PB formation/maintenance<br>• Impair RNA silencing and degradation | • Depletion reduces docking and increases PB number | Aulas et al. (2015) |

(continued)

**Table 8.2** (continued)

| Protein | Function | Potential PB dysfunctions | Evidence | References |
|---|---|---|---|---|
| TRIM21 | E3 ubiquitin ligase | • Impair PB dissolution | • Localizes to PBs | Zheng et al. (2011) |
| UPSP4 | Deubiquitinating enzyme | • Impair PB dissolution | • Localizes to PBs | Zheng et al. (2011) |
| Xrn1 | 5′ to 3′ exonuclease | • Impair RNA degradation | • Repression alters general cellular mRNA | Ingelfinger et al. (2002) Moon et al. (2015) |

## 8.5.1 *Stress Granules and ALS*

TDP-43 is an RBP and a major component of the cytoplasmic inclusions observed in ALS patient neurons (Arai et al. 2006; Neumann et al. 2006). TDP-43 normally localizes to the nucleus, where it plays a key role in RNA metabolism (Lagier-Tourenne et al. 2010; Da Cruz and Cleveland 2011). However in ALS, TDP-43 is depleted from the nucleus and accumulates in large cytoplasmic aggregates. FUS (fused in sarcoma; also referred to as TLS, translocated in sarcoma) is another RBP associated with ALS and is localized in protein inclusions in some familial cases. To date, more than 40 RBPs have been associated with ALS, raising the intriguing possibility that misregulation of RNA processing contributes greatly to the pathology (Kapeli et al. 2017; Chia et al. 2018; Maurel et al. 2018). Indeed, RNA granules have been suggested to be involved in ALS since cytoplasmic TDP-43 inclusions are reported to label with the SG markers TIA-1 and TIAR in post-mortem samples from ALS and FTD patients (Liu-Yesucevitz et al. 2010; Wolozin 2012). This has been furthered by similar findings in cultured cells and in primary neurons responding to cellular stress (Ayala et al. 2008; Dormann et al. 2010). Indeed, overexpression studies of ALS-related TDP-43 and FUS mutant proteins demonstrate enhanced cytoplasmic aggregation following stress compared to their physiological counterparts, ultimately triggering an increased association with SGs (Bosco et al. 2010; Liu-Yesucevitz et al. 2010; Dewey et al. 2011). Interestingly, cells expressing mutant TDP-43 appear to either exhibit attenuated SG formation or are prone to abnormal SG assembly, while FUS mutants are more prone to localize to SGs and increase the size and number of granules (Bosco et al. 2010; Gal et al. 2010; Liu-Yesucevitz et al. 2010; McDonald et al. 2011; Sun et al. 2011). Thus, TDP-43 SG-related pathology has been suggested to be linked to the impairment of SG assembly, while FUS is linked to abnormal interactions with SGs leading to SG dysregulation (Bosco et al. 2010; Sun et al. 2011; Aulas and Vande Velde 2015).

Interestingly, some of the most aggressive cases of ALS are associated with two FUS mutations. These mutations, namely, P525L and R495X, are linked to rare and

**Table 8.3** Summary of transport granule dysfunction caused by neurodegenerative disease-related proteins

| Protein | Expression | Transport granule-related dysfunctions | Models | References |
|---|---|---|---|---|
| C9orf72 | RNA foci overexpression | • TDP-43 sequestration to soma | Rat cortical neurons | Ishiguro et al. (2016) |
| TDP-43 | Mutant overexpression (A315T and Q343R) | • Formation of larger granules sparingly distributed in the processes<br>• Reduced mobility<br>• Reduced granule localization to dendrites upon depolarization | Rat hippocampal neurons | Liu-Yesucevitz et al. (2014) |
| | Mutant basal expression (G298S) | • Insoluble aggregates | IPS-derived motor neurons from fALS patient | Liu-Yesucevitz et al. (2014) |
| | Mutant basal expression (M337V) | • Higher levels of soluble and detergent resistant TDP-43 | IPS-derived motor neurons from ALS patient | Alami et al. (2014) |
| | Mutants overexpression (M337V and A315T) | • Impaired retrograde movement of granules<br>• Cytoplasmic accumulation of TDP-43 and deficiency at the NMJ | *Drosophila* | Alami et al. (2014) |
| | Mutant overexpression (M337V and A315T) | • Increased retrograde movement<br>• Defective Neurofilament-L axonal trafficking | Mouse cortical neurons | Alami et al. (2014) |
| | Basal expression | • Accumulation of MAP1B in cell body | Neurons of the lumbar spinal cord and hippocampus of ALS patients with TDP-43 pathology | Coyne et al. (2014) |
| | Mutant overexpression (M337V) | • Impaired G4-containing mRNA transport | Rat cortical neurons | Ishiguro et al. (2016) |
| | – | • Impaired neuronal RNA metabolism<br>• Impaired synaptic formation<br>• Impaired neurotransmitter release | – | Proposed from Wang et al. (2008a, b) |
| FUS | Mutant overexpression (R521C, R495X, and P525L) | • Spontaneous granule formation<br>• Translation of mRNA that should be targeted to the dendrites or the axon in the soma | NIH/3T3 | Yasuda et al. (2013) |

(continued)

**Table 8.3** (continued)

| Protein | Expression | Transport granule-related dysfunctions | Models | References |
|---|---|---|---|---|
| | – | • Impaired targeting of RNA to axons and dendrites<br>• Impaired synaptic localization<br>• Impaired synaptic changes upon depolarization | – | Proposed from Belly et al. (2005) and Fujii et al. (2005) |
| G3BP1/IMP1 | Wild-type overexpression | • Shift between tau HMW and LMW forms | PC12 stable cell lines | Moschner et al. (2014) |
| | Deletion of 2 LCDs | • Impaired granule formation<br>• Inhibition of sprouting | | |
| | – | • Impaired tau mRNA transportation | – | Proposed from Moschner et al. (2014) |
| APP/tau | NFTs/plaques | • Sequestration of neuronal granules | Brain from AD patients | Proposed from Ginsberg et al. (1999, 1997) |
| APP | Recombinant proteins | • Sequestration of Staufen-1 neuronal granules | In vitro | Proposed from Yu et al. (2015) |

severe forms of juvenile ALS (Conte et al. 2012; Leblond et al. 2016). In these cases, the mutant FUS protein lacks the nuclear localization signal resulting in increased FUS cytoplasmic localization, which is accentuated by environmental stress (Dormann et al. 2010; Zhang and Chook 2012). This suggests that ALS is closely linked to abnormal RBP cytoplasmic localization. Furthermore, in *C. elegans*, FUS mutant mRNPs sequester other RBPs, including Staufen 1, SMN, and TIAR-1, suggesting an impairment of RNA granules with the expression of these mutants (Murakami et al. 2015; Patel et al. 2015). Finally, in vitro, some FUS mutants form membrane-less droplets that transition into a fibrillary solid form at a more rapid rate than the wild-type protein (Patel et al. 2015). If we link these very interesting observations to SGs, FUS mutations are involved in ALS pathobiology through either the transformation of SGs into pathological aggregates or an increase in cytoplasmic localization, thus promoting aggregation and the possible sequestration of specific (regular partners) and non-specific (other RBPs with LCDs) RBPs. Note that these possibilities are not mutually exclusive.

As, previously mentioned, mutations in other RBPs are causative of ALS. Recently, mutations in the LCD of TIA-1, a key SG nucleating protein, have been identified (Mackenzie et al. 2017). The newly described mutations delay SG disassembly and enhance immobile SGs. Very interestingly, TDP-43 recruited into these dysfunctional SGs becomes insoluble, reminiscent of the inclusions observed in patients (Mackenzie et al. 2017). This mutation is therefore suggested to be directly causative of the inclusions observed in patients although TIA-1 itself is curiously not observed in the inclusions (Hirsch-Reinshagen et al. 2017).

Mutations in hnRNP A1 and hnRNP A2/B1 are associated with familial and sporadic forms of ALS as well as the complex disorder inclusion body myopathy with frontotemporal dementia, Paget's disease of bone, with ALS (IBMPFD/ALS) (Kim et al. 2013a). In cellular models, mutant proteins are recruited into SGs at a higher rate than their wild-type counterparts (Kim et al. 2013a). This observation led to the conclusion that the ALS-related mutations induce aggregation-prone species. Indeed, these mutations have been further suggested to induce protein misfolding and/or fibril formation (Sawaya et al. 2007; Kim et al. 2013a). This suggests that SG dynamics are impaired by mutations in hnRNP A1 and hnRNP A2/B1.

Recently, a hexanucleotide repeat expansion in a non-coding region of the *C9ORF72* gene has been described in the majority of ALS/FTD cases (Bigio 2011; Dejesus-Hernandez et al. 2011; Renton et al. 2011; Ratti et al. 2012; Simon-Sanchez et al. 2012). Three pathogenic mechanisms are proposed for ALS/FTD caused by *C9ORF72* expansions. The first mechanism involves the decreased expression (haploinsufficiency) of C9ORF72 protein. This is supported by the observation that expression of *C9ORF72* is decreased in brain tissues from ALS and FTD patients (DeJesus-Hernandez et al. 2011). However, genetic deletion of *C9ORF72* in mice yields macrophage dysfunction, not motor neuron degeneration (Koppers et al. 2015; O'Rourke et al. 2016). Therefore, the contribution of the loss of C9ORF72 function is currently inconclusive in ALS/FTD pathogenesis. Relevant here, the *C9ORF72* protein is reported as required for SG formation in mammalian cells (Maharjan et al. 2017). Furthermore, the protein and transcript levels of key SG proteins TIA-1, HuR, and G3BP1 are reduced in C9ORF72-null cells (Maharjan et al. 2017), providing an explanation for its role in SG formation.

The second mechanism suggests that neurodegeneration is due to the abnormal aggregation of intranuclear RBPs with RNA containing the expanded repeat, leading to RNA toxicity (Mizielinska et al. 2013; Zu et al. 2013). The GGGGCC sequence motif is predicted to bind several RBPs, including hnRNP A1, hnRNP A2/B1, HuR, and FUS, all of which are SG-related proteins (Cartegni et al. 1996; Smith et al. 2006; Sofola et al. 2007; Mori et al. 2013a). hnRNP A2/B1 dysregulation via sequestration into GC-rich RNA repeats is already known to be associated with Fragile X syndrome (Sofola et al. 2007), and hnRNP A3, another protein of the hnRNP A/B family, is localized to *C9ORF72* RNA foci in patients bearing an expanded *C9ORF72* allele (Mori et al. 2013a). Moreover, RNA foci in iPSC-derived motor neurons generated from *C9ORF72*-expanded ALS patients also colabel with hnRNP A1 (Sareen et al. 2013). Therefore, hnRNP A2/B1, HuR, and FUS could also be sequestered within *C9ORF72* RNA foci, potentially impairing SG dynamics.

Finally, the third mechanism suggests that neuronal degeneration is due to the synthesis and aggregation of dipeptide repeat proteins (DPRs) through repeat-associated non-ATG translation (Ash et al. 2013; Mori et al. 2013b). These DPRs include those encoded by sense transcripts (polyGA and polyGR), by antisense transcripts (polyPA and polyPR), and by both (polyGP). Interestingly, in mammalian cells and *Drosophila*, expression of polyGR and polyPR results in interaction with multiple RBPs including the SG components TDP-43, FUS, hnRNP A1 and Ataxin-2 and the formation of insoluble inclusions (Lee et al. 2013, 2016; Lin et al. 2016; Lopez-Gonzalez et al. 2016). Moreover, expression of polyGR and polyPR in U2OS cells induces spontaneous assembly of SGs with reduced dynamism (Lee et al. 2016). This impairment is suggested to be due to the DPRs directly binding and altering the physical properties of hnRNP A1 and TIA-1, rendering them susceptible to phase separation at a lower concentration (Lee et al. 2016). Altogether, these studies show that DPRs rich in arginine impair SG dynamics, without impacting SG formation, and it is presumably due either to DPR interactions with the LCDs of relevant RBPs and/or components of the nucleocytoplasmic transport machinery that traps them in SGs, thereby perturbing nuclear-cytoplasmic movement of RBPs (Lee et al. 2016; Zhang et al. 2018).

The molecular mechanisms by which ALS-related RBPs contribute to disease pathogenesis is not yet fully understood; however, a clear link to SGs has been established. Two main hypotheses, which are not mutually exclusive, may possibly synergize to varying degrees, and have been proposed to explain the role of RBPs and SG biology in ALS/FTD, are gain-of-function and loss-of-function hypotheses. The gain-of-function hypothesis suggests that SG-associated proteins linked to ALS gain pathological functions that impede SG physiological functions and RNA homeostasis (Lee et al. 2011). One example contributing to this hypothesis is the fact that insoluble FUS and TDP-43 mutant proteins form cytoplasmic aggregates which alter SG dynamics (Parker et al. 2012), resulting in SG persistence and increased cell death (Bosco et al. 2010; Dormann et al. 2010; Liu-Yesucevitz et al. 2010, 2011; Dewey et al. 2011; Wolozin 2012). Thus, it is suggested that SG persistence (i.e., failure to resolve) may lead to the sequestration of a number of mRNAs and subsequent perturbation of RNA metabolism and eventual cellular death. Some have also suggested that a persistence of SGs, after stress removal, is due to a gain of function in important regulators and can generate the pathological inclusions observed in patient neurons. Indeed, the fact that in vitro, FUS liquid-like droplets can strikingly transition into a fibril state provides a clue as to how SGs could potentially become pathological inclusions (Molliex et al. 2015). However, to date, it remains to be demonstrated that this dramatic transition occurs in vivo.

The loss-of-function hypothesis, on the other hand, posits that ALS-related proteins lose their essential functions, thus affecting SG physiology. Indeed, TDP-43 has been repeatedly demonstrated as an important regulator of SG dynamics since its depletion reduces SG size and alters SG morphology (McDonald et al. 2011; Aulas et al. 2012). Also, ALS-related RBPs have been implicated in numerous RNA processes including pre-mRNA splicing, RNA stability, and transcriptional regulation (Nussbacher et al. 2015). This has led to the suggestion that the loss of function of ALS-linked RBPs negatively impacts essential SG proteins via defective mRNA processing. Indeed, TDP-43 regulates the expression of key SG components, such as G3BP1 and TIA-1 (McDonald et al. 2011). TDP-43 depletion causes a loss of

G3BP1 protein and mRNA and correlates with dysfunctional SG dynamics (and docking with PBs) which can be rescued by restoration of G3BP1 levels (Aulas et al. 2012, 2015). The exact molecular mechanism governing this regulation remains to be elucidated. Several ALS-related genes are linked to protein homeostasis and SG clearance. Indeed, it is hypothesized that loss-of-function mutations in *VCP* are linked to ALS and IBMPFD (Johnson et al. 2010; Shaw 2010; Miller 2012). The VCP mutation R155H is associated with increased ubiquitination, ER stress, impaired proteasomal activity, and increased cell death (Gitcho et al. 2009), supporting a loss of function for this mutation. Interestingly, VCP mutations are clustered within a region predicted to be necessary for protein-protein interaction (Majcher et al. 2015). Moreover, expression of VCP mutations induces TDP-43 cytoplasmic localization in SK-N-SH cells and in *Drosophila* (Gitcho et al. 2009; Ritson et al. 2010). Furthermore, in flies, the overexpression of the wild-type VCP ortholog (*ter94*) rescues the synaptic and locomotive defects in *caz* mutant flies (FUS-null), while loss-of-function *ter94* mutants do not (Azuma et al. 2014). Collectively, these results suggest that VCP loss of function is relevant to TDP-43 and FUS function and possibly SG dynamics.

Mutations in *UBQLN2*, a gene relevant to the regulation of the UPS and autophagy pathways which are themselves linked to SG clearance, are causative for ALS (Deng et al. 2011; Williams et al. 2012; Zhang et al. 2014b). Interestingly, Ubiquilin-2 interacts with VCP in order to send substrates to the proteasome (Lim et al. 2009; Brown and Kaganovich 2016; Le et al. 2016; Osaka et al. 2016). The ALS-related Ubiquilin-2 mutations P497S and P506T induce an age-dependent aggregation of the protein in the brain and spinal cord of transgenic mice, as well as motor deficits and TDP-43 cytoplasmic inclusions (Le et al. 2016). Interestingly, ALS-linked mutations have longer half-lives than the wild-type protein and show a slower degradation of a substrate (Myc) due to defective proteasome binding (Chang and Monteiro 2015). Altogether, these results suggest that Ubiquilin-2 ALS-linked mutations are associated with a loss of function and thus may impact SG clearance.

### 8.5.2 *Stress Granules and Alzheimer's Disease and Tauopathies*

Tau is primarily known as an axonal microtubule binding protein (Vanderweyde et al. 2016), yet its role in the somatodentric compartment is not fully understood. In the cytosol, tau can bind RNA with a preference for tRNAs (Zhang et al. 2017). Very interestingly, RNA stimulates tau aggregation, and a number of transcripts are found in NFTs (Kampers et al. 1996; Ginsberg et al. 1997; Dinkel et al. 2015). Recently, similar to FUS and hnRNP A1, tau has been shown to phase separate in vitro (Ambadipudi et al. 2017; Wegmann et al. 2018). Phosphorylated, FTD-related mutants (P301L, P301S, and A152T) and high molecular weight soluble phospho-tau from AD patients were found to also phase separate and transition over time toward aggregates that can seed tau aggregation, as is seen in AD. Moreover, tau interacts with RBPs in physiological conditions and these interactions are influenced by the expression of the FTD-related mutation P301L (Gunawardana et al. 2015). Taken together, these findings suggest that tau dysfunction may influence mRNPs and RNA granules and vice versa.

In the soma, tau facilitates SG formation in a TIA-1-dependent manner (Vanderweyde et al. 2016). Tau overexpression in mouse hippocampal neurons accelerates the rate of SG formation, suggesting that it is involved in SG dynamics (Vanderweyde et al. 2016). Interestingly, cells expressing the FTD mutant tau P301L assemble SGs faster and have larger granules compared to cells expressing its wild-type counterpart. Thus, these data imply that dysfunction in SG dynamics is potentially also involved in the pathogenesis of tauopathies. Interestingly, tau interacts with both SG nucleating proteins G3BP1 and TIA-1 (Atlas et al. 2007; Moschner et al. 2014; Vanderweyde et al. 2016). While G3BP1 can interact with tau mRNA in transport granules (to be discussed later), TIA-1 and tau modulate each other's function. Specifically, tau depletion from mouse brain eliminates the interaction of TIA-1 with proteins involved in RNA metabolism and SG-related proteins PABPC1 and SYNCRIP (Vanderweyde et al. 2016). Therefore, in AD and other tauopathies, loss of tau function may impair the formation of functional SGs, rendering neurons more susceptible to stress and cellular death. Moreover, it is important to note that TIA-1 overexpression increases tau insolubility through the modulation of tau misfolding (Vanderweyde et al. 2016). Indeed, it is suggested that this change of physical state may be key to the formation of the NFTs which characterize tauopathies. Recently, tau has been demonstrated to phase separate, similar to FUS and hnRNP A1 (Ambadipudi et al. 2017). Thus, tau has the same capacity as FUS, either upon mutation, with aging, or via interaction with RBPs, to transition from droplet-like granules into fibrils.

In the P301L tau transgenic mouse model, SG number and TIA-1 protein levels increase with the disease course (Vanderweyde et al. 2012). While TIA-1-positive granules are co-labeled with tau in this transgenic mouse model, as well as in AD and FTD patient brains, G3BP1 aggregation is not found in neurons bearing tau aggregates (Vanderweyde et al. 2012). This differential pattern suggests different mechanisms for TIA-1 and G3BP1 granules. A possibility is that in these tauopathies, the increased levels of TIA-1 protein leads to the formation of defective SGs that become trapped in tau aggregates, possibly as a means to provide protection from these toxic aggregates. Since tau modulates the TIA-1 interactome, it is also possible that under stress, tau impairs the interaction of TIA-1 with G3BP1, leading to "incomplete" SG formation. In this scenario, neurons which can form functional SGs, containing TIA-1 and G3BP1, do not have tau aggregation. Another hypothesis is that tau aggregates sequester TIA-1 mRNP complexes before they can interact with G3BP1 and form SGs, while cells without tau aggregates can properly form SGs. In both cases, the constitutive sequestration of functional proteins into tau aggregates or SGs is detrimental for the cell. Interestingly, tau can also be secreted. This secreted tau can be internalized and recruited to SGs in a TIA-1-dependent manner, suggesting that in tauopathies, SG-related pathobiology is contingent on TIA-1 (Brunello et al. 2016).

In AD, microglia remove Aβ plaques via phagocytosis (Ghosh and Geahlen 2015). However, chronic stress induced by Aβ induces the formation of large SGs positive for both TIA-1 and G3BP1 which persist even after the removal of Aβ (Ghosh and Geahlen 2015; Brunello et al. 2016). Long-lasting SGs are associated with sensitization upon encounter with a second stress exposure; thus, Aβ stress presumably renders microglia less resistant to a second stress. These defective SGs sequester important microglial factors such as SYK, a tyrosine kinase essential for both microglial phagocytosis and inflammatory responses (Crowley et al. 1997; Kiefer et al. 1998; Ghosh and Geahlen 2015). SYK sequestration into SGs impairs phagocytosis and drives chronic generation of both reactive oxygen and nitrogen species (Ghosh and Geahlen 2015). This chronic inflammatory response generates constitutive stress conditions that leads to the maintenance of SGs in microglia, which can be deleterious for neighboring cells. Interestingly, SYK is localized to G3BP1-positive puncta in the microglia of mild and severe AD cases, strongly suggesting that the latter mechanism may play a role in disease (Ghosh and Geahlen 2015).

### *8.5.3 Processing Bodies and ALS*

So far, little is known about PB dysfunction in neurodegenerative disease. This is likely due to the fact that no genetic mutations have yet been directly linked to these diseases. However, functions related to PBs have been associated to neuronal loss. Indeed, genetic deletion of key proteins of the RNA silencing pathway results in PB dysfunction as well as neurodegeneration, impaired nerve regeneration, and motor dysfunction (Eulalio et al. 2007c; Haramati et al. 2010; Chen and Wichterle 2012; Wu et al. 2012). These findings suggest that PB-specific functions, such as decay, storage, or miRNA processing, could be impaired in these diseases. Moreover, RNA granules are recognized to interact with each other; therefore, dysfunction in one granule subtype could potentially impair the function of the other granules (Anderson and Kedersha 2009; Decker and Parker 2012). PB dysfunction in neurodegenerative disorders could also be related to their interaction with SGs or transport granules.

Transport granules, SGs and PBs are intrinsically related. First, they share a number of essential components (Anderson and Kedersha 2006). For example, TIA-1 is shared by all three of the granule subtypes. Thus, it is expected that the TIA-1 mutations reported in ALS could affect more than just SGs. If we extrapolate from the observations of TIA-1 mutations on SG dynamics, one may expect that PBs could be similarly impaired in ALS. As PBs are essential to physiological function, this would be deleterious in basal conditions. Immobile and insoluble granules do not efficiently exchange RNAs or proteins with the cytosol. Thus, it is hypothesized that mRNA decay/storage may be impaired in ALS; however, this has not yet been directly demonstrated. Recent evidence indicates that mRNAs are sent to PBs in a context-dependent manner with mitochondrial-related mRNAs being particularly enriched in PBs under stress, while they are excluded under normal conditions (Wang et al. 2018). This specific composition suggests that mitochondrial dysfunction is anticipated to be associated with mutant TIA-1-related PB dysfunction (Hubstenberger et al. 2017; Wang et al. 2018). Interestingly, mitochondrial dysfunction is reported in ALS (Manfredi and Xu 2005; Pickles et al. 2013). Intriguingly, it has also been reported that PBs associate with mitochondria; however, the nature and purpose of this association remain to be determined (Huang et al. 2011).

As discussed previously, PBs and SGs physically interact during docking. So far, little is known about the functional relevance of this process but it is suggested to be necessary for the degradation of specific mRNAs (Anderson and Kedersha 2006). Depletion of TDP-43 reduces the frequency of docking events, a process which is governed by the size of individual SGs (Anderson and Kedersha 2006; Aulas et al. 2015). TDP-43 depletion also causes an increased number of PBs, suggesting an important link between TDP-43 function and PB formation (Aulas et al. 2015). Finally, transport granules and PBs have also been reported to interact (Zeitelhofer et al. 2008), although just as with SGs, the nature of this interaction is not understood.

PBs are involved in miRNA synthesis. Thus, it is reasonable to hypothesize that impaired PB function may also lead to significant changes in miRNA expression and, by extension, transcript stability. Furthermore, miRNAs are considered to be essential to the control of localized translation at the synapse, which is associated with synaptic defects that can progress into synaptic loss. Indeed, miRNA profile changes are reported in ALS (De Felice et al. 2012; Campos-Melo et al. 2013; Pegoraro et al. 2017; Rinchetti et al. 2018), and loss of the specialized synapse, the neuromuscular junction, is one of the first features of ALS pathology.

### *8.5.4 Processing Bodies and Alzheimer's Disease and Tauopathies*

Some of the same PB-related processes identified as possible contributors to ALS pathogenesis may also be relevant to AD. Namely, miRNA profile changes have long been studied in AD, some of which can be directly linked to NFT and Aβ plaque formation. For example, the β-secretase BACE1 which triggers the release of the Aβ peptide is regulated by select miRNAs. Specifically, miR-107 is decreased in the temporal cortex of AD patients (Wang et al. 2008b). Similarly, the miR-29 family is decreased in the anterior temporal cortex of sporadic AD patients (Hebert et al. 2008). Decreased expression of these miRNAs is associated with increased levels of BACE1 and, consequently, increased Aβ peptide levels (Hebert et al. 2008; Wang et al. 2008b; Nelson and Wang 2010). Interestingly, miR-29c overexpression in mouse brain is associated with memory improvement, while depletion of miR-29a is associated with increased neuronal death (Hebert et al. 2008; Wang et al. 2011; Roshan et al. 2014; Yang et al. 2015). Similarly, reduced levels of miR-188-3p are observed in the 5XFAD mouse model. Rescue of miR-188-3p levels via overexpression diminishes BACE1 levels and Aβ peptide burden. This correlated with reduced neuroinflammation and improved synaptic plasticity and memory (Zhang et al. 2014a).

Aβ plaques are generated by the oligomerization of Aβ peptides, which are generated by the consecutive cleavage of the amyloid precursor protein (APP) by the β-secretase BACE1 and the γ-secretase. Several miRNAs regulate APP synthesis and thus the generation of Aβ plaques (Basavaraju and de Lencastre 2016). For example, miR-101 and miR-16 both target APP mRNA and reduce APP production, respectively, in rat hippocampal neurons expressing APP and senescence-accelerated mouse prone 8 (SAMP8) (Vilardo et al. 2010; Liu et al. 2012; Zhang et al. 2015). miRNAs can also indirectly influence Aβ formation. Depletion of miR-137, miR-181c, miR-9, and miR-29a/b-1 causes an increase in the expression of serine palmitoyltransferase, the first rate-limiting enzyme of ceramide synthesis. Ceramide levels are elevated in sporadic AD patients and are suggested to be an important risk factor (Geekiyanage and Chan 2011; Schonrock et al. 2012). An increase of this enzyme is also observed in the AD mouse model TgCRND8 which also features Aβ plaques (Geekiyanage and Chan 2011).

Finally, miRNAs have been shown to regulate tau and its aggregation. Specifically, low levels of the miR-132/122 cluster are reported in tauopathies (Wanet et al. 2012; Smith et al. 2015) and are associated with the suppression of tau expression (Smith et al. 2015). Interestingly, treatment with miR-132 mimics partially restores tau levels and improves cognitive function in the 3×Tg-AD mouse model (Smith et al. 2015). In addition, overexpression of both miR-125b and miR-138 increases tau phosphorylation and, thus, influences NFT formation (Banzhaf-Strathmann et al. 2014; Wang et al. 2015; Yin et al. 2015). Collectively, these results, coupled with the fact that it remains unclear why or how miRNA levels change in AD, demonstrate that miRNA levels can influence AD pathobiology and thereby suggest that PB dysfunction may be associated with disease. Indeed, PB aggregation is associated with increased sequestration of certain RBPs such as TIA-1 (and its interacting partners). This could contribute to an impairment of PB functions and miRNA synthesis. Interestingly, tauopathies are highly associated with TIA-1 loss of function (Vanderweyde et al. 2012, 2016). Lastly, it remains unclear why selected miRNAs are upregulated in the disease and others are not. It is possible that RBPs which facilitate the incorporation of certain families of pre-miRNAs into the RISC complex or the presentation of mRNAs to the miRNAs are sequestered in aggregates, thereby interfering with miRNA-mediated silencing. More studies are needed in order to fully comprehend the role of PBs in AD.

### 8.5.5 Transport RNA Granules in ALS

Most studies have focused on TDP-43 functions in RNA processing. However, TDP-43 is also involved in a number of neuronal mechanisms, including the transport of mRNA from the soma to the dendrites (Sephton et al. 2011; Alami et al. 2014). In primary neurons, TDP-43 is localized in neuronal RNA granules that also contain the SG protein TIA-1; these granules are present in the soma and the dendrites. In contrast, TDP-43 granules containing both TIA-1 and G3BP1 are generally confined to the soma. Interestingly, granules containing TIA-1/G3BP1/TDP-43 are larger, supporting previous work that G3BP1 is responsible for the assembly of SGs (McDonald et al. 2011; Aulas et al. 2012). TDP-43-enriched granules do not colocalize with the PB marker DCP1a but, instead, are localized at the proximity of the granules (Liu-Yesucevitz et al. 2014). This closeness suggests that TDP-43-positive transport granules and PBs interact, similar to SGs and PBs, and thus may exchange mRNA components. Thus, PB functions are potentially essential to transport granules. The ALS-related TDP-43 mutations A315T and Q343R negatively impact the size and distribution of transport granules. Specifically, mutant TDP-43 induced larger RNA granules which were more sparingly distributed in neuronal processes (Liu-Yesucevitz et al. 2014). These larger granules have reduced mobility compared to their wild-type counterparts, suggesting that ALS-related mutations impair transport granule function (Liu-Yesucevitz et al. 2014). This has been confirmed, as the TDP-43 mutations examined both reduced RNA granule trafficking and TDP-43 localization to dendritic arbors following neuronal depolarization (Wang et al. 2008a; Liu-Yesucevitz et al. 2014). These observations suggest that TDP-43 dysfunction stems from the acquisition of aggregation-prone properties either from mutations or from cytoplasmic localization (Conicella et al. 2016). This would impair TDP-43 granule mobility and disassembly, leading to synaptic deficiencies from loss of transport granule function (Dewey et al. 2012; Fallini et al. 2012; Pascual et al. 2012; Wolozin 2012; Alami et al. 2014). Moreover, in rat brain, TDP-43 mRNPs target RNA metabolism, synaptic formation, and neurotransmitter-related transcripts (Sephton et al. 2011). This includes neuroligins and neurexins which are essential for synaptic connectivity and are linked to cognitive dysfunction (Südhof 2008). Therefore, impairing TDP-43-bearing transport granules is deleterious for neuronal processes. In ALS, it is suggested that defects in synaptic processes happen first. Deregulation of RNA metabolism at the synapse may become so impaired that it results in massive synaptic degeneration that may precede clinical symptoms by years.

FUS is also associated with neuronal transport granules and synaptic localization upon depolarization (Belly et al. 2005; Fujii et al. 2005). Indeed, in mouse hippocampal neurons, FUS localizes to dendrites. After activation of mGluR5, FUS translocates to excitatory dendritic spines. Functionally, FUS is reportedly transported to spines in order to control the localization and anchoring of mRNAs at the synapse (Sephton et al. 2014). Therefore, depolarization induces FUS transport in order to properly drive the translational changes required at the spines (Belly et al. 2005; Fujii et al. 2005). The necessity of FUS at the spines is demonstrated by the abnormal spine morphology observed in FUS knockout mice (Fujii et al. 2005). FUS also associates with the tumor-suppressor protein adenomatous polyposis coli (APC) in APC-RNPs which function to target mRNAs to cell protrusions and is required for efficient localized translation at these sites (Yasuda et al. 2013). However, ALS-associated FUS mutants R521C, R495X, and P525L enhance protein recruitment to APC-RNP complexes, generating spontaneous granules and causing bound transcripts to be translated in the cytoplasm rather than at the protrusions (Yasuda et al. 2013). In neurons, loss of the correct localization for translation impairs dendritic spine function resulting in neurodegeneration. This suggests that FUS mutants in ALS are associated with a loss of function with regards to transport granules. mRNAs translated at the wrong place can involve two non-mutually exclusive pathogenic consequences: the incorrectly translated proteins will not be able to exert their functions at their target location and they may exert their activity at the site where it has been translated. If the site is not appropriate, then these mistranslated peptides can potentially confer a gain of toxicity. Both scenarios can be deleterious for neurons.

### *8.5.6 Transport Granules and Alzheimer's Disease*

The axonal localization of tau mRNA to the proximal end of the axon is essential to promote microtubule assembly (Elie et al. 2015; Kadavath et al. 2015). As discussed earlier, tau is highly regulated by G3BP1 (Moschner et al. 2014). G3BP1 is implicated in the transport of tau mRNA in G3BP1-IMP1-positive transport granules (Atlas et al. 2007; Moschner et al. 2014). The formation of G3BP1-IMP1 granules is associated with a shift in the ratio of tau isoforms, increasing the high molecular weight (HMW) forms compared to the low molecular weight (LMW) forms at both the protein and mRNA levels (Moschner et al. 2014). Interestingly, high levels of HMW tau are detected in the CSF of AD patients, as well as in post-mortem AD brains (Takeda et al. 2015, 2016). Moreover, these HMW forms can be absorbed by neurons and seed aggregates, supporting their importance in AD pathogenesis (Takeda et al. 2015).

The relationship of transport granules to AD pathogenesis is also demonstrated by their composition. IMP1 granules have a distinct composition, compared to other neuronal RNA granules (Jonson et al. 2007). Instead of being principally enriched in RBPs, these granules are rich in factors involved in protein secretion, including APP (Jonson et al. 2007). Very interestingly, dysfunction in IMP1 granules may impact both AD-related proteins Aβ and tau. Functionally, G3BP1-IMP1 granules are involved in neuronal sprouting in a G3BP1-dependent manner. G3BP1 contains one RRM and four LCDs (Moschner et al. 2014; Reineke et al. 2015). Removal of two LCDs impairs granule formation and abrogates sprouting, even after treatment with nerve growth factor (NGF) (Moschner et al. 2014). In AD, a massive increase in somatodendritic sprouting is noted: hence, an increase in G3BP1-IMP1 granules is potentially relevant in AD.

Finally, both NFTs and plaques contain transcripts coding for the kainate receptors which are essential for synaptic activity (Ginsberg et al. 1997, 1999). Interestingly, the kainate receptor GluR1 is co-repressed in dendrites by TDP-43 and FRMP to regulate spinogenesis (Majumder et al. 2016). It is therefore possible that NFTs and amyloid plaques sequester neuronal RNA granules, thus impairing their function. In support of this is that TDP-43 immunoreactivity is observed in NFTs in 20% of AD cases in the absence of TDP-43 nuclear depletion, suggesting that nuclear TDP-43 functions are retained (Amador-Ortiz et al. 2007). As tau is rich in disordered domains, it is actually not surprising that dendrite localized TDP-43 can be sequestered in NFTs. Also, using a yeast two-hybrid system, APP has been found to interact with Staufen 1; however, the functional relevance of this interaction remains to be determined (Yu et al. 2015). Nonetheless, it is tempting to speculate that this interaction disrupts the dendritic transport of associated mRNAs and possibly contributes to neuronal dysfunction.

## 8.6 Concluding Remarks

Perturbations in RNA granules are associated with many neurodegenerative diseases. These dysfunctions are critical as they compromise RNA metabolism and contribute to cell death/neuronal loss. In the majority of cases, RNA granule dysfunction is the result of disturbed RBP functions. Thus, it is essential to establish the composition of each RNA granule subtype and understand the role of their components. This will facilitate the study of RNA granule dynamics. Indeed, neurodegenerative diseases are associated with hypo- and hyper-assembly of mRNPs (Shukla and Parker 2016). Comprehending how RNA granules assemble and disassemble will not only shed light on new molecular mechanisms, it will also help understand the pathological processes relevant to these diseases and subsequent therapeutic and/or biomarker development. Furthermore, our understanding of disease-related aggregate formation and whether RBP-containing aggregates are pathogenic or protective will also be advanced.

## References

Alami NH, Smith RB, Carrasco MA, Williams LA, Winborn CS, Han SSW, Kiskinis E, Winborn B, Freibaum BD, Kanagaraj A, Clare AJ, Badders NM, Bilican B, Chaum E, Chandran S, Shaw CE, Eggan KC, Maniatis T, Taylor JP (2014) Axonal transport of TDP-43 mRNA granules is impaired by ALS-causing mutations. Neuron 81(3):536–543

Alberti S, Mateju D, Mediani L, Carra S (2017) Granulostasis: protein quality control of RNP granules. Front Mol Neurosci 10:84

Al-Chalabi A, Hardiman O (2013) The epidemiology of ALS: a conspiracy of genes, environment and time. Nat Rev Neurol 9(11):617–628

Amador-Ortiz C, Lin WL, Ahmed Z, Personett D, Davies P, Duara R, Graff-Radford NR, Hutton ML, Dickson DW (2007) TDP-43 immunoreactivity in hippocampal sclerosis and Alzheimer's disease. Ann Neurol 61(5):435–445

Ambadipudi S, Biernat J, Riedel D, Mandelkow E, Zweckstetter M (2017) Liquid-liquid phase separation of the microtubule-binding repeats of the Alzheimer-related protein tau. Nat Commun 8(1):275

Anderson P, Kedersha N (2002a) Stressful initiations. J Cell Sci 115(Pt 16):3227–3234

Anderson P, Kedersha N (2002b) Visibly stressed: the role of eIF2, TIA-1, and stress granules in protein translation. Cell Stress Chaperones 7(2):213–221

Anderson P, Kedersha N (2006) RNA granules. J Cell Biol 172(6):803–808

Anderson P, Kedersha N (2008) Stress granules: the Tao of RNA triage. Trends Biochem Sci 33 (3):141–150

Anderson P, Kedersha N (2009) RNA granules: post-transcriptional and epigenetic modulators of gene expression. Nat Rev Mol Cell Biol 10(6):430–436

Arai T, Hasegawa M, Akiyama H, Ikeda K, Nonaka T, Mori H, Mann D, Tsuchiya K, Yoshida M, Hashizume Y, Oda T (2006) TDP-43 is a component of ubiquitin-positive tau-negative inclusions in frontotemporal lobar degeneration and amyotrophic lateral sclerosis. Biochem Biophys Res Commun 351(3):602–611

Ash PE, Bieniek KF, Gendron TF, Caulfield T, Lin WL, Dejesus-Hernandez M, van Blitterswijk MM, Jansen-West K, Paul JW 3rd, Rademakers R, Boylan KB, Dickson DW, Petrucelli L (2013) Unconventional translation of C9ORF72 GGGGCC expansion generates insoluble polypeptides specific to c9FTD/ALS. Neuron 77(4):639–646

Ash PEA, Vanderweyde TE, Youmans KL, Apicco DJ, Wolozin B (2014) Pathological stress granules in Alzheimer's disease. Brain Res 1584:52–58

Atlas R, Behar L, Sapoznik S, Ginzburg I (2007) Dynamic association with polysomes during P19 neuronal differentiation and an untranslated-region-dependent translation regulation of the tau mRNA by the tau mRNA-associated proteins IMP1, HuD, and G3BP1. J Neurosci Res 85 (1):173–183

Aulas A, Vande Velde C (2015) Alterations in stress granule dynamics driven by TDP-43 and FUS: a link to pathological inclusions in ALS? Front Cell Neurosci 9:423

Aulas A, Stabile S, Vande Velde C (2012) Endogenous TDP-43, but not FUS, contributes to stress granule assembly via G3BP. Mol Neurodegener 7:54

Aulas A, Caron G, Gkogkas CG, Mohamed NV, Destroismaisons L, Sonenberg N, Leclerc N, Parker JA, Vande Velde C (2015) G3BP1 promotes stress-induced RNA granule interactions to preserve polyadenylated mRNA. J Cell Biol 209(1):73–84

Ayache J, Benard M, Ernoult-Lange M, Minshall N, Standart N, Kress M, Weil D (2015) P-body assembly requires DDX6 repression complexes rather than decay or Ataxin2/2L complexes. Mol Biol Cell 26(14):2579–2595

Ayala YM, Zago P, D'Ambrogio A, Xu YF, Petrucelli L, Buratti E, Baralle FE (2008) Structural determinants of the cellular localization and shuttling of TDP-43. J Cell Sci 121 (Pt 22):3778–3785

Azuma Y, Tokuda T, Shimamura M, Kyotani A, Sasayama H, Yoshida T, Mizuta I, Mizuno T, Nakagawa M, Fujikake N, Ueyama M, Nagai Y, Yamaguchi M (2014) Identification of ter94, Drosophila VCP, as a strong modulator of motor neuron degeneration induced by knockdown of Caz, Drosophila FUS. Hum Mol Genet 23(13):3467–3480

Banzhaf-Strathmann J, Benito E, May S, Arzberger T, Tahirovic S, Kretzschmar H, Fischer A, Edbauer D (2014) MicroRNA-125b induces tau hyperphosphorylation and cognitive deficits in Alzheimer's disease. EMBO J 33(15):1667–1680

Basavaraju M, de Lencastre A (2016) Alzheimer's disease: presence and role of microRNAs. Biomol Concepts 7(4):241–252

Bashkirov VI, Scherthan H, Solinger JA, Buerstedde J-M, Heyer W-D (1997) A mouse cytoplasmic Exoribonuclease (mXRN1p) with preference for G4 Tetraplex substrates. J Cell Biol 136 (4):761–773

Bassell GJ, Warren ST (2008) Fragile X syndrome: loss of local mRNA regulation alters synaptic development and function. Neuron 60(2):201–214

Belly A, Moreau-Gachelin F, Sadoul R, Goldberg Y (2005) Delocalization of the multifunctional RNA splicing factor TLS/FUS in hippocampal neurones: exclusion from the nucleus and accumulation in dendritic granules and spine heads. Neurosci Lett 379(3):152–157

Bentmann E, Haass C, Dormann D (2013) Stress granules in neurodegeneration--lessons learnt from TAR DNA binding protein of 43 kDa and fused in sarcoma. FEBS J 280(18):4348–4370

Bigio EH (2011) C9ORF72, the new gene on the block, causes C9FTD/ALS: new insights provided by neuropathology. Acta Neuropathol 122(6):653–655

Bosco DA, Lemay N, Ko HK, Zhou H, Burke C, Kwiatkowski TJ Jr, Sapp P, McKenna-Yasek D, Brown RH Jr, Hayward LJ (2010) Mutant FUS proteins that cause amyotrophic lateral sclerosis incorporate into stress granules. Hum Mol Genet 19(21):4160–4175

Boyault C, Zhang Y, Fritah S, Caron C, Gilquin B, Kwon SH, Garrido C, Yao TP, Vourc'h C, Matthias P, Khochbin S (2007) HDAC6 controls major cell response pathways to cytotoxic accumulation of protein aggregates. Genes Dev 21(17):2172–2181

Brengues M, Teixeira D, Parker R (2005) Movement of eukaryotic mRNAs between polysomes and cytoplasmic processing bodies. Science 310(5747):486–489

Brown R, Kaganovich D (2016) Look out autophagy, Ubiquilin UPS its game. Cell 166 (4):797–799

Brunello CA, Yan X, Huttunen HJ (2016) Internalized tau sensitizes cells to stress by promoting formation and stability of stress granules. Sci Rep 6:30498

Buchan JR, Kolaitis RM, Taylor JP, Parker R (2013) Eukaryotic stress granules are cleared by autophagy and Cdc48/VCP function. Cell 153(7):1461–1474

Buratti E, Baralle FE (2009) The molecular links between TDP-43 dysfunction and neurodegeneration. Adv Genet 66:1–34

Campos-Melo D, Droppelmann CA, He Z, Volkening K, Strong MJ (2013) Altered microRNA expression profile in amyotrophic lateral sclerosis: a role in the regulation of NFL mRNA levels. Mol Brain 6:26

Cartegni L, Maconi M, Morandi E, Cobianchi F, Riva S, Biamonti G (1996) hnRNP A1 selectively interacts through its Gly-rich domain with different RNA-binding proteins. J Mol Biol 259 (3):337–348

Cestra G, Rossi S, Di Salvio M, Cozzolino M (2017) Control of mRNA translation in ALS Proteinopathy. Front Mol Neurosci 10:85

Chang L, Monteiro MJ (2015) Defective proteasome delivery of Polyubiquitinated proteins by Ubiquilin-2 proteins containing ALS mutations. PLoS One 10(6):e0130162

Chatel-Chaix L, Clement JF, Martel C, Beriault V, Gatignol A, DesGroseillers L, Mouland AJ (2004) Identification of Staufen in the human immunodeficiency virus type 1 gag ribonucleoprotein complex and a role in generating infectious viral particles. Mol Cell Biol 24 (7):2637–2648

Chen J-A, Wichterle H (2012) Apoptosis of limb innervating motor neurons and erosion of motor pool identity upon lineage specific dicer inactivation. Front Neurosci 6:69–69

Chia R, Chio A, Traynor BJ (2018) Novel genes associated with amyotrophic lateral sclerosis: diagnostic and clinical implications. Lancet Neurol 17(1):94–102

Chitiprolu M, Jagow C, Tremblay V, Bondy-Chorney E, Paris G, Savard A, Palidwor G, Barry FA, Zinman L, Keith J, Rogaeva E, Robertson J, Lavallée-Adam M, Woulfe J, Couture J-F, Côté J, Gibbings D (2018) A complex of C9ORF72 and p62 uses arginine methylation to eliminate stress granules by autophagy. Nat Commun 9(1):2794

Chu CY, Rana TM (2006) Translation repression in human cells by microRNA-induced gene silencing requires RCK/p54. PLoS Biol 4(7):e210

Ciryam P, Kundra R, Morimoto RI, Dobson CM, Vendruscolo M (2015) Supersaturation is a major driving force for protein aggregation in neurodegenerative diseases. Trends Pharmacol Sci 36 (2):72–77

Collier NC, Schlesinger MJ (1986) The dynamic state of heat shock proteins in chicken embryo fibroblasts. J Cell Biol 103(4):1495–1507

Conicella AE, Zerze GH, Mittal J, Fawzi NL (2016) ALS mutations disrupt phase separation mediated by alpha-helical structure in the TDP-43 low-complexity C-terminal domain. Structure 24(9):1537–1549

Conte A, Lattante S, Zollino M, Marangi G, Luigetti M, Del Grande A, Servidei S, Trombetta F, Sabatelli M (2012) P525L FUS mutation is consistently associated with a severe form of juvenile amyotrophic lateral sclerosis. Neuromuscul Disord 22(1):73–75

Cougot N, Babajko S, Seraphin B (2004) Cytoplasmic foci are sites of mRNA decay in human cells. J Cell Biol 165(1):31–40

Courchaine EM, Lu A, Neugebauer KM (2016) Droplet organelles? EMBO J 35(15):1603–1612

Coyne AN, Siddegowda BB, Estes PS, Johannesmeyer J, Kovalik T, Daniel SG, Pearson A, Bowser R, Zarnescu DC (2014) Futsch/MAP1B mRNA is a translational target of TDP-43 and is neuroprotective in a Drosophila model of amyotrophic lateral sclerosis. J Neurosci 34 (48):15962–15974

Crowley MT, Costello PS, Fitzer-Attas CJ, Turner M, Meng F, Lowell C, Tybulewicz VLJ, DeFranco AL (1997) A critical role for Syk in signal transduction and phagocytosis mediated by Fcγ receptors on macrophages. J Exp Med 186(7):1027–1039

Da Cruz S, Cleveland DW (2011) Understanding the role of TDP-43 and FUS/TLS in ALS and beyond. Curr Opin Neurobiol 21(6):904–919

Dang Y, Kedersha N, Low WK, Romo D, Gorospe M, Kaufman R, Anderson P, Liu JO (2006) Eukaryotic initiation factor 2alpha-independent pathway of stress granule induction by the natural product pateamine A. J Biol Chem 281(43):32870–32878

David Gerecht PS, Taylor MA, Port JD (2010) Intracellular localization and interaction of mRNA binding proteins as detected by FRET. BMC Cell Biol 11(1):69

De Conti L, Baralle M, Buratti E (2017) Neurodegeneration and RNA-binding proteins. Wiley Interdiscip Rev RNA 8(2):e1394-n/a

De Felice B, Guida M, Guida M, Coppola C, De Mieri G, Cotrufo R (2012) A miRNA signature in leukocytes from sporadic amyotrophic lateral sclerosis. Gene 508(1):35–40

Decker CJ, Parker R (2012) P-bodies and stress granules: possible roles in the control of translation and mRNA degradation. Cold Spring Harb Perspect Biol 4(9):a012286

Decker CJ, Teixeira D, Parker R (2007) Edc3p and a glutamine/asparagine-rich domain of Lsm4p function in processing body assembly in *Saccharomyces cerevisiae*. J Cell Biol 179(3):437–449

Dejesus-Hernandez M, Mackenzie IR, Boeve BF, Boxer AL, Baker M, Rutherford NJ, Nicholson AM, Finch NA, Flynn H, Adamson J, Kouri N, Wojtas A, Sengdy P, Hsiung GY, Karydas A, Seeley WW, Josephs KA, Coppola G, Geschwind DH, Wszolek ZK, Feldman H, Knopman DS, Petersen RC, Miller BL, Dickson DW, Boylan KB, Graff-Radford NR, Rademakers R (2011) Expanded GGGGCC hexanucleotide repeat in noncoding region of C9ORF72 causes chromosome 9p-linked FTD and ALS. Neuron 72(2):245–256

Deng HX, Chen W, Hong ST, Boycott KM, Gorrie GH, Siddique N, Yang Y, Fecto F, Shi Y, Zhai H, Jiang H, Hirano M, Rampersaud E, Jansen GH, Donkervoort S, Bigio EH, Brooks BR, Ajroud K, Sufit RL, Haines JL, Mugnaini E, Pericak-Vance MA, Siddique T (2011) Mutations in UBQLN2 cause dominant X-linked juvenile and adult-onset ALS and ALS/dementia. Nature 477(7363):211–215

Dewey CM, Cenik B, Sephton CF, Dries DR, Mayer P 3rd, Good SK, Johnson BA, Herz J, Yu G (2011) TDP-43 is directed to stress granules by sorbitol, a novel physiological osmotic and oxidative stressor. Mol Cell Biol 31(5):1098–1108

Dewey CM, Cenik B, Sephton CF, Johnson BA, Herz J, Yu G (2012) TDP-43 aggregation in neurodegeneration: are stress granules the key? Brain Res 1462:16–25

Dictenberg JB, Swanger SA, Antar LN, Singer RH, Bassell GJ (2008) A direct role for FMRP in activity-dependent dendritic mRNA transport links filopodial-spine morphogenesis to fragile X syndrome. Dev Cell 14(6):926–939

Dinkel PD, Holden MR, Matin N, Margittai M (2015) RNA binds to tau fibrils and sustains template-assisted growth. Biochemistry 54(30):4731–4740

Dormann D, Rodde R, Edbauer D, Bentmann E, Fischer I, Hruscha A, Than ME, Mackenzie IR, Capell A, Schmid B, Neumann M, Haass C (2010) ALS-associated fused in sarcoma (FUS) mutations disrupt Transportin-mediated nuclear import. EMBO J 29(16):2841–2857

Doyle M, Kiebler MA (2011) Mechanisms of dendritic mRNA transport and its role in synaptic tagging. EMBO J 30(17):3540–3552

Doyle M, Kiebler MA (2012) A zipcode unzipped. Genes Dev 26(2):110–113

Dreyfuss G, Kim VN, Kataoka N (2002) Messenger-RNA-binding proteins and the messages they carry. Nat Rev Mol Cell Biol 3(3):195–205

Edupuganti RR, Geiger S, Lindeboom RGH, Shi H, Hsu PJ, Lu Z, Wang S-Y, Baltissen MPA, Jansen PWTC, Rossa M, Müller M, Stunnenberg HG, He C, Carell T, Vermeulen M (2017) N6-methyladenosine (m6A) recruits and repels proteins to regulate mRNA homeostasis. Nat Struct Mol Biol 24:870

Elie A, Prezel E, Guerin C, Denarier E, Ramirez-Rios S, Serre L, Andrieux A, Fourest-Lieuvin A, Blanchoin L, Arnal I (2015) Tau co-organizes dynamic microtubule and actin networks. Sci Rep 5:9964

Eom T, Antar LN, Singer RH, Bassell GJ (2003) Localization of a beta-actin messenger ribonucleoprotein complex with zipcode-binding protein modulates the density of dendritic filopodia and filopodial synapses. J Neurosci 23(32):10433–10444

Eulalio A, Behm-Ansmant I, Izaurralde E (2007a) P bodies: at the crossroads of post-transcriptional pathways. Nat Rev Mol Cell Biol 8(1):9–22

Eulalio A, Behm-Ansmant I, Schweizer D, Izaurralde E (2007b) P-body formation is a consequence, not the cause, of RNA-mediated gene silencing. Mol Cell Biol 27(11):3970–3981

Eulalio A, Rehwinkel J, Stricker M, Huntzinger E, Yang S-F, Doerks T, Dorner S, Bork P, Boutros M, Izaurralde E (2007c) Target-specific requirements for enhancers of decapping in miRNA-mediated gene silencing. Genes Dev 21(20):2558–2570

Eystathioy T, Jakymiw A, Chan EK, Séraphin B, Cougot N, Fritzler MJ (2003) The GW182 protein colocalizes with mRNA degradation associated proteins hDcp1 and hLSm4 in cytoplasmic GW bodies. RNA 9(10):1171–1173

Fallini C, Bassell GJ, Rossoll W (2012) The ALS disease protein TDP-43 is actively transported in motor neuron axons and regulates axon outgrowth. Hum Mol Genet 21(16):3703–3718

Farina KL, Huttelmaier S, Musunuru K, Darnell R, Singer RH (2003) Two ZBP1 KH domains facilitate beta-actin mRNA localization, granule formation, and cytoskeletal attachment. J Cell Biol 160(1):77–87

Fenger-Gron M, Fillman C, Norrild B, Lykke-Andersen J (2005) Multiple processing body factors and the ARE binding protein TTP activate mRNA decapping. Mol Cell 20(6):905–915

Frankel LB, Lubas M, Lund AH (2017) Emerging connections between RNA and autophagy. Autophagy 13(1):3–23

Franks TM, Lykke-Andersen J (2008) The control of mRNA decapping and P-body formation. Mol Cell 32(5):605–615

Fromm SA, Kamenz J, Noldeke ER, Neu A, Zocher G, Sprangers R (2014) In vitro reconstitution of a cellular phase-transition process that involves the mRNA decapping machinery. Angew Chem Int Ed Engl 53(28):7354–7359

Fujii R, Okabe S, Urushido T, Inoue K, Yoshimura A, Tachibana T, Nishikawa T, Hicks GG, Takumi T (2005) The RNA binding protein TLS is translocated to dendritic spines by mGluR5 activation and regulates spine morphology. Curr Biol 15(6):587–593

Gal J, Zhang J, Kwinter DM, Zhai J, Jia H, Jia J, Zhu H (2010) Nuclear localization sequence of FUS and induction of stress granules by ALS mutants. Neurobiol Aging 32(12):2323.e2327–2323.e2340

Ganassi M, Mateju D, Bigi I, Mediani L, Poser I, Lee HO, Seguin SJ, Morelli FF, Vinet J, Leo G, Pansarasa O, Cereda C, Poletti A, Alberti S, Carra S (2016) A surveillance function of the HSPB8-BAG3-HSP70 chaperone complex ensures stress granule integrity and dynamism. Mol Cell 63(5):796–810

Geekiyanage H, Chan C (2011) MicroRNA-137/181c regulates serine palmitoyltransferase and in turn amyloid beta, novel targets in sporadic Alzheimer's disease. J Neurosci 31 (41):14820–14830

Ghosh S, Geahlen RL (2015) Stress granules modulate SYK to cause microglial cell dysfunction in Alzheimer's disease. EBioMedicine 2(11):1785–1798

Ginsberg SD, Crino PB, Lee VM, Eberwine JH, Trojanowski JQ (1997) Sequestration of RNA in Alzheimer's disease neurofibrillary tangles and senile plaques. Ann Neurol 41(2):200–209

Ginsberg SD, Crino PB, Hemby SE, Weingarten JA, Lee VM, Eberwine JH, Trojanowski JQ (1999) Predominance of neuronal mRNAs in individual Alzheimer's disease senile plaques. Ann Neurol 45(2):174–181

Gitcho MA, Strider J, Carter D, Taylor-Reinwald L, Forman MS, Goate AM, Cairns NJ (2009) VCP mutations causing frontotemporal lobar degeneration disrupt localization of TDP-43 and induce cell death. J Biol Chem 284(18):12384–12398

Glisovic T, Bachorik JL, Yong J, Dreyfuss G (2008) RNA-binding proteins and post-transcriptional gene regulation. FEBS Lett 582(14):1977–1986

Gong C, Maquat LE (2011) lncRNAs transactivate STAU1-mediated mRNA decay by duplexing with 3′ UTRs via Alu elements. Nature 470(7333):284–288

Guerreiro R, Bras J, Hardy J (2015) SnapShot: genetics of ALS and FTD. Cell 160(4):798. e791

Gunawardana CG, Mehrabian M, Wang X, Mueller I, Lubambo IB, Jonkman JE, Wang H, Schmitt-Ulms G (2015) The human tau interactome: binding to the ribonucleoproteome, and impaired binding of the proline-to-leucine mutant at position 301 (P301L) to chaperones and the proteasome. Mol Cell Proteomics 14(11):3000–3014

Haramati S, Chapnik E, Sztainberg Y, Eilam R, Zwang R, Gershoni N, McGlinn E, Heiser PW, Wills AM, Wirguin I, Rubin LL, Misawa H, Tabin CJ, Brown R Jr, Chen A, Hornstein E (2010) miRNA malfunction causes spinal motor neuron disease. Proc Natl Acad Sci U S A 107 (29):13111–13116

Harrison AF, Shorter J (2017) RNA-binding proteins with prion-like domains in health and disease. Biochem J 474(8):1417–1438

Hebert LE, Scherr PA, Bienias JL, Bennett DA, Evans DA (2003) Alzheimer disease in the US population: prevalence estimates using the 2000 census. Arch Neurol 60(8):1119–1122

Hebert SS, Horre K, Nicolai L, Papadopoulou AS, Mandemakers W, Silahtaroglu AN, Kauppinen S, Delacourte A, De Strooper B (2008) Loss of microRNA cluster miR-29a/b-1 in sporadic Alzheimer's disease correlates with increased BACE1/beta-secretase expression. Proc Natl Acad Sci U S A 105(17):6415–6420

Hirokawa N (2006) mRNA transport in dendrites: RNA granules, motors, and tracks. J Neurosci 26 (27):7139–7142

Hirsch-Reinshagen V, Pottier C, Nicholson AM, Baker M, Hsiung G-YR, Krieger C, Sengdy P, Boylan KB, Dickson DW, Mesulam M, Weintraub S, Bigio E, Zinman L, Keith J, Rogaeva E, Zivkovic SA, Lacomis D, Taylor JP, Rademakers R, Mackenzie IRAJANC (2017) Clinical and neuropathological features of ALS/FTD with TIA1 mutations. Acta Neuropathol Commun 5 (1):96

Holmes LEA, Campbell SG, De Long SK, Sachs AB, Ashe MP (2004) Loss of translational control in yeast compromised for the major mRNA decay pathway. Mol Cell Biol 24(7):2998–3010

Hua Y, Zhou J (2004) Survival motor neuron protein facilitates assembly of stress granules. FEBS Lett 572(1–3):69–74

Huang L, Mollet S, Souquere S, Le Roy F, Ernoult-Lange M, Pierron G, Dautry F, Weil D (2011) Mitochondria associate with P-bodies and modulate microRNA-mediated RNA interference. J Biol Chem 286(27):24219–24230

Hubstenberger A, Courel M, Bénard M, Souquere S, Ernoult-Lange M, Chouaib R, Yi Z, Morlot J-B, Munier A, Fradet M, Daunesse M, Bertrand E, Pierron G, Mozziconacci J, Kress M, Weil D (2017) P-body purification reveals the condensation of repressed mRNA regulons. Molecular Cell 68(1):144–157.e145

Huttelmaier S, Zenklusen D, Lederer M, Dictenberg J, Lorenz M, Meng X, Bassell GJ, Condeelis J, Singer RH (2005) Spatial regulation of beta-actin translation by Src-dependent phosphorylation of ZBP1. Nature 438(7067):512–515

Ingelfinger D, Arndt-Jovin DJ, Lührmann R, Achsel T (2002) The human LSm1-7 proteins colocalize with the mRNA-degrading enzymes Dcp1/2 and Xrnl in distinct cytoplasmic foci. RNA 8(12):1489–1501

Ishiguro A, Kimura N, Watanabe Y, Watanabe S, Ishihama A (2016) TDP-43 binds and transports G-quadruplex-containing mRNAs into neurites for local translation. Genes Cells 21(5):466–481

Jain S, Wheeler JR, Walters RW, Agrawal A, Barsic A, Parker R (2016) ATPase-modulated stress granules contain a diverse proteome and substructure. Cell 164(3):487–498

Jakymiw A, Lian S, Eystathioy T, Li S, Satoh M, Hamel JC, Fritzler MJ, Chan EK (2005) Disruption of GW bodies impairs mammalian RNA interference. Nat Cell Biol 7 (12):1267–1274

Johnson JO, Mandrioli J, Benatar M, Abramzon Y, Van Deerlin VM, Trojanowski JQ, Gibbs JR, Brunetti M, Gronka S, Wuu J, Ding J, McCluskey L, Martinez-Lage M, Falcone D, Hernandez DG, Arepalli S, Chong S, Schymick JC, Rothstein J, Landi F, Wang YD, Calvo A, Mora G, Sabatelli M, Monsurro MR, Battistini S, Salvi F, Spataro R, Sola P, Borghero G, Galassi G, Scholz SW, Taylor JP, Restagno G, Chio A, Traynor BJ (2010) Exome sequencing reveals VCP mutations as a cause of familial ALS. Neuron 68(5):857–864

Jonson L, Vikesaa J, Krogh A, Nielsen LK, Hansen T, Borup R, Johnsen AH, Christiansen J, Nielsen FC (2007) Molecular composition of IMP1 ribonucleoprotein granules. Mol Cell Proteomics 6(5):798–811

Ju JS, Fuentealba RA, Miller SE, Jackson E, Piwnica-Worms D, Baloh RH, Weihl CC (2009) Valosin-containing protein (VCP) is required for autophagy and is disrupted in VCP disease. J Cell Biol 187(6):875–888

Kadavath H, Hofele RV, Biernat J, Kumar S, Tepper K, Urlaub H, Mandelkow E, Zweckstetter M (2015) Tau stabilizes microtubules by binding at the interface between tubulin heterodimers. Proc Natl Acad Sci U S A 112(24):7501–7506

Kampers T, Friedhoff P, Biernat J, Mandelkow EM, Mandelkow E (1996) RNA stimulates aggregation of microtubule-associated protein tau into Alzheimer-like paired helical filaments. FEBS Lett 399(3):344–349

Kapeli K, Martinez FJ, Yeo GW (2017) Genetic mutations in RNA-binding proteins and their roles in ALS. Hum Genet 136(9):1193–1214

Kawaguchi Y, Kovacs JJ, McLaurin A, Vance JM, Ito A, Yao T-P (2003) The deacetylase HDAC6 regulates aggresome formation and cell viability in response to misfolded protein stress. Cell 115(6):727–738

Kedersha N, Anderson P (2002) Stress granules: sites of mRNA triage that regulate mRNA stability and translatability. Biochem Soc Trans 30(Pt 6):963–969

Kedersha NL, Gupta M, Li W, Miller I, Anderson P (1999) RNA-binding proteins TIA-1 and TIAR link the phosphorylation of eIF-2 alpha to the assembly of mammalian stress granules. J Cell Biol 147(7):1431–1442

Kedersha N, Chen S, Gilks N, Li W, Miller IJ, Stahl J, Anderson P (2002) Evidence that ternary complex (eIF2-GTP-tRNA(i)(met))-deficient preinitiation complexes are core constituents of mammalian stress granules. Mol Biol Cell 13(1):195–210

Kedersha N, Stoecklin G, Ayodele M, Yacono P, Lykke-Andersen J, Fritzler MJ, Scheuner D, Kaufman RJ, Golan DE, Anderson P (2005) Stress granules and processing bodies are dynamically linked sites of mRNP remodeling. J Cell Biol 169(6):871–884

Khong A, Jain S, Matheny T, Wheeler JR, Parker R (2017a) Isolation of mammalian stress granule cores for RNA-Seq analysis. Methods 137(137):49–54

Khong A, Kerr CH, Yeung CH, Keatings K, Nayak A, Allan DW, Jan E (2017b) Disruption of stress granule formation by the multifunctional cricket paralysis virus 1A protein. J Virol 91(5): e01779-16

Khong A, Matheny T, Jain S, Mitchell SF, Wheeler JR, Parker R (2017c) The stress granule transcriptome reveals principles of mRNA accumulation in stress granules. Mol Cell 68 (4):808–820. e805

Kiebler MA, Bassell GJ (2006) Neuronal RNA granules: movers and makers. Neuron 51 (6):685–690

Kiefer F, Brumell J, Al-Alawi N, Latour S, Cheng A, Veillette A, Grinstein S, Pawson T (1998) The Syk protein tyrosine kinase is essential for Fcgamma receptor signaling in macrophages and neutrophils. Mol Cell Biol 18(7):4209–4220

Kim HJ, Kim NC, Wang YD, Scarborough EA, Moore J, Diaz Z, MacLea KS, Freibaum B, Li S, Molliex A, Kanagaraj AP, Carter R, Boylan KB, Wojtas AM, Rademakers R, Pinkus JL, Greenberg SA, Trojanowski JQ, Traynor BJ, Smith BN, Topp S, Gkazi AS, Miller J, Shaw CE, Kottlors M, Kirschner J, Pestronk A, Li YR, Ford AF, Gitler AD, Benatar M, King OD, Kimonis VE, Ross ED, Weihl CC, Shorter J, Taylor JP (2013a) Mutations in prion-like domains in hnRNPA2B1 and hnRNPA1 cause multisystem proteinopathy and ALS. Nature 495 (7442):467–473

Kim YE, Hipp MS, Bracher A, Hayer-Hartl M, Hartl FU (2013b) Molecular chaperone functions in protein folding and proteostasis. Annu Rev Biochem 82(1):323–355

Kim HH, Lee SJ, Gardiner AS, Perrone-Bizzozero NI, Yoo S (2015) Different motif requirements for the localization zipcode element of beta-actin mRNA binding by HuD and ZBP1. Nucleic Acids Res 43(15):7432–7446

Kitami MI, Kitami T, Nagahama M, Tagaya M, Hori S, Kakizuka A, Mizuno Y, Hattori N (2006) Dominant-negative effect of mutant valosin-containing protein in aggresome formation. FEBS Lett 580(2):474–478

Klann E, Dever TE (2004) Biochemical mechanisms for translational regulation in synaptic plasticity. Nat Rev Neurosci 5(12):931–942

Knowles RB, Sabry JH, Martone ME, Deerinck TJ, Ellisman MH, Bassell GJ, Kosik KS (1996) Translocation of RNA granules in living neurons. J Neurosci 16(24):7812–7820

Köhrmann M, Luo M, Kaether C, DesGroseillers L, Dotti CG, Kiebler MA (1999) Microtubule-dependent recruitment of staufen-green fluorescent protein into large RNA-containing granules and subsequent dendritic transport in living hippocampal neurons. Mol Biol Cell 10 (9):2945–2953

Koppers M, Blokhuis AM, Westeneng HJ, Terpstra ML, Zundel CA, Vieira de Sa R, Schellevis RD, Waite AJ, Blake DJ, Veldink JH, van den Berg LH, Pasterkamp RJ (2015) C9orf72 ablation in mice does not cause motor neuron degeneration or motor deficits. Ann Neurol 78 (3):426–438

Krichevsky AM, Kosik KS (2001) Neuronal RNA granules: a link between RNA localization and stimulation-dependent translation. Neuron 32(4):683–696

Krisenko MO, Higgins RL, Ghosh S, Zhou Q, Trybula JS, Wang WH, Geahlen RL (2015) Syk is recruited to stress granules and promotes their clearance through autophagy. J Biol Chem 290 (46):27803–27815

Kroschwald S, Maharana S, Mateju D, Malinovska L, Nüske E, Poser I, Richter D, Alberti S (2015) Promiscuous interactions and protein disaggregases determine the material state of stress-inducible RNP granules. Elife 4:e06807

Kuriakose T, Kanneganti TD (2017) ZBP1: innate sensor regulating cell death and inflammation. Trends Immunol 39(2):123–134

Kwon S, Zhang Y, Matthias P (2007) The deacetylase HDAC6 is a novel critical component of stress granules involved in the stress response. Genes Dev 21(24):3381–3394

Labbadia J, Morimoto RI (2015) The biology of proteostasis in aging and disease. Annu Rev Biochem 84:435–464

Lagier-Tourenne C, Polymenidou M, Cleveland DW (2010) TDP-43 and FUS/TLS: emerging roles in RNA processing and neurodegeneration. Hum Mol Genet 19(R1):R46–R64

Le NT, Chang L, Kovlyagina I, Georgiou P, Safren N, Braunstein KE, Kvarta MD, Van Dyke AM, LeGates TA, Philips T, Morrison BM, Thompson SM, Puche AC, Gould TD, Rothstein JD, Wong PC, Monteiro MJ (2016) Motor neuron disease, TDP-43 pathology, and memory deficits in mice expressing ALS-FTD-linked UBQLN2 mutations. Proc Natl Acad Sci U S A 113(47): E7580–E7589

Lebeau G, Maher-Laporte M, Topolnik L, Laurent CE, Sossin W, Desgroseillers L, Lacaille JC (2008) Staufen1 regulation of protein synthesis-dependent long-term potentiation and synaptic function in hippocampal pyramidal cells. Mol Cell Biol 28(9):2896–2907

Lebeau G, DesGroseillers L, Sossin W, Lacaille JC (2011a) mRNA binding protein staufen 1-dependent regulation of pyramidal cell spine morphology via NMDA receptor-mediated synaptic plasticity. Mol Brain 4:22

Lebeau G, Miller LC, Tartas M, McAdam R, Laplante I, Badeaux F, DesGroseillers L, Sossin WS, Lacaille JC (2011b) Staufen 2 regulates mGluR long-term depression and Map1b mRNA distribution in hippocampal neurons. Learn Mem 18(5):314–326

Leblond CS, Webber A, Gan-Or Z, Moore F, Dagher A, Dion PA, Rouleau GA (2016) De novo FUS P525L mutation in juvenile amyotrophic lateral sclerosis with dysphonia and diplopia. Neurol Genet 2(2):e63

Lechler MC, David DC (2017) More stressed out with age? Check your RNA granule aggregation. Prion 11(5):313–322

Lee EB, Lee VM, Trojanowski JQ (2011) Gains or losses: molecular mechanisms of TDP43-mediated neurodegeneration. Nat Rev Neurosci 13(1):38–50

Lee YB, Chen HJ, Peres JN, Gomez-Deza J, Attig J, Stalekar M, Troakes C, Nishimura AL, Scotter EL, Vance C, Adachi Y, Sardone V, Miller JW, Smith BN, Gallo JM, Ule J, Hirth F, Rogelj B, Houart C, Shaw CE (2013) Hexanucleotide repeats in ALS/FTD form length-dependent RNA foci, sequester RNA binding proteins, and are neurotoxic. Cell Rep 5(5):1178–1186

Lee KH, Zhang P, Kim HJ, Mitrea DM, Sarkar M, Freibaum BD, Cika J, Coughlin M, Messing J, Molliex A, Maxwell BA, Kim NC, Temirov J, Moore J, Kolaitis RM, Shaw TI, Bai B, Peng J, Kriwacki RW, Taylor JP (2016) C9orf72 dipeptide repeats impair the assembly, dynamics, and function of membrane-less organelles. Cell 167(3):774–788. e717

Liberek K, Lewandowska A, Ziętkiewicz S (2008) Chaperones in control of protein disaggregation. EMBO J 27(2):328–335

Lim PJ, Danner R, Liang J, Doong H, Harman C, Srinivasan D, Rothenberg C, Wang H, Ye Y, Fang S, Monteiro MJ (2009) Ubiquilin and p97/VCP bind erasin, forming a complex involved in ERAD. J Cell Biol 187(2):201–217

Lin AC, Holt CE (2007) Local translation and directional steering in axons. EMBO J 26 (16):3729–3736

Lin Y, Protter DS, Rosen MK, Parker R (2015) Formation and maturation of phase-separated liquid droplets by RNA-binding proteins. Mol Cell 60(2):208–219

Lin Y, Mori E, Kato M, Xiang S, Wu L, Kwon I, McKnight SL (2016) Toxic PR poly-dipeptides encoded by the C9orf72 repeat expansion target LC domain polymers. Cell 167(3):789–802. e712

Ling SHM, Decker CJ, Walsh MA, She M, Parker R, Song H (2008) Crystal structure of human Edc3 and its functional implications. Mol Cell Biol 28(19):5965

Ling SC, Polymenidou M, Cleveland DW (2013) Converging mechanisms in ALS and FTD: disrupted RNA and protein homeostasis. Neuron 79(3):416–438

Liu J, Rivas FV, Wohlschlegel J, Yates JR 3rd, Parker R, Hannon GJ (2005a) A role for the P-body component GW182 in microRNA function. Nat Cell Biol 7(12):1261–1266

Liu J, Valencia-Sanchez MA, Hannon GJ, Parker R (2005b) MicroRNA-dependent localization of targeted mRNAs to mammalian P-bodies. Nat Cell Biol 7(7):719–723

Liu W, Liu C, Zhu J, Shu P, Yin B, Gong Y, Qiang B, Yuan J, Peng X (2012) MicroRNA-16 targets amyloid precursor protein to potentially modulate Alzheimer's-associated pathogenesis in SAMP8 mice. Neurobiol Aging 33(3):522–534

Liu-Yesucevitz L, Bilgutay A, Zhang YJ, Vanderweyde T, Citro A, Mehta T, Zaarur N, McKee A, Bowser R, Sherman M, Petrucelli L, Wolozin B (2010) Tar DNA binding protein-43 (TDP-43) associates with stress granules: analysis of cultured cells and pathological brain tissue. PLoS One 5(10):e13250

Liu-Yesucevitz L, Bassell GJ, Gitler AD, Hart AC, Klann E, Richter JD, Warren ST, Wolozin B (2011) Local RNA translation at the synapse and in disease. J Neurosci 31(45):16086–16093

Liu-Yesucevitz L, Lin AY, Ebata A, Boon JY, Reid W, Xu YF, Kobrin K, Murphy GJ, Petrucelli L, Wolozin B (2014) ALS-linked mutations enlarge TDP-43-enriched neuronal RNA granules in the dendritic arbor. J Neurosci 34(12):4167–4174

Lopez-Gonzalez R, Lu Y, Gendron TF, Karydas A, Tran H, Yang D, Petrucelli L, Miller BL, Almeida S, Gao FB (2016) Poly(GR) in C9ORF72-related ALS/FTD compromises mitochondrial function and increases oxidative stress and DNA damage in iPSC-derived motor neurons. Neuron 92(2):383–391

Lukavsky PJ, Daujotyte D, Tollervey JR, Ule J, Stuani C, Buratti E, Baralle FE, Damberger FF, Allain FH (2013) Molecular basis of UG-rich RNA recognition by the human splicing factor TDP-43. Nat Struct Mol Biol 20(12):1443–1449

Lunde BM, Moore C, Varani G (2007) RNA-binding proteins: modular design for efficient function. Nat Rev Mol Cell Biol 8(6):479–490

Lykke-Andersen J (2002) Identification of a human decapping complex associated with hUpf proteins in nonsense-mediated decay. Mol Cell Biol 22(23):8114–8121

Mackenzie IR, Nicholson AM, Sarkar M, Messing J, Purice MD, Pottier C, Annu K, Baker M, Perkerson RB, Kurti A, Matchett BJ, Mittag T, Temirov J, Hsiung GR, Krieger C, Murray ME, Kato M, Fryer JD, Petrucelli L, Zinman L, Weintraub S, Mesulam M, Keith J, Zivkovic SA, Hirsch-Reinshagen V, Roos RP, Zuchner S, Graff-Radford NR, Petersen RC, Caselli RJ, Wszolek ZK, Finger E, Lippa C, Lacomis D, Stewart H, Dickson DW, Kim HJ, Rogaeva E, Bigio E, Boylan KB, Taylor JP, Rademakers R (2017) TIA1 mutations in amyotrophic lateral sclerosis and frontotemporal dementia promote phase separation and alter stress granule dynamics. Neuron 95(4):808–816. e809

Maharjan N, Kunzli C, Buthey K, Saxena S (2017) C9ORF72 regulates stress granule formation and its deficiency impairs stress granule assembly, hypersensitizing cells to stress. Mol Neurobiol 54(4):3062–3077

Majcher V, Goode A, James V, Layfield R (2015) Autophagy receptor defects and ALS-FTLD. Mol Cell Neurosci 66(Pt A):43–52

Majumder P, Chu JF, Chatterjee B, Swamy KB, Shen CJ (2016) Co-regulation of mRNA translation by TDP-43 and fragile X syndrome protein FMRP. Acta Neuropathol 132(5):721–738

Mallardo M, Deitinghoff A, Muller J, Goetze B, Macchi P, Peters C, Kiebler MA (2003) Isolation and characterization of staufen-containing ribonucleoprotein particles from rat brain. Proc Natl Acad Sci U S A 100(4):2100–2105

Manfredi G, Xu Z (2005) Mitochondrial dysfunction and its role in motor neuron degeneration in ALS. Mitochondrion 5(2):77–87

Maniecka Z, Polymenidou M (2015) From nucleation to widespread propagation: a prion-like concept for ALS. Virus Res 207(Supplement C):94–105

Masuda K, Marasa B, Martindale JL, Halushka MK, Gorospe M (2009) Tissue- and age-dependent expression of RNA-binding proteins that influence mRNA turnover and translation. Aging (Albany NY) 1(8):681–698

Mateju D, Franzmann TM, Patel A, Kopach A, Boczek EE, Maharana S, Lee HO, Carra S, Hyman AA, Alberti S (2017) An aberrant phase transition of stress granules triggered by misfolded protein and prevented by chaperone function. EMBO J 36(12):1669–1687

Mathys H, Basquin J, Ozgur S, Czarnocki-Cieciura M, Bonneau F, Aartse A, Dziembowski A, Nowotny M, Conti E, Filipowicz W (2014) Structural and biochemical insights to the role of the CCR4-NOT complex and DDX6 ATPase in microRNA repression. Mol Cell 54(5):751–765

Maurel C, Dangoumau A, Marouillat S, Brulard C, Chami A, Hergesheimer R, Corcia P, Blasco H, Andres CR, Vourc'h P (2018) Causative genes in amyotrophic lateral sclerosis and protein degradation pathways: a link to neurodegeneration. Mol Neurobiol 55(8):6480–6499

Maziuk B, Ballance HI, Wolozin B (2017) Dysregulation of RNA binding protein aggregation in neurodegenerative disorders. Front Mol Neurosci 10:89

Mazroui R, Di Marco S, Kaufman RJ, Gallouzi IE (2007) Inhibition of the ubiquitin-proteasome system induces stress granule formation. Mol Biol Cell 18(7):2603–2618

McDonald KK, Aulas A, Destroismaisons L, Pickles S, Beleac E, Camu W, Rouleau GA, Vande Velde C (2011) TAR DNA-binding protein 43 (TDP-43) regulates stress granule dynamics via differential regulation of G3BP and TIA-1. Hum Mol Genet 20(7):1400–1410

McKee AE, Minet E, Stern C, Riahi S, Stiles CD, Silver PA (2005) A genome-wide in situ hybridization map of RNA-binding proteins reveals anatomically restricted expression in the developing mouse brain. BMC Dev Biol 5:14

Meister G, Landthaler M, Peters L, Chen PY, Urlaub H, Luhrmann R, Tuschl T (2005) Identification of novel argonaute-associated proteins. Curr Biol 15(23):2149–2155

Meyer H, Weihl CC (2014) The VCP/p97 system at a glance: connecting cellular function to disease pathogenesis. J Cell Sci 127(Pt 18):3877–3883

Meyer KD, Saletore Y, Zumbo P, Elemento O, Mason CE, Jaffrey SR (2012) Comprehensive analysis of mRNA methylation reveals enrichment in 3′ UTRs and near stop codons. Cell 149 (7):1635–1646

Miller JW, Smith BN, Topp SD, Al-Chalabi A, Shaw CE, Vance C (2012) Mutation analysis of VCP in British familial and sporadic amyotrophic lateral sclerosis patients. Neurobiol Aging 33 (11):2721. e2721–2722

Mizielinska S, Lashley T, Norona FE, Clayton EL, Ridler CE, Fratta P, Isaacs AM (2013) C9orf72 frontotemporal lobar degeneration is characterised by frequent neuronal sense and antisense RNA foci. Acta Neuropathol 126(6):845–857

Molliex A, Temirov J, Lee J, Coughlin M, Kanagaraj AP, Kim HJ, Mittag T, Taylor JP (2015) Phase separation by low complexity domains promotes stress granule assembly and drives pathological fibrillization. Cell 163(1):123–133

Monshausen M, Gehring NH, Kosik KS (2004) The mammalian RNA-binding protein Staufen2 links nuclear and cytoplasmic RNA processing pathways in neurons. NeuroMolecular Med 6 (2–3):127–144

Moon SL, Blackinton JG, Anderson JR, Dozier MK, Dodd BJT, Keene JD, Wilusz CJ, Bradrick SS, Wilusz J (2015) XRN1 stalling in the 5′ UTR of hepatitis C virus and bovine viral diarrhea virus is associated with dysregulated host mRNA stability. PLoS Pathog 11(3):e1004708

Mori K, Lammich S, Mackenzie IR, Forne I, Zilow S, Kretzschmar H, Edbauer D, Janssens J, Kleinberger G, Cruts M, Herms J, Neumann M, Van Broeckhoven C, Arzberger T, Haass C (2013a) hnRNP A3 binds to GGGGCC repeats and is a constituent of p62-positive/TDP43-negative inclusions in the hippocampus of patients with C9orf72 mutations. Acta Neuropathol 125(3):413–423

Mori K, Weng SM, Arzberger T, May S, Rentzsch K, Kremmer E, Schmid B, Kretzschmar HA, Cruts M, Van Broeckhoven C, Haass C, Edbauer D (2013b) The C9orf72 GGGGCC repeat is translated into aggregating dipeptide-repeat proteins in FTLD/ALS. Science 339 (6125):1335–1338

Moschner K, Sundermann F, Meyer H, da Graca AP, Appel N, Paululat A, Bakota L, Brandt R (2014) RNA protein granules modulate tau isoform expression and induce neuronal sprouting. J Biol Chem 289(24):16814–16825

Moujaber O, Mahboubi H, Kodiha M, Bouttier M, Bednarz K, Bakshi R, White J, Larose L, Colmegna I, Stochaj U (2017) Dissecting the molecular mechanisms that impair stress granule formation in aging cells. Biochim Biophys Acta Mol Cell Res 1864(3):475–486

Murakami T, Qamar S, Lin JQ, Schierle GS, Rees E, Miyashita A, Costa AR, Dodd RB, Chan FT, Michel CH, Kronenberg-Versteeg D, Li Y, Yang SP, Wakutani Y, Meadows W, Ferry RR, Dong L, Tartaglia GG, Favrin G, Lin WL, Dickson DW, Zhen M, Ron D, Schmitt-Ulms G, Fraser PE, Shneider NA, Holt C, Vendruscolo M, Kaminski CF, St George-Hyslop P (2015) ALS/FTD mutation-induced phase transition of FUS liquid droplets and reversible hydrogels into irreversible hydrogels impairs RNP granule function. Neuron 88(4):678–690

Neary D, Snowden J, Mann D (2005) Frontotemporal dementia. Lancet Neurol 4(11):771–780
Nelson PT, Wang WX (2010) MiR-107 is reduced in Alzheimer's disease brain neocortex: validation study. J Alzheimers Dis 21(1):75–79
Neumann M, Sampathu DM, Kwong LK, Truax AC, Micsenyi MC, Chou TT, Bruce J, Schuck T, Grossman M, Clark CM, McCluskey LF, Miller BL, Masliah E, Mackenzie IR, Feldman H, Feiden W, Kretzschmar HA, Trojanowski JQ, Lee VM (2006) Ubiquitinated TDP-43 in frontotemporal lobar degeneration and amyotrophic lateral sclerosis. Science 314 (5796):130–133
Nonhoff U, Ralser M, Welzel F, Piccini I, Balzereit D, Yaspo ML, Lehrach H, Krobitsch S (2007) Ataxin-2 interacts with the DEAD/H-box RNA helicase DDX6 and interferes with P-bodies and stress granules. Mol Biol Cell 18(4):1385–1396
Nott TJ, Petsalaki E, Farber P, Jervis D, Fussner E, Plochowietz A, Craggs TD, Bazett-Jones DP, Pawson T, Forman-Kay JD, Baldwin AJ (2015) Phase transition of a disordered nuage protein generates environmentally responsive membraneless organelles. Mol Cell 57(5):936–947
Noubissi FK, Elcheva I, Bhatia N, Shakoori A, Ougolkov A, Liu J, Minamoto T, Ross J, Fuchs SY, Spiegelman VS (2006) CRD-BP mediates stabilization of betaTrCP1 and c-myc mRNA in response to beta-catenin signalling. Nature 441(7095):898–901
Nussbacher JK, Batra R, Lagier-Tourenne C, Yeo GW (2015) RNA-binding proteins in neurodegeneration: seq and you shall receive. Trends Neurosci 38(4):226–236
O'Rourke JG, Bogdanik L, Yanez A, Lall D, Wolf AJ, Muhammad AK, Ho R, Carmona S, Vit JP, Zarrow J, Kim KJ, Bell S, Harms MB, Miller TM, Dangler CA, Underhill DM, Goodridge HS, Lutz CM, Baloh RH (2016) C9orf72 is required for proper macrophage and microglial function in mice. Science 351(6279):1324–1329
Osaka M, Ito D, Suzuki N (2016) Disturbance of proteasomal and autophagic protein degradation pathways by amyotrophic lateral sclerosis-linked mutations in ubiquilin 2. Biochem Biophys Res Commun 472(2):324–331
Parker R, Sheth U (2007) P bodies and the control of mRNA translation and degradation. Mol Cell 25(5):635–646
Parker SJ, Meyerowitz J, James JL, Liddell JR, Crouch PJ, Kanninen KM, White AR (2012) Endogenous TDP-43 localized to stress granules can subsequently form protein aggregates. Neurochem Int 60(4):415–424
Pascual ML, Luchelli L, Habif M, Boccaccio GL (2012) Synaptic activity regulated mRNA-silencing foci for the fine tuning of local protein synthesis at the synapse. Commun Integr Biol 5(4):388–392
Patel A, Lee HO, Jawerth L, Maharana S, Jahnel M, Hein MY, Stoynov S, Mahamid J, Saha S, Franzmann TM, Pozniakovski A, Poser I, Maghelli N, Royer LA, Weigert M, Myers EW, Grill S, Drechsel D, Hyman AA, Alberti S (2015) A liquid-to-solid phase transition of the ALS protein FUS accelerated by disease mutation. Cell 162(5):1066–1077
Paul S, Dansithong W, Figueroa KP, Scoles DR, Pulst SM (2018) Staufen1 links RNA stress granules and autophagy in a model of neurodegeneration. Nat Commun 9(1):3648–3648
Pegoraro V, Merico A, Angelini C (2017) Micro-RNAs in ALS muscle: differences in gender, age at onset and disease duration. J Neurol Sci 380:58–63
Perez-Pepe M, Fernández-Alvarez AJ, Boccaccio GL (2018) Life and work of stress granules and processing bodies: new insights into their formation and function. Biochemistry 57 (17):2488–2498
Pickles S, Destroismaisons L, Peyrard SL, Cadot S, Rouleau GA, Brown RH Jr, Julien JP, Arbour N, Vande Velde C (2013) Mitochondrial damage revealed by immunoselection for ALS-linked misfolded SOD1. Hum Mol Genet 22(19):3947–3959
Radhakrishnan A, Green R (2016) Connections underlying translation and mRNA stability. J Mol Biol 428(18):3558–3564
Radó-Trilla N, Albà M (2012) Dissecting the role of low-complexity regions in the evolution of vertebrate proteins. BMC Evol Biol 12:155–155

Ratti A, Corrado L, Castellotti B, Del Bo R, Fogh I, Cereda C, Tiloca C, D'Ascenzo C, Bagarotti A, Pensato V, Ranieri M, Gagliardi S, Calini D, Mazzini L, Taroni F, Corti S, Ceroni M, Oggioni GD, Lin K, Powell JF, Soraru G, Ticozzi N, Comi GP, D'Alfonso S, Gellera C, Silani V (2012) C9ORF72 repeat expansion in a large Italian ALS cohort: evidence of a founder effect. Neurobiol Aging 33(10):2528. e2527–2514

Reijns MA, Alexander RD, Spiller MP, Beggs JD (2008) A role for Q/N-rich aggregation-prone regions in P-body localization. J Cell Sci 121(Pt 15):2463–2472

Reineke LC, Kedersha N, Langereis MA, van Kuppeveld FJ, Lloyd RE (2015) Stress granules regulate double-stranded RNA-dependent protein kinase activation through a complex containing G3BP1 and Caprin1. MBio 6(2):e02486

Renton AE, Majounie E, Waite A, Simon-Sanchez J, Rollinson S, Gibbs JR, Schymick JC, Laaksovirta H, van Swieten JC, Myllykangas L, Kalimo H, Paetau A, Abramzon Y, Remes AM, Kaganovich A, Scholz SW, Duckworth J, Ding J, Harmer DW, Hernandez DG, Johnson JO, Mok K, Ryten M, Trabzuni D, Guerreiro RJ, Orrell RW, Neal J, Murray A, Pearson J, Jansen IE, Sondervan D, Seelaar H, Blake D, Young K, Halliwell N, Callister JB, Toulson G, Richardson A, Gerhard A, Snowden J, Mann D, Neary D, Nalls MA, Peuralinna T, Jansson L, Isoviita VM, Kaivorinne AL, Holtta-Vuori M, Ikonen E, Sulkava R, Benatar M, Wuu J, Chio A, Restagno G, Borghero G, Sabatelli M, Heckerman D, Rogaeva E, Zinman L, Rothstein JD, Sendtner M, Drepper C, Eichler EE, Alkan C, Abdullaev Z, Pack SD, Dutra A, Pak E, Hardy J, Singleton A, Williams NM, Heutink P, Pickering-Brown S, Morris HR, Tienari PJ, Traynor BJ (2011) A hexanucleotide repeat expansion in C9ORF72 is the cause of chromosome 9p21-linked ALS-FTD. Neuron 72(2):257–268

Rinchetti P, Rizzuti M, Faravelli I, Corti S (2018) MicroRNA metabolism and dysregulation in amyotrophic lateral sclerosis. Mol Neurobiol 55(3):2617–2630

Ritson GP, Custer SK, Freibaum BD, Guinto JB, Geffel D, Moore J, Tang W, Winton MJ, Neumann M, Trojanowski JQ, Lee VM, Forman MS, Taylor JP (2010) TDP-43 mediates degeneration in a novel drosophila model of disease caused by mutations in VCP/p97. J Neurosci 30(22):7729–7739

Rodriguez-Ortiz CJ, Hoshino H, Cheng D, Liu-Yescevitz L, Blurton-Jones M, Wolozin B, LaFerla FM, Kitazawa M (2013) Neuronal-specific overexpression of a mutant valosin-containing protein associated with IBMPFD promotes aberrant ubiquitin and TDP-43 accumulation and cognitive dysfunction in transgenic mice. Am J Pathol 183(2):504–515

Roshan R, Shridhar S, Sarangdhar MA, Banik A, Chawla M, Garg M, Singh VP, Pillai B (2014) Brain-specific knockdown of miR-29 results in neuronal cell death and ataxia in mice. RNA 20 (8):1287–1297

Sareen D, O'Rourke JG, Meera P, Muhammad AK, Grant S, Simpkinson M, Bell S, Carmona S, Ornelas L, Sahabian A, Gendron T, Petrucelli L, Baughn M, Ravits J, Harms MB, Rigo F, Bennett CF, Otis TS, Svendsen CN, Baloh RH (2013) Targeting RNA foci in iPSC-derived motor neurons from ALS patients with a C9ORF72 repeat expansion. Sci Transl Med 5 (208):208ra149

Savas JN, Makusky A, Ottosen S, Baillat D, Then F, Krainc D, Shiekhattar R, Markey SP, Tanese N (2008) Huntington's disease protein contributes to RNA-mediated gene silencing through association with Argonaute and P bodies. Proc Natl Acad Sci U S A 105(31):10820

Savas JN, Ma B, Deinhardt K, Culver BP, Restituito S, Wu L, Belasco JG, Chao MV, Tanese N (2010) A role for Huntington disease protein in dendritic RNA granules. J Biol Chem 285 (17):13142–13153

Sawaya MR, Sambashivan S, Nelson R, Ivanova MI, Sievers SA, Apostol MI, Thompson MJ, Balbirnie M, Wiltzius JJ, McFarlane HT, Madsen AO, Riekel C, Eisenberg D (2007) Atomic structures of amyloid cross-beta spines reveal varied steric zippers. Nature 447(7143):453–457

Scharf KD, Heider H, Hohfeld I, Lyck R, Schmidt E, Nover L (1998) The tomato Hsf system: HsfA2 needs interaction with HsfA1 for efficient nuclear import and may be localized in cytoplasmic heat stress granules. Mol Cell Biol 18(4):2240–2251

Schmidt HB, Rohatgi R (2016) In vivo formation of vacuolated multi-phase compartments lacking membranes. Cell Rep 16(5):1228–1236

Schonrock N, Matamales M, Ittner LM, Gotz J (2012) MicroRNA networks surrounding APP and amyloid-beta metabolism--implications for Alzheimer's disease. Exp Neurol 235(2):447–454

Sen GL, Blau HM (2005) Argonaute 2/RISC resides in sites of mammalian mRNA decay known as cytoplasmic bodies. Nat Cell Biol 7(6):633–636

Sephton CF, Cenik C, Kucukural A, Dammer EB, Cenik B, Han Y, Dewey CM, Roth FP, Herz J, Peng J, Moore MJ, Yu G (2011) Identification of neuronal RNA targets of TDP-43-containing ribonucleoprotein complexes. J Biol Chem 286(2):1204–1215

Sephton CF, Tang AA, Kulkarni A, West J, Brooks M, Stubblefield JJ, Liu Y, Zhang MQ, Green CB, Huber KM, Huang EJ, Herz J, Yu G (2014) Activity-dependent FUS dysregulation disrupts synaptic homeostasis. Proc Natl Acad Sci U S A 111(44):E4769–E4778

Shaw CE (2010) Capturing VCP: another molecular piece in the ALS jigsaw puzzle. Neuron 68 (5):812–814

Sheth U, Parker R (2003) Decapping and decay of messenger RNA occur in cytoplasmic processing bodies. Science 300(5620):805–808

Shukla S, Parker R (2016) Hypo- and hyper-assembly diseases of RNA-protein complexes. Trends Mol Med 22(7):615–628

Simon-Sanchez J, Dopper EG, Cohn-Hokke PE, Hukema RK, Nicolaou N, Seelaar H, de Graaf JR, de Koning I, van Schoor NM, Deeg DJ, Smits M, Raaphorst J, van den Berg LH, Schelhaas HJ, De Die-Smulders CE, Majoor-Krakauer D, Rozemuller AJ, Willemsen R, Pijnenburg YA, Heutink P, van Swieten JC (2012) The clinical and pathological phenotype of C9ORF72 hexanucleotide repeat expansions. Brain 135(Pt 3):723–735

Singh N, Blobel G, Shi H (2015) Hooking She3p onto She2p for myosin-mediated cytoplasmic mRNA transport. Proc Natl Acad Sci U S A 112(1):142–147

Smith RA, Miller TM, Yamanaka K, Monia BP, Condon TP, Hung G, Lobsiger CS, Ward CM, McAlonis-Downes M, Wei H, Wancewicz EV, Bennett CF, Cleveland DW (2006) Antisense oligonucleotide therapy for neurodegenerative disease. J Clin Invest 116(8):2290–2296

Smith PY, Hernandez-Rapp J, Jolivette F, Lecours C, Bisht K, Goupil C, Dorval V, Parsi S, Morin F, Planel E, Bennett DA, Fernandez-Gomez FJ, Sergeant N, Buee L, Tremblay ME, Calon F, Hebert SS (2015) miR-132/212 deficiency impairs tau metabolism and promotes pathological aggregation in vivo. Hum Mol Genet 24(23):6721–6735

Smith J, Calidas D, Schmidt H, Lu T, Rasoloson D, Seydoux G (2016) Spatial patterning of P granules by RNA-induced phase separation of the intrinsically-disordered protein MEG-3. Elife 5:e21337

Sofola OA, Jin P, Qin Y, Duan R, Liu H, de Haro M, Nelson DL, Botas J (2007) RNA-binding proteins hnRNP A2/B1 and CUGBP1 suppress fragile X CGG premutation repeat-induced neurodegeneration in a Drosophila model of FXTAS. Neuron 55(4):565–571

Steward O, Levy WB (1982) Preferential localization of polyribosomes under the base of dendritic spines in granule cells of the dentate gyrus. J Neurosci 2(3):284–291

Stöhr N, Lederer M, Reinke C, Meyer S, Hatzfeld M, Singer RH, Hüttelmaier S (2006) ZBP1 regulates mRNA stability during cellular stress. J Cell Biol 175(4):527–534

Südhof TC (2008) Neuroligins and Neurexins link synaptic function to cognitive disease. Nature 455(7215):903–911

Sun Z, Diaz Z, Fang X, Hart MP, Chesi A, Shorter J, Gitler AD (2011) Molecular determinants and genetic modifiers of aggregation and toxicity for the ALS disease protein FUS/TLS. PLoS Biol 9(4):e1000614

Tada T, Sheng M (2006) Molecular mechanisms of dendritic spine morphogenesis. Curr Opin Neurobiol 16(1):95–101

Takeda S, Wegmann S, Cho H, DeVos SL, Commins C, Roe AD, Nicholls SB, Carlson GA, Pitstick R, Nobuhara CK, Costantino I, Frosch MP, Muller DJ, Irimia D, Hyman BT (2015) Neuronal uptake and propagation of a rare phosphorylated high-molecular-weight tau derived from Alzheimer's disease brain. Nat Commun 6:8490

Takeda S, Commins C, DeVos SL, Nobuhara CK, Wegmann S, Roe AD, Costantino I, Fan Z, Nicholls SB, Sherman AE, Trisini Lipsanopoulos AT, Scherzer CR, Carlson GA, Pitstick R, Peskind ER, Raskind MA, Li G, Montine TJ, Frosch MP, Hyman BT (2016) Seed-competent high-molecular-weight tau species accumulates in the cerebrospinal fluid of Alzheimer's disease mouse model and human patients. Ann Neurol 80(3):355–367

Teixeira D, Parker R (2007) Analysis of P-body assembly in *Saccharomyces cerevisiae*. Mol Biol Cell 18(6):2274–2287

Teixeira D, Sheth U, Valencia-Sanchez MA, Brengues M, Parker R (2005) Processing bodies require RNA for assembly and contain nontranslating mRNAs. RNA 11(4):371–382

Tiraboschi P, Hansen LA, Thal LJ, Corey-Bloom J (2004) The importance of neuritic plaques and tangles to the development and evolution of AD. Neurology 62(11):1984–1989

Tiruchinapalli DM, Oleynikov Y, Kelic S, Shenoy SM, Hartley A, Stanton PK, Singer RH, Bassell GJ (2003) Activity-dependent trafficking and dynamic localization of zipcode binding protein 1 and beta-actin mRNA in dendrites and spines of hippocampal neurons. J Neurosci 23 (8):3251–3261

Tourriere H, Chebli K, Zekri L, Courselaud B, Blanchard JM, Bertrand E, Tazi J (2003) The RasGAP-associated endoribonuclease G3BP assembles stress granules. J Cell Biol 160 (6):823–831

van Dijk E, Cougot N, Meyer S, Babajko S, Wahle E, Séraphin B (2002) Human Dcp2: a catalytically active mRNA decapping enzyme located in specific cytoplasmic structures. EMBO J 21(24):6915–6924

Van Treeck B, Parker R (2018) Emerging roles for intermolecular RNA-RNA interactions in RNP assemblies. Cell 174(4):791–802

Vanderweyde T, Yu H, Varnum M, Liu-Yesucevitz L, Citro A, Ikezu T, Duff K, Wolozin B (2012) Contrasting pathology of the stress granule proteins TIA-1 and G3BP in tauopathies. J Neurosci 32(24):8270–8283

Vanderweyde T, Apicco DJ, Youmans-Kidder K, Ash PEA, Cook C, da Rocha EL, Jansen-West K, Frame AA, Citro A, Leszyk JD, Ivanov P, Abisambra JF, Steffen M, Li H, Petrucelli L, Wolozin B (2016) Interaction of tau with the RNA-binding protein TIA1 regulates tau pathophysiology and toxicity. Cell Rep 15(7):1455–1466

Velikkakath AK, Nishimura T, Oita E, Ishihara N, Mizushima N (2012) Mammalian Atg2 proteins are essential for autophagosome formation and important for regulation of size and distribution of lipid droplets. Mol Biol Cell 23(5):896–909

Vessey JP, Macchi P, Stein JM, Mikl M, Hawker KN, Vogelsang P, Wieczorek K, Vendra G, Riefler J, Tubing F, Aparicio SA, Abel T, Kiebler MA (2008) A loss of function allele for murine Staufen1 leads to impairment of dendritic Staufen1-RNP delivery and dendritic spine morphogenesis. Proc Natl Acad Sci U S A 105(42):16374–16379

Vikesaa J, Hansen TV, Jonson L, Borup R, Wewer UM, Christiansen J, Nielsen FC (2006) RNA-binding IMPs promote cell adhesion and invadopodia formation. EMBO J 25 (7):1456–1468

Vilardo E, Barbato C, Ciotti M, Cogoni C, Ruberti F (2010) MicroRNA-101 regulates amyloid precursor protein expression in hippocampal neurons. J Biol Chem 285(24):18344–18351

Walters RW, Muhlrad D, Garcia J, Parker R (2015) Differential effects of Ydj1 and Sis1 on Hsp70-mediated clearance of stress granules in *Saccharomyces cerevisiae*. RNA 21(9):1660–1671

Wanet A, Tacheny A, Arnould T, Renard P (2012) miR-212/132 expression and functions: within and beyond the neuronal compartment. Nucleic Acids Res 40(11):4742–4753

Wang IF, Wu LS, Chang HY, Shen CK (2008a) TDP-43, the signature protein of FTLD-U, is a neuronal activity-responsive factor. J Neurochem 105(3):797–806

Wang WX, Rajeev BW, Stromberg AJ, Ren N, Tang G, Huang Q, Rigoutsos I, Nelson PT (2008b) The expression of microRNA miR-107 decreases early in Alzheimer's disease and may accelerate disease progression through regulation of beta-site amyloid precursor protein-cleaving enzyme 1. J Neurosci 28(5):1213–1223

Wang WX, Huang Q, Hu Y, Stromberg AJ, Nelson PT (2011) Patterns of microRNA expression in normal and early Alzheimer's disease human temporal cortex: white matter versus gray matter. Acta Neuropathol 121(2):193–205

Wang X, Tan L, Lu Y, Peng J, Zhu Y, Zhang Y, Sun Z (2015) MicroRNA-138 promotes tau phosphorylation by targeting retinoic acid receptor alpha. FEBS Lett 589(6):726–729

Wang C, Schmich F, Srivatsa S, Weidner J, Beerenwinkel N, Spang A (2018) Context-dependent deposition and regulation of mRNAs in P-bodies. Elife 7:e29815

Wegmann S, Eftekharzadeh B, Tepper K, Zoltowska KM, Bennett RE, Dujardin S, Laskowski PR, MacKenzie D, Kamath T, Commins C, Vanderburg C, Roe AD, Fan Z, Molliex AM, Hernandez-Vega A, Muller D, Hyman AA, Mandelkow E, Taylor JP, Hyman BT (2018) Tau protein liquid-liquid phase separation can initiate tau aggregation. EMBO J 37(7):e98049

Welshhans K, Bassell GJ (2011) Netrin-1-induced local beta-actin synthesis and growth cone guidance requires zipcode binding protein 1. J Neurosci 31(27):9800–9813

Wenk GL (2003) Neuropathologic changes in Alzheimer's disease. J Clin Psychiatry 64(Suppl 9):7–10

Wheeler JR, Matheny T, Jain S, Abrisch R, Parker R (2016) Distinct stages in stress granule assembly and disassembly. Elife 5:e18413

Wheeler JR, Jain S, Khong A, Parker R (2017) Isolation of yeast and mammalian stress granule cores. Methods 126:12–17

Williams KL, Warraich ST, Yang S, Solski JA, Fernando R, Rouleau GA, Nicholson GA, Blair IP (2012) UBQLN2/ubiquilin 2 mutation and pathology in familial amyotrophic lateral sclerosis. Neurobiol Aging 33(10):2527.e2523–2510

Willis D, Li KW, Zheng JQ, Chang JH, Smit A, Kelly T, Merianda TT, Sylvester J, van Minnen J, Twiss JL (2005) Differential transport and local translation of cytoskeletal, injury-response, and neurodegeneration protein mRNAs in axons. J Neurosci 25(4):778–791

Wolozin B (2012) Regulated protein aggregation: stress granules and neurodegeneration. Mol Neurodegener 7:56–56

Wolozin B, Apicco D (2015) RNA binding proteins and the genesis of neurodegenerative diseases. Adv Exp Med Biol 822:11–15

Wu D, Raafat A, Pak E, Clemens S, Murashov AK (2012) Dicer-microRNA pathway is critical for peripheral nerve regeneration and functional recovery in vivo and regenerative axonogenesis in vitro. Exp Neurol 233(1):555–565

Wu Y, Zhu J, Huang X, Du Z (2016) Crystal structure of a dimerization domain of human Caprin-1: insights into the assembly of an evolutionarily conserved ribonucleoprotein complex consisting of Caprin-1, FMRP and G3BP1. Acta Crystallogr Sect D 72(6):718–727

Yang G, Song Y, Zhou X, Deng Y, Liu T, Weng G, Yu D, Pan S (2015) MicroRNA-29c targets beta-site amyloid precursor protein-cleaving enzyme 1 and has a neuroprotective role in vitro and in vivo. Mol Med Rep 12(2):3081–3088

Yasuda K, Zhang H, Loiselle D, Haystead T, Macara IG, Mili S (2013) The RNA-binding protein Fus directs translation of localized mRNAs in APC-RNP granules. J Cell Biol 203(5):737–746

Yin H, Sun Y, Wang X, Park J, Zhang Y, Li M, Yin J, Liu Q, Wei M (2015) Progress on the relationship between miR-125 family and tumorigenesis. Exp Cell Res 339(2):252–260

Yu X, Caltagarone J, Smith MA, Bowser R (2005) DNA damage induces cdk2 protein levels and histone H2B phosphorylation in SH-SY5Y neuroblastoma cells. J Alzheimers Dis 8(1):7–21

Yu Y, Li Y, Zhang Y (2015) Screening of APP interaction proteins by DUALmembrane yeast two-hybrid system. Int J Clin Exp Pathol 8(3):2802–2808

Zalfa F, Achsel T, Bagni C (2006) mRNPs, polysomes or granules: FMRP in neuronal protein synthesis. Curr Opin Neurobiol 16(3):265–269

Zeitelhofer M, Karra D, Macchi P, Tolino M, Thomas S, Schwarz M, Kiebler M, Dahm R (2008) Dynamic interaction between P-bodies and transport ribonucleoprotein particles in dendrites of mature hippocampal neurons. J Neurosci 28(30):7555–7562

Zhang ZC, Chook YM (2012) Structural and energetic basis of ALS-causing mutations in the atypical proline-tyrosine nuclear localization signal of the fused in sarcoma protein (FUS). Proc Natl Acad Sci U S A 109(30):12017–12021

Zhang HL, Eom T, Oleynikov Y, Shenoy SM, Liebelt DA, Dictenberg JB, Singer RH, Bassell GJ (2001) Neurotrophin-induced transport of a beta-actin mRNP complex increases beta-actin levels and stimulates growth cone motility. Neuron 31(2):261–275

Zhang J, Hu M, Teng Z, Tang YP, Chen C (2014a) Synaptic and cognitive improvements by inhibition of 2-AG metabolism are through upregulation of microRNA-188-3p in a mouse model of Alzheimer's disease. J Neurosci 34(45):14919–14933

Zhang KY, Yang S, Warraich ST, Blair IP (2014b) Ubiquilin 2: a component of the ubiquitin-proteasome system with an emerging role in neurodegeneration. Int J Biochem Cell Biol 50:123–126

Zhang B, Chen CF, Wang AH, Lin QF (2015) MiR-16 regulates cell death in Alzheimer's disease by targeting amyloid precursor protein. Eur Rev Med Pharmacol Sci 19(21):4020–4027

Zhang X, Lin Y, Eschmann NA, Zhou H, Rauch JN, Hernandez I, Guzman E, Kosik KS, Han S (2017) RNA stores tau reversibly in complex coacervates. PLoS Biol 15(7):e2002183

Zhang K, Daigle JG, Cunningham KM, Coyne AN, Ruan K, Grima JC, Bowen KE, Wadhwa H, Yang P, Rigo F, Taylor JP, Gitler AD, Rothstein JD, Lloyd TE (2018) Stress granule assembly disrupts nucleocytoplasmic transport. Cell 173(4):958–971.e917

Zheng D, Chen C-YA, Shyu A-B (2011) Unraveling regulation and new components of human P-bodies through a protein interaction framework and experimental validation. RNA 17 (9):1619–1634

Zu T, Liu Y, Banez-Coronel M, Reid T, Pletnikova O, Lewis J, Miller TM, Harms MB, Falchook AE, Subramony SH, Ostrow LW, Rothstein JD, Troncoso JC, Ranum LP (2013) RAN proteins and RNA foci from antisense transcripts in C9ORF72 ALS and frontotemporal dementia. Proc Natl Acad Sci U S A 110(51):E4968–E4977

# Chapter 9
# Lessons from (pre-)mRNA Imaging

**Srivathsan Adivarahan and Daniel Zenklusen**

**Abstract** Cells are complex assemblies of molecules organized into organelles and membraneless compartments, each playing important roles in ensuring cellular homeostasis. The different steps of the gene expression pathway take place within these various cellular compartments, and studying gene regulation and RNA metabolism requires incorporating the spatial as well as temporal separation and progression of these processes. Microscopy has been a valuable tool to study RNA metabolism, as it allows the study of biomolecules in the context of intact individual cells, embryos or tissues, preserving cellular context often lost in experimental approaches that require the collection and lysis of cells in large numbers to obtain sufficient material for different types of assays. Indeed, from the first detection of RNAs and ribosomes in cells to today's ability to study the behaviour of single RNA molecules in living cells, or the expression profile and localization of hundreds of mRNA simultaneously in cells, constant effort in developing tools for microscopy has extensively contributed to our understanding of gene regulation. In this chapter, we will describe the role various microscopy approaches have played in shaping our current understanding of mRNA metabolism and outline how continuous development of new approaches might help in finding answers to outstanding questions or help to look at old dogmas through a new lens.

**Keywords** mRNA · mRNPs · Electron microscopy · In situ hybridization · smFISH · Polysomes · RNA imaging · Single molecule microscopy · Gene expression

---

S. Adivarahan · D. Zenklusen (✉)
Département de Biochimie et médecine moléculaire, Université de Montréal, Montréal, Québec, Canada
e-mail: srivathsan.adivarahan@umontreal.ca; daniel.r.zenklusen@umontreal.ca



M. Oeffinger, D. Zenklusen (eds.), *The Biology of mRNA: Structure and Function*, Advances in Experimental Medicine and Biology 1203,
https://doi.org/10.1007/978-3-030-31434-7_9

## 9.1 Tools for RNA Visualization at Different Scales

The dynamic regulation of gene expression is critical for cells and organisms to develop and maintain homeostasis. mRNAs play a central role in this process, acting as messenger molecules that connect the information stored in the genome and the machineries translating them into proteins. However, despite their often-short-lived role as templates for proteins synthesis, controlling mRNA metabolism is among the most complex cellular processes composed of numerous steps, many of which are subject to regulation and quality control, involving hundreds of proteins. Therefore, a long-standing and critical effort has been made towards studying gene regulation and mRNA metabolism, as mis-regulation in any step can lead to a wide range of diseases (Cooper et al. 2009). Microscopy approaches have been critical tools in this effort, as they allow to study these processes in the context of the native environment of the cell.

Many imaging-based approaches have been developed to observe mRNA and messenger ribonucleoproteins (mRNPs) in cells, either in a fixed cell or a living cell context, and each of these approaches has its own strengths and limitations. While fixation prior to any kind of labelling for RNA detection comes with the benefit of allowing complex labelling protocols and long exposures during image acquisition that is often required for robust detection and multiplexing, the dynamics of interactions within the cell is lost and can only be captured using live-cell approaches. Light microscopy techniques are most commonly used to study mRNAs, but they are limited in resolution; however, recently developed single molecule localization and super-resolution microscopy approaches have helped overcome this barrier, but their usage in understanding the behaviour of RNAs in cells is still in its infancy . Electron microscopy, on the other hand, is superior in terms of resolution and has been used in combination with different staining protocols or combined with immunolabelling to detect RNPs or target specific RNA binding proteins in cells or in vitro; however, it is limited in terms of labelling efficiencies for specific RNAs and multiplexing. In this chapter, we will describe the most commonly used techniques for visualizing mRNPs both in fixed and live cells, at low and high resolution, before discussing in more detail how RNA imaging has contributed to the current understanding of mRNA metabolism.

### *9.1.1 RNA Detection in Fixed Cells and Tissues*

Most studies involving mRNA detection using microscopy have been performed in fixed cells and tissues. The main reason to work in a fixed cell environment is largely technical, allowing access to a wider range of tools and methodologies that are easier to implement and requiring less sophisticated microscopy setups to image, making them a preferable choice over live-cell imaging. Moreover, due to the crudeness of sample preparation and the destructive nature of high-energy electron beams, EM studies have to be restricted to fixed cells. Two approaches are generally used to visualize mRNPs—either through direct targeting of the mRNA or indirect targeting of associated RNA-binding proteins within the mRNA-protein complex. When using RNA as the target for mRNA imaging, the most common tools use antisense probes, most often DNA probes of various lengths, that hybridize specifically to an mRNA of interest. These probes can be coupled with labels that can be recognized using either electron microscopy or tomography (EM-in situ hybridization), fluorescent in situ hybridization (FISH) or, in the early days of RNA detection, radioactivity. The use of fluorescent dyes instead of heavy metals in EM or radioactive materials is advantageous as it allows for multiplexing using probes labelled with spectrally differentiable fluorophores. Early RNA studies were often limited to the detection of either highly abundant mRNAs or mRNAs that show high local concentrations within specific cellular compartments, such as localized RNAs, largely due to the limited sensitivity and low signal-to-noise ratio of RNA FISH when using single and often long (>1 kB) fluorescent probes. The development of methods that allowed for detection of single mRNA molecules in cells, independent of their abundance, represented a milestone in RNA imaging and opened the door for more quantitative approaches to mRNA imaging in cells. However, adoption of the technique as the standard tool for cellular mRNA imaging was a slow process. The development of single molecule resolution RNA FISH (smFISH) by the Singer laboratory in 1998 was the first of many crucial steps towards this process (Femino et al. 1998). In a seminal paper by Femino et al., multiple DNA oligonucleotides probes ~50 nt in length were targeted to hybridize with the beta- and gamma-actin mRNAs allowing for the simultaneous detection of single mRNA molecules of multiple transcripts within the same cell (Fig. 9.1a top). However, the limited availability of sensitive cameras and high-end imaging equipment, combined with the need for custom synthesis of densely labelled probes that were both expensive and harder to generate for laboratories that did not had access to a DNA synthesizer, limited the adoption of the technique. Over the last decade, however, various modifications to the initial approach have been made that have made single mRNA detection much more accessible. The approach that is currently most widely adopted uses 35–50 DNA oligonucleotides, each 20 nt in length and coupled to a single fluorescent dye. The probes are hybridized in fixed cells in low formamide concentrations, resulting in robust single molecule detection (Fig. 9.1a bottom) (Raj et al. 2008). The high signal-to-noise ratio observed for single mRNAs has seen its wide adoption for mRNAs imaging in many organisms, cells and tissues. A more cost-efficient adaptation of this approach, termed single molecule inexpensive FISH (smiFISH), has also been developed that uses target-specific probes containing a transcript-specific sequence as well as an overhang that can hybridize with a common set of fluorescently labelled antisense probes (Fig. 9.1b) (Tsanov et al. 2016). Additionally, alternative approaches have been successfully implemented, using either branched probes (Sinnamon and Czaplinski 2014; Wang et al. 2012), rolling circle amplification (Larsson et al. 2010) or click chemistry to padlock probes to the target mRNA or probes hybridized to the target mRNA (Rouhanifard et al. 2019) to increase signal amplification (Fig. 9.1c–f). Furthermore, to overcome the

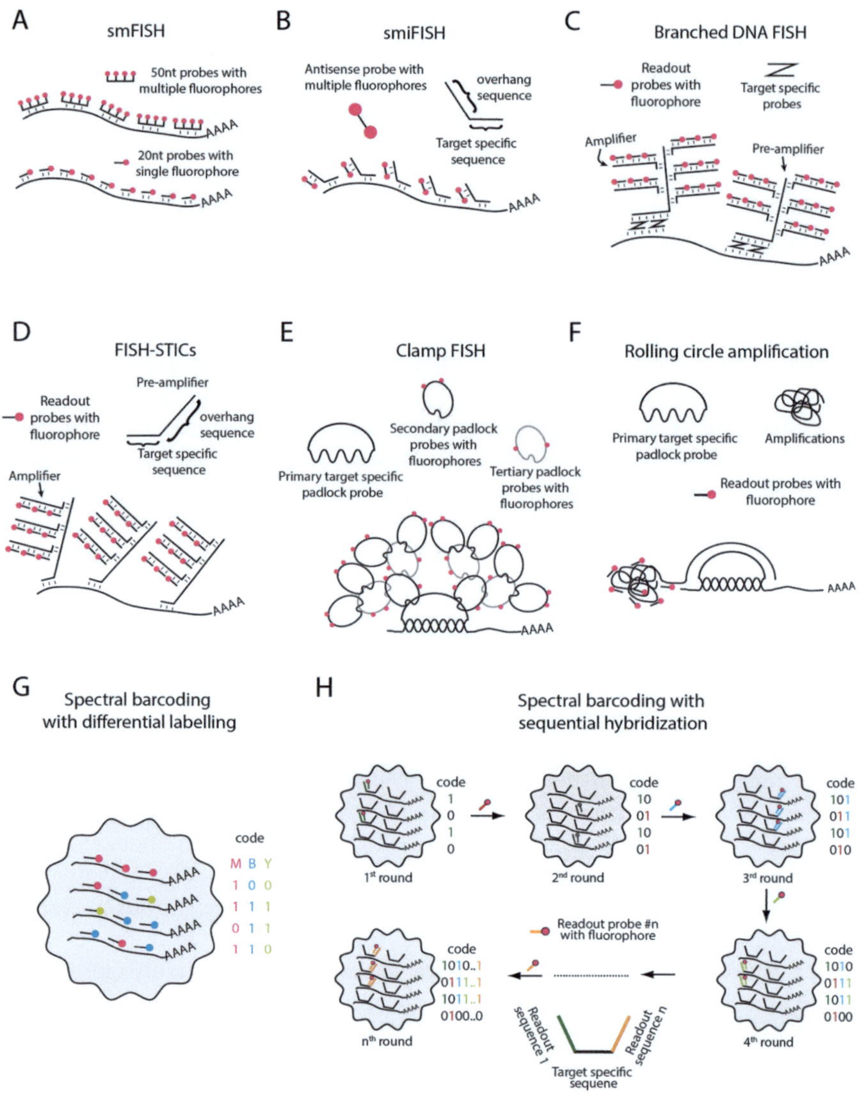


**Fig. 9.1** Methods used to visualize mRNA in fixed cells. (**a**) Single molecule RNA in situ hybridization (smFISH) uses either multiple 50 nt ssDNA oligos labelled with multiple dyes or 20 nt ssDNA oligos labelled at a single position. (**b**) smiFISH uses a 20–35 nt target-specific sequence plus a 28 nt overhang which can hybridize to an antisense FLAP probe coupled to fluorescent dyes. (**c**) Branched DNA FISH requires hybridization of two gene-specific probes to allow the hybridization of a preamplifier and subsequent amplifier probes that are then detected with dye-labelled readout probes. (**d**) FISH-STICS is similar to branched DNA FISH, with the preamplifier sequence present as an overhang to the gene-specific probe. (**e**) and (**f**) Padlock-based systems for detection rely on single-stranded target-specific probes with ligatable ends. ClampFISH uses multiple round of hybridization with padlock probes with each round amplifying the signal.

low multiplexing capability of RNA FISH due to the limited availability of spectrally differentiable fluorophores, spectral barcoding has often been used, either by using probes labelled with specific combinations of dyes for specific RNAs or through sequential rounds of hybridization with a subset of probes followed by rounds of imaging and stripping of hybridized probes, and has allowed for the detection of tens to hundreds of RNAs in the same cell (Fig. 9.1g, h) (Chen et al. 2015; Codeluppi et al. 2018; Eng et al. 2017, 2019; Jakt et al. 2013; Levsky et al. 2002; Lubeck and Cai 2012; Lubeck et al. 2014; Wang et al. 2018).

mRNAs in cells are part of mRNPs, and mRNAs can also be visualized indirectly by visualizing protein bound to mRNAs, either using antibodies to specific RNA-binding proteins (RBPs) or using fluorescent protein fusions. Similar to FISH probes, antibodies can be conjugated either with fluorescent dyes or heavy metals to be imaged using either fluorescence or electron microscopy, respectively. However, there are important differences to direct RNA detection as most RBPs bind to many different mRNAs and, in addition, exist in cells in RNA-bound as well as in free fractions. Imaging RBPs, therefore, reveals a different kind of information than the RNA-centric information obtained from hybridization approaches that target specific transcripts. Nevertheless, combining RBP imaging and FISH is a powerful tool to study regulatory mechanisms acting on mRNAs.

For electron microscopy and tomography studies, mRNPs can be labelled using heavy metal salt solutions such as uranyl acetate and lead citrate. These salts can react with cellular structures including RNA and RNA-binding protein to increase their contrast when imaging by electron microscopy (Bozzola and Russell 1999). This methodology can either be used alone or combined with EM-ISH or antibody-based targeting of RNA-binding proteins to further increase the labelling of mRNPs or identification of specific proteins as part of the mRNP complex. More recently, the advent of cryo-electron microscopy/tomography has made it possible to determine structures of different RNA-protein complexes in vitro without the need for crystallization (Kühlbrandt 2014); however, its usage in imaging mRNPs in cells might be limited as mRNPs are heterogeneous both in protein and mRNA composition. Moreover, the crowded environment of the cell combined with the low contrast while imaging has limited the usage of cryo-electron tomography to specific regions of the cell, where it is possible to spatially separate RNPs (Mahamid et al. 2016).

**Fig. 9.1** (continued) Rolling circle amplification uses one padlock probe from which the signal can be amplified. (**g**) and (**h**) Spectral barcoding approaches to detect multiple mRNA targets either through differential labelling (**g**) or multiple rounds of hybridization and stripping to generate unique barcodes for specific mRNAs (h). See text for more details

## 9.1.2 RNA Detection in Living Cells

Cellular processes are dynamic and the various steps within the gene expression pathway involve spatial progression through different cellular structures and compartments. Investigating such dynamic processes is limited when using approaches that rely on cell fixation, and various imaging techniques have been developed to visualize mRNPs in living cells (Fig. 9.2). Early approaches relied on hybridizing probes labelled with single fluorophore to target mRNAs. These probes contained either a single fluorescent dye, were labeled with an acceptor or donor FRET pair that only fluoresce when bound to the target RNA or were designed such as to contain a dye and quencher on the same oligonucleotide sequence, where the dye is quenched when the probe is not hybridized to its target (Fig. 9.2a, b) (Bao et al. 2009; Molenaar et al. 2001; Santangelo et al. 2004; Tyagi and Kramer 1996). However, the usage of these probes for visualizing mRNAs in live cells was challenging and often limited due to their low signal-to-noise ratio, fast degradation and difficulty to

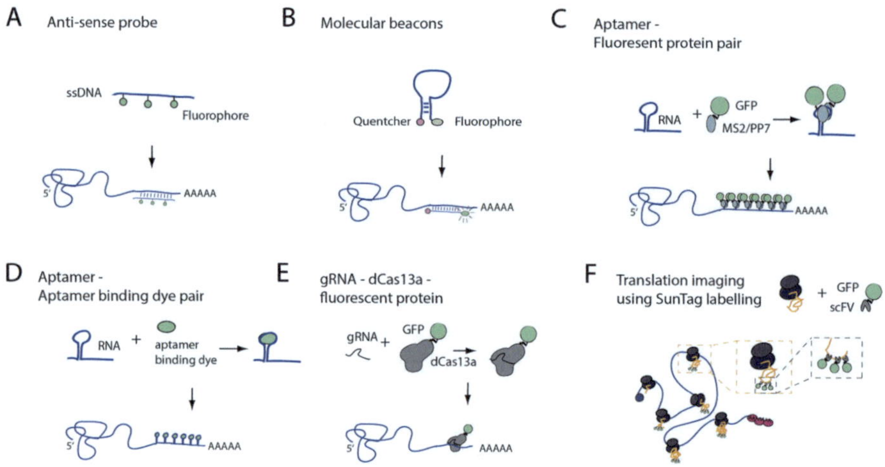


**Fig. 9.2** Tools for life cell RNA imaging. Cartoons illustrating different methods to visualize mRNA in living cells. (**a**) Antisense probes. Single-stranded DNA or RNA probes labelled with fluorescent dyes can be inserted into cells using different transfection or injection strategies where they hybridize to specific mRNAs. (**b**) Molecular beacons change their fluorescent properties when binding to a target mRNA, thereby reducing background. (**c**) Aptamer-fluorescent protein pair. RNA stem-loops bound with high affinity and specificity by RNA-binding proteins such as the capsid proteins from the bacteriophages MS2 and PP7, which when fused to a fluorescent protein result in a fluorescent labelled RNA. (**d**) Aptamer-aptamer binding dye pair. Molecules designed to bind to RNA aptamers such as Spinach or Mango can result in fluorescent RNAs. (**e**) Cas13a RNA imaging. Cas13a binds RNA specifically, mediated by a guide RNA (gRNA). Co-expressing a gRNA and a catalytic dead mutant termed dCas13a fused to a fluorescent protein results in a fluorescent labelled mRNA. Multiple gRNAs to an mRNA are used to enhance the signal. (**f**) Translation imaging using the SunTag peptide labelling system. An antibody fused to GFP (scFv-GFP) that recognizes a short multimerized peptide sequence at the N-terminus of a nascent protein allows imaging of translating mRNAs. See text for more details

permeate through the cell membrane, requiring the use of delivery methods such as microinjection, electroporation, cell membrane permeabilization or packaging in cell-penetrating peptides [discussed in Bao et al. (2009)].

To overcome many of these drawbacks, aptamer-based RNA visualization approaches have been developed that allow detection of RNAs either using endogenously expressed fluorescent proteins (aptamer-protein combination) or through membrane-permeable fluorescent dyes (aptamer-dye combination) (Fig. 9.2c, d). The most commonly used aptamer-protein combinations are derived from bacteriophage capsid proteins that bind with high affinity and specificity to short stem-loop RNA structures. Because they are derived from bacteriophages, these proteins do not have endogenous targets in eukaryotic systems. Coat protein/RNA stem-loop combination of the MS2 and PP7 bacteriophages is most frequently used, but other combinations, such as lambda N or U1A, have also been utilized (Bertrand et al. 1998; Brodsky and Silver 2000; Daigle and Ellenberg 2007; Larson et al. 2011; Urbanek et al. 2014). Insertion of a specific aptamer sequence to an RNA of interest and co-expression of a coat protein fused to a fluorescent protein result in a fluorescently labelled RNA. However, insertion of a single stem-loop does not allow for detection of single RNAs, and aptamer sequences need to be multimerized to amplify the signal. To obtain robust single molecule sensitivity, typically 12–24 stem-loops are inserted to an mRNA of interest, often within the 3′ untranslated region. Over the years, many modifications have been made to the system to fine-tune signal-to-noise ratio and to adopt the system for the study of specific processes, either through modifications to the RNA aptamer sequences, dimerization of the proteins or fine-tuning the expression levels of the aptamer binding proteins (Tutucci et al. 2017; Wu et al. 2012, 2015a). Aptamer labelled RNAs are either ectopically expressed or integrated into genomic loci or the aptamer sequence can be integrated to endogenously expressed mRNAs. Common in lower eukaryotes such as *S. cerevisiae* for a long time, genomic integration only recently got adapted in higher eukaryotes using different genome editing approaches such as TALEN (Ochiai et al. 2014) or CRISPR/Cas9 (Spille et al. 2019).

One limitation of the MS2/PP7 systems is that it requires the expression of the aptamer binding proteins fused to a fluorescent protein. To circumvent this problem, dye binding aptamers have been developed such as Mango and Spinach (Dolgosheina et al. 2014; Paige et al. 2011). Aptamer-dye combinations provide a distinct advantage in terms of theoretically stronger signal because of the use of organic dyes that are generally brighter and more photostable than fluorescent proteins and, for dyes that change their fluorescent properties upon binding to the aptamer (fluorgenic dyes), can further reduce background. However, despite having great potential, aptamer-dye pairs have not yet shown to result in robust single molecule detection, possibly due to issues in RNA folding, cell permeability and/or dye binding properties. Aptamer-based imaging systems are discussed more in detail in (Dolgosheina and Unrau 2016; Urbanek et al. 2014). Moreover, the limitations of aptamer-based methods still apply to aptamer-dye combinations requiring genetic manipulations to insert aptamer sequences to the RNA of interest and are therefore laborious for studying endogenously expressed RNAs.

More recently, a CRISPR-Cas9-based method was developed that uses an RNA-targeting Cas9 to recognize RNAs of interest (Nelles et al. 2016), as well as Cas13a, which directly binds to RNA (Abudayyeh et al. 2017) (Fig. 9.2e). Though successfully applied for detecting the population of highly abundant endogenously expressed mRNAs, signal-to-noise ratio sufficient for single molecule detection has not yet been reported.

In addition to new methodologies for mRNP imaging in cells, tools for image analysis have simultaneously been developed with the aim to facilitate detection, localization and tracking of single RNA molecules. Single-particle tracking algorithms initially developed for tracking receptor diffusion on cell surfaces were later utilized to tracking of single molecules in cells with a very high spatial accuracy (~10–20 nm) (Cherry et al. 1998). These algorithms were further developed by Thompson et al. to enable sub-diffraction resolution localization of single particles for a wide range of circumstances (Thompson et al. 2002). To overcome the resolution limit determined by the wavelength of light, the signal emitted from individual spatially distinct particles was fitted to a 2D Gaussian with the centroid of the Gaussian being able to determine molecule localization to a very high precision. This and similar approaches have since been widely adapted to create tools for localization and counting of single mRNPs in fixed cells and localization, as well as counting and tracking of single mRNPs in living cells (Jaqaman et al. 2008; Lionnet et al. 2011; Mueller et al. 2013; Tinevez et al. 2017).

Together, these imaging techniques have been used to study various aspects of mRNP metabolism, starting from transcription to degradation, as well as have been used to study biophysical properties of mRNPs. Below, we will discuss how imaging approaches have contributed to the current understanding of these processes.

## 9.2 Visualizing Nuclear (pre-)mRNPs

The life of an mRNA starts with its synthesis by RNA polymerase II, when mRNAs are produced as precursors that require extensive processing and maturation before being released from chromatin to find their way to the nuclear periphery to be exported to the cytoplasm for translation. Imaging has been an important tool to study all aspects of nuclear RNA metabolism, including mRNA synthesis and processing, and has revealed important aspects of the kinetics and dynamics of these various processes (Fig. 9.3).

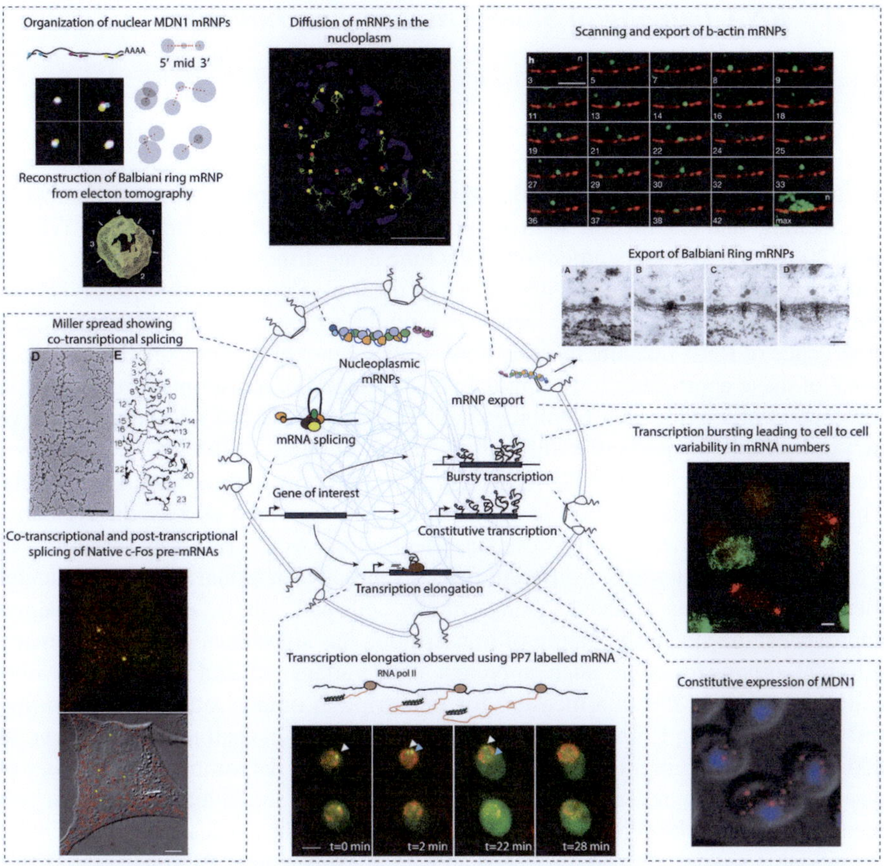


**Fig 9.3** Visualizing nuclear mRNA metabolism. Clockwise starting at the top left. *Organization and dynamics of nuclear mRNPs*. (Top) Conformations of nuclear MDN1 mRNAs in HEK293 cells visualized using smFISH (left), their ball-and-stick representations (right) and cartoon showing regions of probe hybridization (top) from Adivarahan et al. (2018). (Bottom) Reconstruction of BR mRNPs observed with electron tomography. Modified with permission from Mehlin et al. (1992). (Right) Restricted diffusion of mRNPs through interchromatin space visualized using molecular beacons. Blue shows DAPI signal. Modified from Vargas et al. (2005). Copyright (2005) National Academy of Sciences, USA. *mRNP export*. Scanning and export of beta-actin mRNA visualized using MS2 RNA tagging. Figure shows time series of a single β-actin mRNP reaching the nuclear periphery prior to its export. Bottom right panel shows overlay of time series. Nuclear pores are labelled in red. Modified with permission from Grünwald and Singer (2010). Export of BR mRNPs observed using electron microscopy. mRNPs were found to dock to the nuclear basket and remodelled before exiting through the central channel. Modified with permission from Mehlin et al. (1992). *Transcription bursting*. smFISH illustrating transcription bursting of an inducible reporter (green) and the large subunit of RNA polymerase II (red). Modified with permission from Raj et al. (2006). *Constitutive transcription*. Image showing constitutive expression of MDN1 mRNA in *Saccharomyces cerevisiae* using smFISH. Modified from Zenklusen et al. (2008). *Measuring transcription elongation*. Images showing expression of PP7 labelled reporter mRNA (green) in live cells. Nuclear pores are shown in red. Panels represent different time points with arrows pointing sites of active transcription. Modified from Larson et al. (2011). *mRNA processing*.

### *9.2.1 RNA Imaging to Study Transcription Initiation and Elongation*

Transcription regulation is a complex process initiated by recruitment of the pre-initiation complex at the promoter region, a process itself influenced by a multitude of factors including the binding of transcription factors, chromatin remodelling and interaction with regulatory elements such as enhancers (Hager et al. 2009). Extensively studied for a long time using different experimental systems and approaches, including in vitro assays to determine binding affinities of TF and the role of general and specific factors in modulating the transcription reaction, the emergence of RNA imaging to study transcription quickly revealed the limitations of some of these approaches to recapitulate many aspects of transcription regulation in the context of a living cell (Coulon et al. 2013). One factor that made application of in vitro studies in particular difficult to living systems is that many regulatory elements that have to assemble at promoters are present in finite numbers in cells. This implies that transcription in cells can best be described as a stochastic rather than a purely deterministic process, an effect which would result in variability of RNA numbers expressed in different cells, even between clonal cells grown under identical conditions. Indeed, using single cell and single molecule imaging approaches, the stochastic nature of transcription has since been described in many different organisms. One such approach uses variants of smFISH to determine cellular mRNA levels as a measure for transcription output, similar to measuring mRNA levels using RNAseq or qRT-PCR, but at the single cell level. Moreover, in addition to quantifying total RNA, smFISH also allows determining the number of nascent transcripts, revealing transcriptional activity at individual loci. These two measurements can be combined with modelling approaches to describe transcription behaviour in single cells and have been applied in many studies (Bartman et al. 2019; Halpern et al. 2015; Paré et al. 2009; Raj and van Oudenaarden 2008; Raj et al. 2006; Senecal et al. 2014; Zenklusen et al. 2008). Alternatively, transcription can be monitored in real time by inserting aptamer repeats into genes and measuring the intensity of fluorescence signals of nascent mRNAs using fluorescently tagged proteins with high affinity to these aptamers, such as MS2 and PP7 (see above). As each initiation and termination event leads to fluctuation in transcription site intensity, these measurements can reveal transcription dynamics, including initiation frequencies (Chubb et al. 2006; Darzacq et al. 2007; Golding et al. 2005; Larson

---

**Fig 9.3** (continued) (Top) Electron micrograph of a Miller spread chromatin isolated from *Drosophila* embryos and drawing showing tracing of the micrograph. The numbers represent different mRNA templates transcribed. Looping of introns can be observed along with the co-transcriptional splicing of mRNPs identified by the deposition of the spliceosome complexes. Modified with permission from Beyer and Osheim (1988). (Bottom) Co- and post-transcriptional splicing of c-Fox pre-mRNAs visualized by smFISH. Exons shown in red, introns in green. Bottom shows overlay of the localized signals on a bright-field image with yellow spots representing unspliced pre-mRNAs not colocalizing with transcription sites. Modified with permission from Vargas et al. (2011). See text for more details

et al. 2011; Muramoto et al. 2012; Yunger et al. 2010). Such measurements have revealed many important features on how genes are transcribed that could not be obtained using classical approaches, including the stochastic nature of many aspects of transcription regulation. One of the most consequential observations was that most genes are transcribed in a discontinuous manner, where periods of active transcription are interspaced with periods where there is no new initiation by RNA polymerase II (Chubb et al. 2006; Golding et al. 2005; Muramoto et al. 2012; Zenklusen et al. 2008). Thereafter, various studies have showed that both the duration of 'on' and 'off' periods, as well as the initiation frequency during the 'on' time, often described as a transcription burst, are extensively regulated. Factors such as histone modifications, promoter architecture, binding of transcription factors, formation of enhancer-promoter loops, cell volume and position of genes in the genome were all found to regulate transcription bursting which continues to be extensively studied (Bartman et al. 2016; Chen and Larson 2016; Lenstra et al. 2016; Nicolas et al. 2018; Padovan-Merhar et al. 2015; Raj and van Oudenaarden 2008; Raj et al. 2006; Senecal et al. 2014; Suter et al. 2011).

Assays used to study transcription initiation have also been used to measure the speed of an RNA polymerase along the template, either by modelling smFISH data or by correlating signal fluctuations from time traces of aptamer labelled RNAs. These measurements revealed a high amount of variability in the elongation speed of RNA polymerase II, ranging from ~25 nt/s in *E. coli* (Golding and Cox 2004; Golding et al. 2005) to 5–1000 nt/s in eukaryotes (Ben-Ari et al. 2010; Brody et al. 2011; Darzacq et al. 2007; Femino et al. 1998; Hocine et al. 2013; Larson et al. 2011; Maiuri et al. 2011; Wada et al. 2009; Yunger et al. 2010). Moreover, transcription elongation rates were found to vary from cell to cell, with some of the variations linked to the cell cycle (Hocine et al. 2013; Larson et al. 2011). One cause of the variability in elongation rates was attributed to RNA polymerase pausing, previously suggested by ChIP studies that showed non-uniform distribution of RNA polymerase II across genes, with intermittent spikes (Churchman and Weissman 2011; Jonkers et al. 2014; Zeitlinger et al. 2007). This was further confirmed by FRAP studies on the transcription site of MS2-labelled mRNA, revealing that RNA polymerases can stochastically pause during the elongation step, or at the 3′ terminus post the polyadenylation site (Boireau et al. 2007; Darzacq et al. 2007). In addition to stochastic pausing events, ChIP results also indicated the RNA polymerases could pause throughout the body of the gene with particularly enrichment near the promoter (termed promoter-proximal pausing), before nucleosome dyads and at intron-exon junctions, suggesting a wider role for RNA polymerase in regulation of gene expression (Churchman and Weissman 2011; Kwak et al. 2013; Lenstra et al. 2016). However, single molecule FRAP observations on reporter mRNAs did not observe pausing in the body of intron-containing genes (Brody et al. 2011), indicating that pausing might not be a universal for all intron-exon junctions. Overall, single molecule microscopy techniques have provided a platform to image the transcriptional behaviour of genes at the level of single alleles, providing a better spatial resolution as well as sensitivity in comparison to single cell sequencing techniques.

However, despite the progress in determining factors affecting transcriptional bursting, the molecular mechanisms regulating this process are not yet fully understood

### 9.2.2 Pre-mRNA Maturation

mRNAs are first transcribed as precursors, and the process of mRNA maturation involves multiple processing steps, including modification of the 5′, excision of introns, 3′ end cleavage and polyadenylation as well as chemical modification of bases along the length of the mRNA. Many of these processes occur co-transcriptionally, are coupled with each other and are essential to ensure that mRNPs are properly assembled to allow for their export and subsequent translation. Already early into the discovery of RNA processing steps, imaging provided important insights into this complex process. Hybridization approaches combined with electron microscopy were critical for the discovery of introns, when experiments in the Roberts and Sharp laboratories showed DNA segments looping from DNA-RNA R-loop regions, when RNA was hybridized to viral genomic DNA fragments (Berget et al. 1977; Chow et al. 1977). Similarly, electron micrographs of chromatin spreads from *Drosophila melanogaster* showed RNP assemblies at the intron-exon junctions which were later identified as spliceosomes, indicating that splicing might be a co-transcriptional process (Osheim et al. 1985). Co-transcriptional spliceosome assembly was later elucidated by many studies and approaches, including using variants of chromatin immunoprecipitation that allowed cross-linking of snRNPs and splicing factors to chromatin in transcription- and/or splicing-dependent manner (Alpert et al. 2017; Görnemann et al. 2005; Kotovic et al. 2003) as well as using an in vitro TIRF microscopy system with labelled RNAs and spliceosome components (Hoskins et al. 2011). Dynamics of spliceosome association at sites of transcription was further studied using single molecule microscopy approaches using either antibodies against U snRNP proteins or FISH probes against U snRNAs (Brody et al. 2011; Schmidt et al. 2011; Wetterberg et al. 2001). Interestingly, it was shown that recruitment of the U1 snRNP to active transcription sites could occur independent of the presence of introns in the pre-mRNA, indicating an RNA-independent recruitment possibly mediated by RNA pol II (Brody et al. 2011). The same study found that mRNAs with higher number of introns had more spliceosome components recruited to the transcription site, suggesting that multiple spliceosomes could potentially assemble onto the same mRNA. However, the number of spliceosomes acting on the pre-mRNA is likely to vary depending on the strength of the 5′ and 3′ splice sites, the presence of RNA secondary structures and splicing of adjacent introns. Moreover, it is not clear which proteins are recruited as preassembled complexes and which join as individual proteins. Co-transcriptional recruitment to sites of transcription or loading onto pre-mRNPs was also observed for various splicing regulators, including several SR proteins using either immunofluorescence or immuno-EM (Björk et al. 2006, 2009; Brody et al. 2011; Misteli et al. 1998; Wetterberg et al. 1996).

Co-transcriptional assembly of splicing factors resulting in the co-transcriptional splicing was first observed on chromatin Miller spreads from *Drosophila* embryos. These electron micrographs showed nascent pre-mRNA with multiple stages of intron excision with loops of introns 5′ and 3′ in the process of getting excised (Beyer and Osheim 1988). Since then, co-transcriptional splicing has been reported for several intron-containing mRNAs either using dual-colour RNA labelling of specific intron and exon sequences through incorporation of MS2, lambda N or PP7 aptamer sequences to monitor splicing in live cells, or by in situ hybridization in fixed cells (Brody et al. 2011; Coulon et al. 2014; Martin et al. 2013; Schmidt et al. 2011; Vargas et al. 2011). These experiments also showed that not all introns are spliced at the site of transcription, as a small fraction of intron-containing mRNAs was observed to be either retained at the site of transcription, close to the site of transcription or within the nucleoplasm, with some indication that splicing of mRNAs is enhanced post-transcriptionally (Boireau et al. 2007; Brody et al. 2011; Coulon et al. 2014; Vargas et al. 2011; Waks et al. 2011). Co- or post-transcriptional splicing might be defined by many factors, including the position of an intron within a pre-mRNA, the strength of splice sites and possibly other regulatory processes that might facilitate faster or slower splicing of specific introns. Splicing of introns has also been shown to impact elongation rates, and recent studies have provided evidence that splicing of one intron can influence splicing of nearby introns (Blazquez et al. 2018; Boehm et al. 2018). Moreover, intron retention has recently been proposed as a mechanism to regulate mRNA and protein expression, and imaging approaches will likely play an important role in dissecting the mechanisms of this regulatory process (Bahar Halpern et al. 2015; Wegener and Müller-McNicoll 2018).

Upon reaching the 3′ of a gene, RNA polymerases have to terminate transcription, and mRNAs are cleaved and polyadenylated before being released into the nucleoplasm. Studies using various experimental approaches have shown that termination, cleavage and release of the mRNA are possibly linked to other transcriptional process including elongation and splicing (Bentley 2014; Kyburz et al. 2006; Niwa and Berget 1991). Single molecule live-cell imaging has been used to determine the relationship between splicing and release of mRNAs from the transcription site (Coulon et al. 2014; Martins et al. 2011). Using MS2-labelled mRNAs, it was found that inhibition of splicing using spliceostatin A did not result in an increased release time of mRNAs; on the contrary, beta-globin mRNAs were released faster after treatment with the drug, indicating a link between splicing and 3′ end processing possibly through pausing of RNA polymerases at the 3′ end of the gene. Similar experiments were used to measure the post-transcriptional dwell times of transcripts at sites of transcription. These measurements showed varying release times for different genes and cell cycle stages, ranging from 60 s to 8 mins (Boireau et al. 2007; Coulon et al. 2014; Darzacq et al. 2007; Larson et al. 2011). Overall, however, the mechanisms modulating mRNPs release are still poorly understood, and only few models have so far been proposed to explain these observations (Lenstra et al. 2016).

### 9.2.3 Organization and Movement of mRNPs Within the Nucleoplasm

It is thought that nascent mRNAs do not exist in cells as long extended polymers but that pre-RNAs are co-transcriptionally folded and packaged into pre-mRNPs, a process mediated in part by RNA-binding proteins (RBPs), many of which contain homo- or hetero-dimerizing domains (Singh et al. 2015). Very little is known about how (pre-)mRNP formation is achieved, and much of our knowledge comes from electron microscopy (EM) experiments visualizing the long Balbiani ring (BR) mRNPs expressed from polytene chromosomes of the dipteran *Chironomus tentans*. Due to the large size of mRNAs expressed from the *BR1*, *BR2.1*, *BR2.2* and *BR6* loci (between 35 and 40 kB in length), the resulting mRNPs are sufficiently electron dense to be visualized using EM and have been an extremely valuable model system to study different aspects of mRNP metabolism (Björk and Wieslander 2015). These studies showed that mRNA packaging begins sequentially with the formation of a short 19–20 nm-thick fibre which is later packaged into a globular particle of ~50 nm diameter, before being released into the nucleoplasm (Skoglund et al. 1986). Complementing these studies, measuring diffusion characteristics of BR mRNPs labelled using complementary oligonucleotides suggests similar-sized particles (Siebrasse et al. 2008). A somewhat different kind of organization was observed for nuclear 18 kB-long MDN1 mRNPs in human tissue culture cells using smFISH and super-resolution microscopy. These mRNPs were found to have a more linear architecture and were similar to purified nuclear mRNPs from *S. cerevisiae*, which showed elongated, rodlike structures with variable length but a constant width when visualized by EM (Batisse et al. 2009; Adivarahan et al. 2018). Such a linear organization is also consistent with data from a recent developed RNA-RNA proximity ligation approach (Metkar et al. 2018). Due to the limited number of studies investigating the organization of nuclear mRNPs, it is still difficult to assess whether there exists a universal mechanism that mediates organization of mRNPs in the nucleus, and the role of different co- and post-transcriptional processes in regulating this process.

Once released from the site of transcription, mRNPs need to reach the nuclear pore to be exported to the cytoplasm. While very early studies suggested that there might be directed movement of mRNPs from the site of transcription to the nuclear pore, as stated by the 'gene gating hypothesis' proposed by (Blobel 1985), various studies using either EM or fluorescent microscopy approaches since then have shown that mRNPs move within the nucleoplasm and reach the nuclear pore through diffusion. First indications for this nondirected movement came once again from visualizing nuclear BR mRNPs which were observed to have a random distribution within the nucleoplasm, suggesting that these mRNPs do not have a defined path from the site of transcription towards nuclear pores, but possibly diffuse throughout the nucleoplasm in a random manner (Singh et al. 1999). However, the first direct measurement of diffusion kinetics of nuclear mRNAs used oligo dT probes labelled with caged carboxyfluorescein that, when allowed to penetrate cells, hybridized to nuclear poly(A) RNA and permitted monitoring diffusion of all poly(A) RNAs in cells. These studies showed that poly(A) RNAs move freely within the nonchromosomal space of the nucleus with properties characteristic of diffusion (Politz et al. 1999). Thereafter, various other studies have visualized mRNP diffusion in the nucleoplasm and found them to have a wide distribution of diffusion coefficients (Calapez et al. 2002; Molenaar et al. 2004; Politz et al. 1998, 1999). Using single-particle imaging approaches such as antisense oligonucleotides targeted to specific mRNA or using the MS2 tagging system, it was then revealed that nuclear mRNP diffusion was, although random in its movement, restricted to the extranucleolar space (Mor et al. 2010; Shav-Tal et al. 2004; Siebrasse et al. 2008; Vargas et al. 2005). Moreover, diffusion was slowed while passing through high-density chromatin, suggesting possible interactions with chromatin, and resolution of this stalling required the presence of ATP (Miralles et al. 2000; Shav-Tal et al. 2004; Vargas et al. 2005).

At least for some mRNPs, the path taken from the site of transcription towards the nuclear pore might be more complex than simple diffusion through the interchromatin space to reach the nuclear periphery. In higher eukaryotes, mRNAs containing inverted Alu repeat elements in their 3′ UTRs have been shown to localize to paraspeckles, membrane-less nuclear subcompartments, from where they can be released upon further processing or binding of specific RNA-binding proteins that promote their export. A first example for such localization was the CTN-RNA, an alternatively processed transcript expressed from the mCAT2 locus that contains alternative 5′ and 3′ UTRs but is otherwise identical to the protein-coding mCAT mRNA. The longer CTN-RNA 3′ UTR contains Alu-like SINE repeats that are A-to-I edited, resulting in the RNA localizing to paraspeckles. Upon stress, the transcript is processed to the mCAT2 mRNA, released and transported to the cytoplasm (Prasanth et al. 2005). Alternatively, paraspeckle localization of different Alu repeat containing mRNAs was shown to be mediated by the binding of the Staufen 2 protein (Elbarbary et al. 2013); however, the mechanism that facilitates this localization is not yet known. While paraspeckles are often located adjacent to nuclear speckles, nuclear domains located in the interchromatin regions and enriched in splicing factors, poly(A) RNAs and noncoding RNAs, the abundance of paraspeckles is much lower than that of nuclear speckles (Galganski et al. 2017; Staněk and Fox 2017), and the dynamics of mRNP localization to paraspeckles remains unclear. It is possible that mRNPs might be transcribed and spliced in or close to nuclear speckles and are subsequently transferred to paraspeckles. Alternatively, it remains possible that mRNAs once released from the transcription site diffuse through the interchromatin space to reach either the nuclear speckles or paraspeckles. In addition to Alu containing mRNAs, recent studies have shown that many other mRNAs are retained within the nucleus and that the process of mRNA retention is regulated. Combining fractionation with RNA sequencing and smFISH, Bahar Halpern et al. found that in different mouse metabolic tissues such as beta cells, liver and gut, many mRNAs exhibited varying retention within the nucleus depending on exposure to different metabolic conditions (Bahar Halpern et al. 2015). Similarly, a high-throughput RNA

FISH study aimed towards determining expression variability of over 900 different mRNAs in HeLa cells suggested that mRNAs can be nuclear retained and that their slow export buffers expression noise in the cytoplasm (Battich et al. 2015). Furthermore, many transcripts can contain retained introns which results in their nuclear retention (Wegener and Müller-McNicoll 2018). However, the mechanistic details on how nuclear retention is achieved and whether these mRNAs are retained in specific subnuclear compartments is not yet known.

### *9.2.4 mRNP Export Through the Nuclear Pore Complex*

Diffusion takes mRNPs to the nuclear periphery where they interact with the nuclear pore complex (NPC) to be exported. The time for mRNPs to reach the nuclear periphery varies widely across organisms and largely depends on the size of the nucleus, taking only a few seconds in lower eukaryotes, such as *S. cerevisiae*, but possibly up to minutes in human cell nuclei (Grünwald and Singer 2010; Mor et al. 2010; Oeffinger and Zenklusen 2012; Saroufim et al. 2015; Shav-Tal et al. 2004; Siebrasse et al. 2012; Smith et al. 2015). Live-cell single molecule fluorescence microscopy has shown that when mRNPs reach the periphery, they often first scan the region possibly making contact with multiple nuclear pore complexes before stably docking onto a nuclear pore for export (Grünwald and Singer 2010; Mor et al. 2010; Saroufim et al. 2015; Siebrasse et al. 2012). At the NPC, mRNPs first interact with the nuclear basket, a structure attached to the central framework of the nuclear pore complex that protrudes towards the nuclear interior (Buchwalter et al. 2018). Docking to the NPC has been shown to be a rate-limiting step for the export of mRNPs. Different single molecule studies found prolonged residency times of mRNPs at NPCs, with some of these studies being able to map prolonged residency at the basket (Grünwald and Singer 2010; Mor et al. 2010; Saroufim et al. 2015; Siebrasse et al. 2012). This increased residency might be a result of mRNPs being rearranged at the basket due to the release and/or binding of specific proteins that could facilitate its interaction with the nuclear basket and/or its translocation through the central channel. Indeed, EM studies of the BR mRNPs showed that the large BR mRNPs are unfolded at the distal ring of the nuclear basket before entering the basket with the 5′ of the mRNA first (Mehlin et al. 1992). However, it is unclear whether such remodelling is required for all mRNPs as most are at least one order magnitude smaller than BR mRNPs. Moreover, as mentioned above, recent single molecule super-resolution microscopy studies suggest that mRNPs in mammalian cells, as well as nuclear mRNPs purified from yeast, show a linear organization which could negate the need for such a reorganization (Batisse et al. 2009; Adivarahan et al. 2018).

Upon accesses of the central framework of the NPC, translocation is a very fast process (Grünwald and Singer 2010; Siebrasse et al. 2012). Using a super-registration approach to follow the translocation process of MS2-labelled beta-actin mRNAs through the NPC, Grünwald and Singer showed that translocation only takes around 20 ms. Moreover, mRNPs can move in either direction within the central channel, suggesting the directionality is not encoded by the central channel but by events at either side of the NPC (Grünwald and Singer 2010). Consistent with such a model, residency times at the cytoplasmic side of the pore are similar to the residency times at the nuclear basket, around 80 ms. Moreover, mRNP rearrangements, in part mediated by RNA helicases, are thought to be required at the cytoplasmic side of the NPC to facilitate the release of mRNPs into the cytoplasm (Alcázar-Román et al. 2006; Smith et al. 2015; Weirich et al. 2006).

## 9.3 Visualizing Cytoplasmic mRNPs

The main function for cytoplasmic mRNAs is to associate with ribosomes for translation. However, following their translocation through the nuclear pore, many mRNAs are first transported to various cytoplasmic compartments before associating with ribosomes and initiating protein synthesis. Mechanisms for mRNA localization are diverse, including diffusion followed by local retention and motor-driven movement, a process best described in neurons (Buxbaum et al. 2015). Moreover, mRNAs can switch between translationally active and repressed states upon certain stimuli such as stress, and this is, at least in part, concurrent with their accumulation in membraneless organelles such a stress granules (SG) (Guzikowski et al. 2019). Similarly, degradation has been linked to membraneless organelles called processing bodies (P-bodies) that contain high concentration of proteins involved in RNA degradation and are distinct from SGs. Imaging has been pivotal in the identification and characterization of all these processes, in particular RNA localization and local translation, with many methods now applied to study mRNA metabolism using microscopy being first developed to study mRNA localization, including smFISH and the MS2 aptamer system (Bertrand et al. 1998; Femino et al. 1998). Moreover, recent developments now allow monitoring translation at the single mRNA level in real time, as well as to study localization and mRNA turnover more directly in their relation to translation regulation and dynamic association with phase separated compartments, such as P-bodies and stress granules.

### *9.3.1 Discovery of mRNA Localization and Local mRNA Translation*

Much like for studying nuclear mRNA metabolism, EM and transcript-specific fluorescent RNA imaging have both played important roles towards today's understanding of many aspects of cytoplasmic RNA metabolism. When researchers initially became aware of the extensive cytoplasmic compartmentalization, the question arose as to how proteins are targeted to these subcellular structures and organelles. One proposed mechanism was that protein targeting could occur through post-translational transport, with mRNAs translated anywhere in the cytoplasm and proteins finding their final location by diffusion or through some active transport mechanism. However, EM images of polysomes, clusters of ribosomes translating a single mRNA, provided a first indication that translation of at least some protein coding mRNAs might occur in a more regulated and localized manner. Polysomes were observed to localize at different cellular structures such as the endoplasmic reticulum (Christensen et al. 1987; Lin and Chang 1975) and mitochondria (Kellems et al. 1975), as well as within dendritic spines (Steward and Levy 1982). Later, in situ hybridization approaches showed the localization of specific cytosolic protein-coding mRNAs, such as the actin-coding mRNAs in *Styela plicata* embryos, has a distinct localization pattern (Jeffery et al. 1983). Localization of few specific examples of mRNAs has since been observed in many organisms such as *S. cerevisiae* (Long et al. 1997), *Xenopus* (Melton 1987; Yisraeli and Melton 1988), *Drosophila* (Akam 1983) and mammalian cells (Lawrence and Singer 1986). For a long time thought to be a process restricted to only few specific transcripts, a high-throughput in situ hybridization study that surveyed localization patterns revealed that 71% of the 3370 transcripts examined preferentially localized to distinct subcellular compartments in *Drosophila* embryos, suggesting that RNA localization, and possibly localized translation, is the rule rather than the exception (Lécuyer et al. 2007). Similarly, recent studies have identified subcellular localization of a large number of mRNAs in specialized cells like neurons, as well as single-cell eukaryotes like yeast, further establishing the role of mRNA localization in regulation of gene expression (Cajigas et al. 2012; Gonsalvez et al. 2005; Jung et al. 2014). mRNA localization and localized translation offer distinct advantages compared to protein targeting through diffusion, in particular in larger cells such neurons but also in dividing cells (Buxbaum et al. 2015). Localized translation can quickly increase the local concentration of proteins, circumventing time and energy that would otherwise be required to transport each individual protein molecule. Moreover, a single mRNA can undergo many rounds of translation and, therefore, allow a fast response to stimuli at the site of localization, by either increasing or decreasing translation. Additionally, localization of mRNAs might help restrict synthesis and hence localization of proteins to subcellular compartments which could be essential in case the proteins either are toxic to the cell or have alternate functions based on their localization.

RNA imaging has also been extensively used as a readout when determining the mechanisms that mediate mRNA localization and localized translation across different transcripts and organisms. mRNA localization commonly depends on the presence of *cis*-acting localization elements, often called ‘zip codes’, and are frequently located within the 3′ UTR of an mRNA. In situ hybridization is most often used as a functional readout for localization, such as during the characterization of one of the first RNA localization elements located within 3′ UTR of the chicken β-actin mRNA that mediates the localization of the mRNA to the leading lamellae of chicken embryo fibroblasts (Kislauskis et al. 1994). The short sequence was shown to be sufficient to mediate localization of the reporter mRNA when isolated from its host RNA context and placed into a reporter RNA, an assay often used to define localization sequences. In addition, mutations to the localization element deterred but did not abolish the localization of mRNAs, suggesting multiple important elements within the sequence (Kislauskis et al. 1994). The same study determined that this 'zip code' was conserved across species both in sequence and function, as replacing the 3′-UTR with one from the human β-actin gene did not alter localization pattern for β-galactosidase mRNAs in chicken cells. Similar localization elements have been found for mRNAs in other organisms targeting mRNAs to various cellular compartments, including the *ASH1* mRNA in *S. cerevisiae* (Bertrand et al. 1998; Long et al. 1997; Takizawa et al. 1997), Vg1 in *Xenopus* (Mowry and Melton 1992), oskar (Kim-Ha et al. 1993), nanos (Gavis and Lehmann 1992; Gavis et al. 1996), bicoid (Macdonald and Struhl 1988) in *Drosophila* and MBP mRNA in neurons (Ainger et al. 1997). In addition to the use of FISH, the RNA aptamer system has been extensively applied to the study of RNA localization in live cells. The MS2 system was first developed to study the localization of the *ASH1* mRNA to the bud tip of the daughter cell in dividing cells in *S. cerevisiae* and enabled the demonstration of a motor-driven localization of this particular mRNA to the daughter cell. This process was found to depend on a protein complex with the myosin protein She3p as a core component, which allowed the mRNA to move along actin cables (Bertrand et al. 1998). Since then, the MS2 and other aptamer systems have become indispensable tools for studying localization dynamics, with many examples showing motor-driven movements, such as in neurons, or diffusion-based localization and retention, as observed in fly oocytes (Becalska and Gavis 2009; Buxbaum et al. 2015; Lee et al. 2016; Wu et al. 2016).

*cis*-RNA localization elements work in conjunction with trans-acting factors, i.e. RNA-binding proteins (RBPs) that bind to these zip-code sequences and are required for transport of mRNAs to their subcellular destination. Many RNA-binding proteins have been implicated in transport of mRNAs, some of which function through interaction with other protein partners that can link them to motor proteins such as myosin, kinesin or dynein (Buxbaum et al. 2015). Examples for *trans*-acting factors include the Staufen protein, required for the localization of oskar and bicoid mRNAs in *Drosophila*, the Imp1/ZBP1 and ZBP2 proteins that are important for localization of β-actin mRNA in mammals and Vera for Vg1 localization in *Xenopus* (Deshler et al. 1997; Farina et al. 2003; Hüttelmaier et al. 2005; Johnston et al. 1991; Martin and Ephrussi 2009). However, the use of imaging in characterizing the role of these proteins in the localization process has been challenging. In comparison to RNA imaging that allows visualizing individual mRNAs, visualizing of single proteins, and in particular their association with mRNAs, is still challenging. RBPs are typically labelled using fluorescent proteins or by immunolabelling, but signals from such stainings are difficult to attribute to specific RNA-protein complexes. Therefore, the readout from RBP imaging is much less direct and can represent both unbound and bound fractions, with the bound fraction possibly representing RBPs associated with multiple mRNA targets. Nevertheless, RBP imaging has been an important tool for studying mRNA localization, as RBPs colocalizing with mRNAs or in transport granules in neurons were shown to have similar localization dynamics to mRNAs and can therefore be used to study

RNA localization mechanisms (Buxbaum et al. 2015). To bridge the gap between the single molecule sensitivity of mRNA imaging and protein imaging, approaches have been developed to measure interactions of RBPs with localized mRNAs, such as fluorescence fluctuation spectroscopy (FFS) which was used by Wu et al. to characterize the interaction and stoichiometry of ZBP1 association with β-actin mRNA in living cells (Wu et al. 2015b). However, further technological development is needed in order to use RBPs as targets to monitor mRNP localization dynamics (see Outlook).

In addition to the sequence-specific 'zip-code' binding proteins, localization of certain mRNAs has been found to depend on proteins deposited during splicing, including proteins that are part of the exon junction complex; however, the mechanism behind the role of these proteins in mRNA localization is less well understood (Martin and Ephrussi 2009). Furthermore, recent studies in *Drosophila* suggest yet another mechanism regulating localization of mRNAs through modulation of local stability of mRNAs within the cell as has been observed for *Hsp83* mRNAs (Bashirullah et al. 2001; Martin and Ephrussi 2009). The different mechanisms regulating mRNA localization are only in the process of being unravelled, with the localization elements determining localization of a vast majority of mRNAs yet to be identified. Moreover, it has been shown mRNA localization is linked to translation, with at least some mRNAs being transported in a translationally silenced form and with translation of these mRNAs only initiated upon reception of specific signals at the site of localization (Buxbaum et al. 2015; Halstead et al. 2015; Hüttelmaier et al. 2005; Yoon et al. 2016). Combining recently developed translation imaging assays with single molecule RNA microscopy and advancements in protein imaging will be essential to dissect these processes in a much more detailed manner (see also below).

### 9.3.2 Dynamics of Translation Initiation and Elongation

Although some mRNAs are translationally repressed after reaching the cytoplasm, many are thought to rapidly associate with ribosomes and start translation. Once again, imaging of BR mRNPs was the first indication for fast translation initiation, with ribosomes shown to assemble on BR mRNPs even before the entire mRNP was fully exported to the cytoplasm (Mehlin et al. 1992). However, due to their large size, translocation for BR mRNPs might be slower than for most mRNPs, and translation might not initiate during or immediately after export for most cellular mRNAs. Nevertheless, using an elegant live-cell imaging approach, Halstead et al. were able to show that translation initiation can be fast for certain reporter mRNAs. They developed a new translation imaging approach, termed 'translating RNA imaging by coat protein knock-off (TRICK)', that allows distinguishing between untranslated mRNAs and mRNAs that have undergone at least one round of translation. The TRICK system consists of an mRNA reporter with two aptamer sequences within the body of the mRNA (MS2 and PP7). While the MS2 sequence is placed in the 3′ untranslated regions, the PP7 sequence, generally incorporated in the

3′ UTR of mRNAs, was instead placed within the open reading frame of the mRNA, and the entire sequence was translated along with an upstream ORF. This required modification of the PP7 repeat sequence to separate the individual stem-loops, so that translating ribosomes could efficiently displace the PCP-GFP proteins during the first round of translation. As the PCP-GFP contains a nuclear localization signal, ribosome displaced PCP-GFP will be transported back to the nucleus depleting their abundance in the cytoplasm to allow rebinding, whereas the MS2 signal is maintained. Therefore, a cytoplasmic mRNA will lose one label but maintain the second label after it has been translated at least once. Using this system, they showed that 94% of cytoplasmic mRNA from their TRICK reporter had undergone translation of least once, suggesting that translation occurs most likely within minutes after export to the cytoplasm, if not faster (Halstead et al. 2015). This system was also used to study localized translation, in particular the role of Oskar protein in osk mRNA localization, and will be a useful tool for determining translation regulation in the future. However, one limitation of this assay is that it does not allow to directly test whether mRNAs are actually associated with ribosomes.

One way of attempting to distinguish between translating and non-translating mRNAs is based on the reasoning that polysomal mRNAs, which are part of much larger assemblies in comparison to non-translating mRNAs, should show altered diffusion characteristics. Tracking labelled ribosomal proteins together with MS2/PP7 labelled β-actin mRNA in living cells revealed that association with ribosomes significantly slowed down mRNA diffusion in a manner that scaled with the ribosome occupancy (Katz et al. 2016; Wu et al. 2015b; and see below). Using this assay, it was shown that mRNAs in focal adhesions exhibited slowed and confined diffusion suggesting that β-actin mRNAs localized to these regions were heavily translating (Katz et al. 2016).

While polysomes can be tracked by labelling ribosomes, association of mRNAs with ribosomes does not imply translation in all cases. In addition, diffusion of mRNAs might become restricted for other reasons than their ribosome association. An early attempt to quantify translation in single cells used a reporter mRNA expressing a protein that contained a tetra-cysteine motif in its N-terminus which can be bound by the biarsenial dyes FlAsH and ReAsH. Using pulse-chase labelling in living cells allowed to visualize newly synthesized proteins and to spatially correlate them with the sites of β-actin mRNA localization (Rodriguez et al. 2006). Although it was possible to visualize sites of localized translation, it did not allow monitoring of translation at a single molecule level. This became possible with the development of protein tagging systems that used multiple epitopes within the N-terminus of the reporter protein, which, upon expression, could amplify the signal of nascent peptides. Two such tags have been used to image translation at the single molecule level, the SunTag and ‘spaghetti monster’ (SM) tag (Tanenbaum et al. 2014; Viswanathan et al. 2015). The SunTag system uses endogenously expressed single-chain antibody fragments (scFV) against a short epitope of the yeast Gcn4p that when fused to GFP can specifically bind to proteins containing, in general, multiples of this epitope (Fig. 9.2f). The SM-tag contains multimerized epitopes recognized by either fluorescently labeled anti-myc or anti-Flag antigen-binding fragment (Fab), introduced into cells by injection or through bead loading. While the signal intensity emitted by an individual nascent protein is not different than for a mature protein, the signal at translating mRNAs is amplified as translation within polysomes results in multiple nascent peptides at a single mRNA, all of which containing epitope sequences. This results in signal intensities at translating mRNAs that are integer multiples compared to the signal of a single proteins and therefore allows to determine ribosome occupancy on individual mRNAs, similar to determining polymerase density and dynamics at a transcription site by determining nascent mRNA signal intensities and fluctuations (Morisaki et al. 2016; Pichon et al. 2016; Wang et al. 2016; Wu et al. 2016; Yan et al. 2016). Moreover, to further facilitate the detection of translation sites, some studies have inserted degradation tags into their reporter proteins, resulting in low background except from nascent peptides (Wu et al. 2016).

The ability to monitor translation in real time and at the single mRNA level revealed important features of translation regulation, with some of them being analogous to observations first made when imaging transcription. Monitoring signal intensities of nascent peptides at individual translating mRNAs revealed that translation, similar to transcription, occurs in bursts, with mRNAs alternating between active and inactive states of translation (Morisaki et al. 2016; Pichon et al. 2016; Wu et al. 2016; Yan et al. 2016). Moreover, similar to RNA polymerase, ribosome stalling was observed at a fraction of mRNAs, even for codon optimized transcripts, and introduction of previously suggested stalling sequences further increased this fraction (Yan et al. 2016). mRNAs in polysomes were also found to have slower diffusion coefficients in comparison to ones that are not translating, as previously observed (Pichon et al. 2016; Wang et al. 2016; Wu et al. 2016; Yan et al. 2016). Monitoring fluctuations in ribosome occupancy also allowed for the calculation of initiation and elongation rates at individual mRNAs. This showed that translation rates vary significantly between mRNAs with different 5′ untranslated regions but also between different molecules of the same transcript within a cell, with initiation rates ranging from every 13 s to every 45 s. Elongation rates also varied significantly between 3 and 18 amino acids per second. These values also allowed for the determination of the average spacing between ribosomes, revealing significant variance across different studies and reporter mRNAs with spacing of around 160–910 nucleotides between individual ribosomes (Pichon et al. 2016; Wang et al. 2016; Wu et al. 2016; Yan et al. 2016).

Expanding the toolbox for translation imaging with the development of additional epitope-scFV combinations or SM-tags has further widened the scope of their usage for imaging of translation dynamics (Boersma et al. 2019; Zhao et al. 2019). Using two different epitope tags translated in different reading frames (SunTag and MoonTag), Boersma and co-workers revealed heterogeneity in start site selection that varied for different genes as well as for a specific mRNA during different stages in its life cycle (Boersma et al. 2019). Furthermore, although no evidence for frameshifting was observed in human mRNAs, viral RNAs seem to exhibit frameshifting allowing for synthesis of multiple proteins from the same RNA template (Boersma et al. 2019; Lyon et al. 2019).

### 9.3.3 Spatial Organization of Translating mRNAs

Translation is regulated by a set of proteins that help recruit the 43S pre-initiation complex to the 5′ end of the mRNA. The cap binding protein eIF4E, the scaffold protein eIF4G and the DEAD box helicase eIF4A, together forming the eIF4F complex, have been shown to have a critical role in regulating translation initiation. Moreover, eIF4G interacts with the poly(A) binding protein PABC1, and this interaction was shown to stimulate translation of mRNAs in vitro and in vivo (Imataka et al. 1998; Tarun and Sachs 1996; Tarun et al. 1997; Wakiyama et al. 2000). The interactions between all these components can be reconstituted in vitro using in vitro transcribed mRNAs and purified proteins, resulting in a closed-loop configuration of the mRNA mediated by PABC1 and eIF4F that can be visualized using electron microscopy (Wells et al. 1998). Together with biochemical evidence, these observations have resulted in a model that suggests that translating mRNAs are present in cells in a closed-loop configuration. This model is also supported, at least in part, by early electron microscopy studies in cells that showed polysome conformations resembling a closed-loop state for ER-associated polysomes (Christensen and Bourne 1999; Christensen et al. 1987). However, in addition to circular conformations, these EM studies have identified polysomes in many different configurations, including spiral, G-spiral and hairpin shapes, questioning whether all translating mRNAs exist in such a closed-loop configuration (Christensen and Bourne 1999; Christensen et al. 1987). Furthermore, a recent cryo-electron tomography study in human glioblastoma cells found the majority of polysomes had a helical organization indicating an open conformation with the ends separate (Brandt et al. 2010). Interestingly, polysome conformations observed in this study were very similar to conformations that had previously been seen in bacteria, suggesting that polysome organization could be evolutionary conserved (Brandt et al. 2009). However, one limitation of using polysome imaging to understand mRNA organization is that the RNA is not visible in these images. Ribosome densities determined via different assays suggest a spacing between ribosomes in the order of several hundreds of nucleotides, making it difficult to ascertain where most of the mRNA is located within polysomes. Furthermore, mRNAs with long 3′ UTRs will have long regions not occupied by ribosomes. In an attempt to obtain a more mRNA-centric view of mRNA organization during translation, recent studies used smFISH to determine the spatial relationship of different regions within mRNAs in human cell lines (Adivarahan et al. 2018; Khong and Parker 2018). These studies did not observe closed-loop conformations for the mRNAs studied but rather suggested that translation results in a decompaction of mRNAs that separates the ends. This could indicate that the interaction between eIF4F and PABC1 only reflects a transient state during translation initiation when mRNAs are still compact or during translationally inactive states as the result of translation bursting. Alternatively, it is also possible that closed-loop translation happens only for specific classes of mRNAs. New assays that allow studying the dynamics of mRNA conformation and compaction in living cells will be required to test these models (Vicens et al. 2018).

### *9.3.4 Spatial Organization of Translationally Inhibited RNAs*

Translation is a highly energy-consuming process, and producing new proteins is one of the main requirements for cells to grow and divide. Upon cellular stress, cells need to conserve energy to ensure their survival, resulting in translational inhibition of most mRNAs. Moreover, various stresses induce the formation of phase-separated compartments termed stress granules (SG) which are composed of poly(A) RNAs and different mRNA-binding proteins, including components of the pre-initiation complex (Decker and Parker 2012). Initially suggested as static structures of mRNA storage sequestering translationally inactive mRNAs, recent studies revealed that association of mRNA and RBPs with SGs can be dynamic. In addition, a combination of RNA-sequencing of purified SGs and smFISH experiments found that translation inhibition by itself is not a sufficient criterion for mRNAs to localize to SGs, with larger mRNAs more frequently found in SG compared to short transcripts (Khong et al. 2017). In situ hybridization and live-cell tracking using MS2-tagged mRNAs and SunTag/SM-tag labelling later showed that formation of stress granules in cells precedes the recruitment of the mRNAs tested, with stable association of mRNAs with SG requiring runoff of all ribosomes from the translating mRNA (Khong and Parker 2018; Moon et al. 2019; Wilbertz et al. 2019). Furthermore, mRNAs were found to have a very compact conformation upon ribosome run-off, and this compact conformation was maintained when mRNAs are associated with SGs (Khong and Parker 2018; Adivarahan et al. 2018). Together, these results suggest that compaction of mRNAs might be a prerequisite for their recruitment to SGs. mRNA recruitment to SGs was also found to be influenced by *cis*-elements such as the presence of a TOP motif, a sequence motif found within the 5′ UTR of many highly translated mRNAs and that TOP motif containing mRNAs more frequently exhibited a stable SG association (Halstead et al. 2015). In addition, single protein tracking of the SG proteins G3BP1 and IMP1 showed dynamic biphasic partition of these proteins within SGs, suggesting that SGs contain relatively immobile nanocores; however, whether static association of all mRNAs within SGs requires localization to these regions is still unclear (Niewidok et al. 2018). Lastly, consistent with the suggested role of SGs as storage compartment during stress, mRNAs localizing to SGs were shown to be capable of resuming translation once the stress was dissolved, with their translation kinetics indistinguishable from non-SG localized cytoplasmic mRNAs (Wilbertz et al. 2019). However, many of the rules that define why some translationally inactive mRNAs accumulate in SGs upon stress whereas others do not still need to be determined.

### *9.3.5 Towards Death of an mRNA*

In the cytoplasm, the processes modulating translation are believed to be in direct competition with the mRNA decay pathway. For example, the eIF4F complex that binds to the 5′ cap and is responsible for translation initiation is thought to compete with the decapping complex for access to the cap (Decker and Parker 2012; Schwartz and Parker 1999, 2000). This competition between mRNA translation and decay factors ultimately determines mRNA stability. The balance between the two processes can further be regulated through regulatory elements like miRNA binding sites or AU-rich elements (Duchaine and Fabian 2019; Grudzien-Nogalska and Kiledjian 2017). In situ hybridization has also proven to be a useful tool for studying mRNA degradation in cells as targeting FISH probes to different regions of the mRNAs allows for the detection of degradation intermediates. Using such an approach, it was shown that decay of mRNAs in yeast could be regulated by the promoter sequence, with the stability of the *SWI5* and *CLB2* mRNAs regulated in a cell cycle-dependent manner (Trcek et al. 2011). Using a similar in situ hybridization approach, a study in trypanosome found that decay of mRNAs was predominantly mediated by 5′-3′ exonuclease Xrn1 (Kramer 2016).

Many of the factors implicated in mRNA degradation were found to accumulate in membraneless organelles called processing bodies (P-bodies) (Decker and Parker 2012). P-bodies exist in most cells under normal growth conditions; however, their size and number are dependent on the pool of non-translating mRNAs and are greatly increased upon exposure to stress (Decker and Parker 2012; Teixeira et al. 2005). The presence of the degradation factors including the decapping factors Dcp1 and Dcp2, as well as many regulatory proteins of the RNA degradation pathway such as Ccr4 and GW182, led to the hypothesis that P-bodies act as degradation factories, where mRNAs are brought at the end of their lives to be degraded. To directly test this model, Horvathova and colleagues developed a reporter that allowed visualization of degradation intermediates in living cells termed '*3*(three)'--RNA end accumulation during turnover' or *TREAT* (Horvathova et al. 2017). The reporter was designed such that it contained pseudo-knots (PKs) between PP7 and MS2 stem-loops (Fig. 9.4). These PK sequences, derived from insect-borne flaviviruses, are resistant to Xrn1 degradation, believed to be the dominant pathway for mRNA degradation in mammals. By positioning the PK sequences between the PP7 and MS2 loops, thus ensures that Xrn1 mediated degradation intermediates will only contain the MS2 signal. This determined that full-length mRNAs, but no degradation intermediates, localized to P-bodies, suggesting that degradation, at least from some mRNAs, does not occur within P-bodies. Consistent with such a model, it was also shown that individual degradation events for these mRNAs occurred within the cytosol rather than in P-bodies (Horvathova et al. 2017).

P-bodies are often found adjacent to SGs, and early models suggested that mRNAs localized within SGs during stress could be translocated to p-bodies for preferntial degradation of certain transcripts. To test this model, Wilbertz et al. followed fluorescently tagged mRNAs during stress conditions in cells labelled with markers for p-bodies and SGs. They observed that although their reporter mRNAs localize to both SGs and P-bodies, exchange between the two compartments, was an extremely rare event (Moon et al. 2019; Wilbertz et al. 2019). Together, these new studies suggest that SGs and p-bodies might function independently and concurrently in regulating mRNA metabolism under conditions of stress.

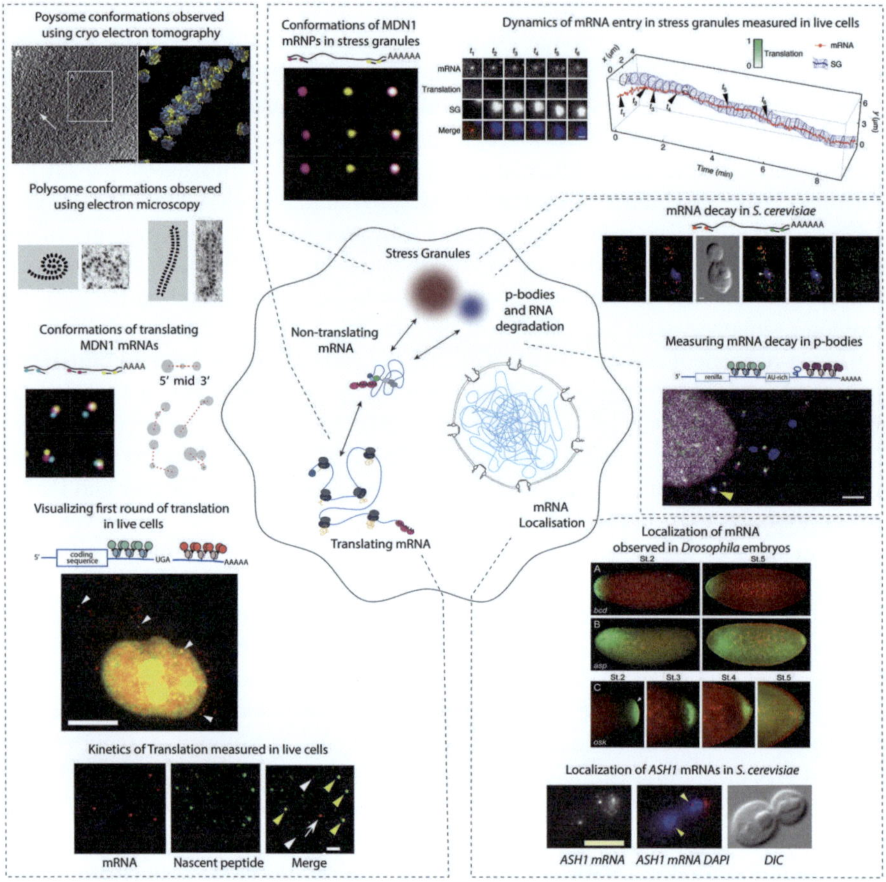


**Fig. 9.4** Visualizing different steps of cytoplasmic mRNA metabolism. *Organization of mRNAs during translation and translation kinetics*. (Top to bottom) Polysome conformations in human glioblastoma cells visualized by cryo-electron tomography. Isosurface model shown on the right representing a helical polysome conformation. Modified with permission from Brandt et al. (2010). Sample images of spiral and hairpin configurations of endoplasmic reticulum-localized polysomes in cultured fibroblasts visualized by electron microscopy. Modified with permission from Christensen and Bourne (1999). Open conformation of cytoplasmic MDN1 mRNAs in HEK293 cells visualized using smFISH (left) and their ball-and-stick representations (right). Modified from Adivarahan et al. (2018). TRICK assay to visualize the first round of translation. Red signals correspond to MS2 signal, green signal to PP7 signal. Overlapping red and green signals, visualized as yellow, represent mRNAs that are yet to undergo first round of translation. Modified with permission from Halstead et al. (2015). Measuring translation kinetics using the SunTag labelling system. The red signals correspond to the MS2-tagged mRNAs, green signals to nascent peptides. Modified with permission from Wu et al. (2016). *Stress granules (SG)*. (From left to right) Conformations of MDN1 mRNAs in stress granules as visualized using smFISH and super-resolution microscopy. Cartoon showing regions of probe hybridization. Modified from Adivarahan et al. (2018). Dynamics of mRNA entry into SGs observed for MS2-tagged mRNAs in living cells. The plot on the right shows the path of an mRNA from initial entry to stable association with SG. Modified with permission from Moon et al. (2019). *RNA degradation*. (Top to bottom) mRNA

It is however, still to determined if the observations made using these reporter mRNAs can be applied for all mRNAs in general. Moreover, the lack of exchange of mRNAs between the two membrane-less compartments could suggest an additional step of regulation with some mRNAs preferentially recruited to one compartment over the other.

## 9.4 Outlook

mRNA imaging has contributed extensively to the current understanding of different aspects of RNA metabolism and has in recent years become an increasingly important tool to study quantitative aspects as well as the dynamics of the different processes along the gene expression pathway. The continuous development and refinement of RNA imaging approaches will further facilitate studying these processes in even more detail, allowing to find answers to new questions or look at old questions with a new set of tools. Single molecule imaging is likely to continue to provide an ideal platform to study different processes that regulate mRNA metabolism, having the advantage over traditional methods in being able to yield high-resolution spatial and temporal information.

One of the main limitations for cellular RNA imaging today is the low-throughput nature of the many of these approaches. While it is relatively straightforward to image hundreds or even thousands of cells using automated image acquisition and image analysis, most approaches still only allow to study one or few mRNAs at a time. However, recent developments in in situ hybridization approaches in fixed cells using sequential hybridization and barcoding have enabled for imaging of thousands of RNAs within the same cell (MERFISH, SeqFISH, seqFISH+) (Chen et al. 2015; Eng et al. 2019; Lubeck et al. 2014; Shah et al. 2018). These approaches have the potential to become complementary, or even more powerful, than (single-cell) RNA sequencing methodologies and may open the doors to a microscopy centric transcriptome analysis that is not limited by expression levels and is able to provide high-resolution spatial information. Moreover, combining these with expansion or clearing protocols might further increase resolution and facilitate the use of such approaches in tissues and animals.

**Fig. 9.4** (continued) decay kinetics in *S. cerevisiae* measured using smFISH. The red signal corresponds to probes hybridizing to the 5′, green probes to the 3′ end. Modified with permission from Trcek et al. (2011). Measuring mRNA decay using the TREAT reporter. The green signal corresponds to the PP7 signals, magenta the MS2 signals and blue represents Dcp1a, a marker for P-bodies. Modified with permission from Horvathova et al. (2017). *RNA localization*. (Top to bottom) Visualizing mRNA localization in *Drosophila* at different stages of development for transcripts (from top to bottom) *bcd*, *asp* and *osk* mRNAs (mRNAs in green, nucleus in red). Modified with permission from Lécuyer et al. (2007). Localization of *ASH1* mRNA observed in *S. cerevisiae* using FISH. Nuclei were stained with DAPI. Arrowheads indicate transcription sites. From Powrie et al. (2011)

The spatial organization of mRNPs is one of the last unexplored topics in mRNA research that is likely to profit from future advances in imaging methods. Recent approaches combining smFISH and super-resolution microscopy have already revealed important new insights into mRNP organization in cells and showed that even long-standing dogmas, such as the closed-loop model for translation, have to be revisited (Pierron and Weil 2018; Adivarahan et al. 2018; Khong and Parker 2018; Vicens et al. 2018). Further adaptations of super-resolution approaches, including STED and dSTORM/PALM and yet to be developed methods, will allow to delve deeper into the structural organization of RNA-protein complexes and its role in mRNA metabolism. Similarly, recent improvements in cryo-EM revolution, which have led to high-resolution structures of many large protein/ RNA-protein complexes, can provide an interesting avenue towards exploring mRNP organization. However, purifying specific mRNPs from heterogeneous mRNP populations in sufficient quantities and homogeneity for cryo-EM analysis will likely remain challenging. In addition, correlative imaging in cells by combining cryo-EM with fluorescence microscopy will allow to combine the strength of specific labelling of fluorescent approaches with the resolution of election microscopy and could be a powerful approach to study mRNPs in cells.

Lastly, further expanding the tools to image mRNPs in living cells will be essential for moving towards the ability to follow mRNP metabolism through its different stages and will require tools that allow imaging single mRNAs as well as its associated proteins through time and space. Current methods allow this only for very short time periods and/or in limited subregions of the cells. Advances in labelling, illumination and image acquisition will be required to move towards this goal. Further improvements of aptamer-based RNA visualization approaches that make use of bright, photostable and membrane-permeable dyes such as the Janelia Fluor dyes are already helping to overcome some of these drawbacks (Grimm et al. 2015, 2016, 2017). Similarly, the use of proteins labelling systems such as Halo-, CLIP- and SNAP-tags which enable single protein imaging will allow us to better investigate the dynamics and stoichiometry of RBPs on mRNA and how this participates in regulating RNA metabolism (Grimm et al. 2015, 2016, 2017; Keppler et al. 2002; Los et al. 2008). However, achieving all this will also require further improvements in the fluorophores for live-cell imaging in terms of photon emission, photostability, cell permeability and labelling efficiencies to their respective tags as well as combining them with less phototoxic imaging methodologies such as light-sheet microscopy or the development of entirely new tools to follow biomolecules in cells.

**Acknowledgements** We thank members of the Zenklusen laboratory for discussion and comments on the manuscript. This work has been supported by Canadian Institutes of Health Research (Project Grant-366682), Le Fonds de recherche du Québec—Santé (Chercheur-boursier Junior 2) and Natural Sciences and Engineering Research Council of Canada for DZ.

# References

Abudayyeh OO, Gootenberg JS, Essletzbichler P, Han S, Joung J, Belanto JJ, Verdine V, Cox DB, Kellner MJ, Regev A et al (2017) RNA targeting with CRISPR–Cas13. Nature *550*:280

Adivarahan S, Livingston N, Nicholson B, Rahman S, Wu B, Rissland OS, Zenklusen D (2018) Spatial organization of single mRNPs at different stages of the gene expression pathway. Mol Cell *72*:727–738.e5

Ainger K, Avossa D, Diana AS, Barry C, Barbarese E, Carson JH (1997) Transport and localization elements in myelin basic protein mRNA. J Cell Biol 138:1077–1087

Akam ME (1983) The location of Ultrabithorax transcripts in Drosophila tissue sections. EMBO J 2:2075–2084

Alcázar-Román AR, Tran EJ, Guo S, Wente SR (2006) Inositol hexakisphosphate and Gle1 activate the DEAD-box protein Dbp5 for nuclear mRNA export. Nat Cell Biol 8:711–716

Alpert T, Herzel L, Neugebauer KM (2017) Perfect timing: splicing and transcription rates in living cells. Wiley Interdiscip Rev RNA 8:1–3

Bahar Halpern K, Caspi I, Lemze D, Levy M, Landen S, Elinav E, Ulitsky I, Itzkovitz S (2015) Nuclear retention of mRNA in mammalian tissues. Cell Rep 13:2653–2662

Bao G, Rhee W, Tsourkas A (2009) Fluorescent probes for live-cell RNA detection. Annu Rev Biomed Eng 11:25–47

Bartman CR, Hsu SC, Hsiung C, Raj A, Blobel GA (2016) Enhancer regulation of transcriptional bursting parameters revealed by forced chromatin looping. Mol Cell 62:237–247

Bartman CR, Hamagami N, Keller CA, Giardine B, Hardison RC, Blobel GA, Raj A (2019) Transcriptional burst initiation and polymerase pause release are key control points of transcriptional regulation. Mol Cell *73*:519–532.e4

Bashirullah A, Cooperstock RL, Lipshitz HD (2001) Spatial and temporal control of RNA stability. Proc Natl Acad Sci U S A 98:7025–7028

Batisse J, Batisse C, Budd A, Böttcher B, Hurt E (2009) Purification of nuclear poly(A)-binding protein Nab2 reveals association with the yeast transcriptome and a messenger ribonucleoprotein core structure. J Biol Chem 284:34911–34917

Battich N, Stoeger T, Pelkmans L (2015) Control of transcript variability in single mammalian cells. Cell 163:1596–1610

Becalska AN, Gavis ER (2009) Lighting up mRNA localization in Drosophila oogenesis. Development 136:2493–2503

Ben-Ari Y, Brody Y, Kinor N, Mor A, Tsukamoto T, Spector DL, Singer RH, Shav-Tal Y (2010) The life of an mRNA in space and time. J Cell Sci 123:1761–1774

Bentley DL (2014) Coupling mRNA processing with transcription in time and space. Nat Rev Genet 15. https://doi.org/10.1038/nrg3662

Berget S, Moore C, Sharp P (1977) Spliced segments at the 5′ terminus of adenovirus 2 late mRNA. Proc Natl Acad Sci U S A 74:3171–3175

Bertrand E, Chartrand P, Schaefer M, Shenoy SM, Singer RH, Long RM (1998) Localization of ASH1 mRNA particles in living yeast. Mol Cell 2:437–445

Beyer A, Osheim Y (1988) Splice site selection, rate of splicing, and alternative splicing on nascent transcripts. Genes Dev 2:754–765

Björk P, Wieslander L (2015) The Balbiani ring story: synthesis, assembly, processing, and transport of specific messenger RNA–protein complexes. Annu Rev Biochem 84:65–92

Björk P, Wetterberg-Strandh I, Baurén G, Wieslander L (2006) Chironomus tentans-repressor splicing factor represses SR protein function locally on pre-mRNA exons and is displaced at correct splice sites. Mol Biol Cell 17:32–42

Björk P, Jin S, Zhao J, Singh O, Persson J-O, Hellman U, Wieslander L (2009) Specific combinations of SR proteins associate with single pre-messenger RNAs in vivo and contribute different functions. J Cell Biol 184:555–568

Blazquez L, Emmett W, Faraway R, Pineda J, Bajew S, Gohr A, Haberman N, Sibley CR, Bradley RK, Irimia M et al (2018) Exon junction complex shapes the transcriptome by repressing recursive splicing. Mol Cell 72:496–509.e9

Blobel G (1985) Gene gating: a hypothesis. Proc Natl Acad Sci U S A 82:8527–8529

Boehm V, Britto-Borges T, Steckelberg A-L, Singh KK, Gerbracht JV, Gueney E, Blazquez L, Altmüller J, Dieterich C, Gehring NH (2018) Exon junction complexes suppress spurious splice sites to safeguard transcriptome integrity. Mol Cell 72:482–495.e7

Boersma S, Khuperkar D, Verhagen BMP, Sonneveld S, Grimm JB, Lavis LD, Tanenbaum ME (2019) Multi-color single molecule imaging uncovers extensive heterogeneity in mRNA decoding. Cell 178:458–472

Boireau S, Maiuri P, Basyuk E, de la Mata M, Knezevich A, Pradet-Balade B, Bäcker V, Kornblihtt A, Marcello A, Bertrand E (2007) The transcriptional cycle of HIV-1 in real-time and live cells. J Cell Biol 179:291–304

Bozzola JJ, Russell L (1999) Electron microscopy: principles and techniques for biologists. Jones and Bartlett, Sudbury, MA

Brandt F, Etchells SA, Ortiz JO, Elcock AH, Hartl UF, Baumeister W (2009) The native 3D organization of bacterial polysomes. Cell 136:261–271

Brandt F, Carlson L-A, Hartl UF, Baumeister W, Grünewald K (2010) The three-dimensional organization of polyribosomes in intact human cells. Mol Cell 39:560–569

Brodsky A, Silver P (2000) Pre-mRNA processing factors are required for nuclear export. RNA New York NY 6:1737–1749

Brody Y, Neufeld N, Bieberstein N, Causse SZ, Böhnlein E-M, Neugebauer KM, Darzacq X, Shav-Tal Y (2011) The in vivo kinetics of RNA polymerase II elongation during co-transcriptional splicing. PLoS Biol 9:e1000573

Buchwalter A, Kaneshiro JM, Hetzer MW (2018) Coaching from the sidelines: the nuclear periphery in genome regulation. Nat Rev Genet 20:1

Buxbaum AR, Haimovich G, Singer RH (2015) In the right place at the right time: visualizing and understanding mRNA localization. Nat Rev Mol Cell Bio 16:95-109

Buxbaum AR, Haimovich G, Singer RH (2015) In the right place at the right time: visualizing and understanding mRNA localization. Nat Rev Mol Cell Biol 16:95–109

Cajigas IJ, Tushev G, Will TJ, tom Dieck S, Fuerst N, Schuman EM (2012) The local transcriptome in the synaptic neuropil revealed by deep sequencing and high-resolution imaging. Neuron 74:453–466

Calapez A, Pereira HM, Calado A, Braga J, Rino J, Carvalho C, Tavanez J, Wahle E, Rosa AC, Carmo-Fonseca M (2002) The intranuclear mobility of messenger RNA binding proteins is ATP dependent and temperature sensitive. J Cell Biol 159:795–805

Chen H, Larson DR (2016) What have single-molecule studies taught us about gene expression? Genes Dev 30:1796–1810

Chen K, Boettiger AN, Moffitt JR, Wang S, Zhuang X (2015) Spatially resolved, highly multiplexed RNA profiling in single cells. Science 348:aaa6090

Cherry RJ, Smith PR, Morrison I, Fernandez N (1998) Mobility of cell surface receptors: a re-evaluation. FEBS Lett 430:88–91

Chow LT, Roberts JM, Lewis JB, Broker TR (1977) A map of cytoplasmic RNA transcripts from lytic adenovirus type 2, determined by electron microscopy of RNA:DNA hybrids. Cell 11:819–836

Christensen KA, Bourne CM (1999) Shape of large bound polysomes in cultured fibroblasts and thyroid epithelial cells. Anat Rec 255:116–129

Christensen KA, Kahn LE, Bourne CM (1987) Circular polysomes predominate on the rough endoplasmic reticulum of somatotropes and mammotropes in the rat anterior pituitary. Am J Anat 178:1–10

Chubb JR, Trcek T, Shenoy SM, Singer RH (2006) Transcriptional pulsing of a developmental gene. Curr Biol 16:1018–1025

Churchman SL, Weissman JS (2011) Nascent transcript sequencing visualizes transcription at nucleotide resolution. Nature 469:368

Codeluppi S, Borm LE, Zeisel A, Manno G, van Lunteren JA, Svensson CI, Linnarsson S (2018) Spatial organization of the somatosensory cortex revealed by osmFISH. Nat Methods 15:932–935

Cooper TA, Wan L, Dreyfuss G (2009) RNA and disease. Cell 136:777–793

Coulon A, Chow CC, Singer RH, Larson DR (2013) Eukaryotic transcriptional dynamics: from single molecules to cell populations. Nat Rev Genet 14:572–584

Coulon A, Ferguson ML, de Turris V, Palangat M, Chow CC, Larson DR (2014) Kinetic competition during the transcription cycle results in stochastic RNA processing. Elife 3:e03939

Daigle N, Ellenberg J (2007) λN-GFP: an RNA reporter system for live-cell imaging. Nat Methods 4. https://doi.org/10.1038/nmeth1065

Darzacq X, Shav-Tal Y, de Turris V, Brody Y, Shenoy S, Phair RD, Singer RH (2007) In vivo dynamics of RNA polymerase II transcription. Nat Struct Mol Biol 14:796–806

Decker CJ, Parker R (2012) P-bodies and stress granules: possible roles in the control of translation and mRNA degradation. CSH Perspect Biol 4:a012286

Deshler JO, Highett MI, Schnapp BJ (1997) Localization of Xenopus Vg1 mRNA by Vera protein and the endoplasmic reticulum. Science 276:1128–1131

Dolgosheina EV, Unrau PJ (2016) Fluorophore-binding RNA aptamers and their applications. Wiley Interdiscip Rev RNA 7:843–851

Dolgosheina EV, Jeng SC, Panchapakesan SS, Cojocaru R, Chen PS, Wilson PD, Hawkins N, Wiggins PA, Unrau PJ (2014) RNA mango aptamer-fluorophore: a bright, high-affinity complex for RNA labeling and tracking. ACS Chem Biol 9:2412–2420

Duchaine TF, Fabian MR (2019) Mechanistic insights into MicroRNA-mediated gene silencing. CSH Perspect Biol 11:a032771

Elbarbary RA, Li W, Tian B, Maquat LE (2013) STAU1 binding 3′ UTR IRAlus complements nuclear retention to protect cells from PKR-mediated translational shutdown. Genes Dev 27:1495–1510

Eng C-H, Shah S, Thomassie J, Cai L (2017) Profiling the transcriptome with RNA SPOTs. Nat Methods 14:1153–1155

Eng C-H, Lawson M, Zhu Q, Dries R, Koulena N, Takei Y, Yun J, Cronin C, Karp C, Yuan G-C et al (2019) Transcriptome-scale super-resolved imaging in tissues by RNA seqFISH+. Nature 568:235–239

Farina KL, Hüttelmaier S, Musunuru K, Darnell R, Singer RH (2003) Two ZBP1 KH domains facilitate β-actin mRNA localization, granule formation, and cytoskeletal attachment. J Cell Biol 160:77–87

Femino AM, Fay FS, Fogarty K, Singer RH (1998) Visualization of single RNA transcripts in situ. Science 280:585–590

Galganski L, Urbanek MO, Krzyzosiak WJ (2017) Nuclear speckles: molecular organization, biological function and role in disease. Nucleic Acids Res. https://doi.org/10.1093/nar/gkx759

Gavis ER, Lehmann R (1992) Localization of nanos RNA controls embryonic polarity. Cell 71:301–313

Gavis ER, Curtis D, Lehmann R (1996) Identification of cis-acting sequences that control nanos RNA localization. Dev Biol 176:36–50

Golding I, Cox EC (2004) RNA dynamics in live *Escherichia coli* cells. Proc Natl Acad Sci U S A 101:11310–11315

Golding I, Paulsson J, Zawilski SM, Cox EC (2005) Real-time kinetics of gene activity in individual bacteria. Cell 123:1025–1036

Gonsalvez GB, Urbinati CR, Long RM (2005) RNA localization in yeast: moving towards a mechanism. Biol Cell 97:75–86

Görnemann J, Kotovic KM, Hujer K, Neugebauer KM (2005) Cotranscriptional spliceosome assembly occurs in a stepwise fashion and requires the cap binding complex. Mol Cell 19:53–63

Grimm JB, English BP, Chen J, Slaughter JP, Zhang Z, Revyakin A, Patel R, Macklin JJ, Normanno D, Singer RH et al (2015) A general method to improve fluorophores for live-cell and single-molecule microscopy. Nat Methods 12. https://doi.org/10.1038/nmeth.3256

Grimm JB, English BP, Choi H, Muthusamy AK, Mehl BP, Dong P, Brown TA, Lippincott-Schwartz J, Liu Z, Lionnet T et al (2016) Bright photoactivatable fluorophores for single-molecule imaging. Nat Methods 13. https://doi.org/10.1038/nmeth.4034

Grimm JB, Muthusamy AK, Liang Y, Brown TA, Lemon WC, Patel R, Lu R, Macklin JJ, Keller PJ, Ji N et al (2017) A general method to fine-tune fluorophores for live-cell and in vivo imaging. Nat Methods 14. https://doi.org/10.1038/nmeth.4403

Grudzien-Nogalska E, Kiledjian M (2017) New insights into decapping enzymes and selective mRNA decay. Wiley Interdiscip Rev RNA 8:e1379

Grünwald D, Singer RH (2010) In vivo imaging of labelled endogenous β-actin mRNA during nucleocytoplasmic transport. Nature 467:604

Guzikowski AR, Chen YS, Zid BM (2019) Stress-induced mRNP granules: form and function of processing bodies and stress granules. Wiley Interdiscip Rev RNA 10(3):e1524

Hager GL, McNally JG, Misteli T (2009) Transcription dynamics. Mol Cell 35:741–753

Halpern K, Tanami S, Landen S, Chapal M, Szlak L, Hutzler A, Nizhberg A, Itzkovitz S (2015) Bursty gene expression in the intact mammalian liver. Mol Cell 58:147–156

Halstead JM, Lionnet T, Wilbertz JH, Wippich F, Ephrussi A, Singer RH, Chao JA (2015) An RNA biosensor for imaging the first round of translation from single cells to living animals. Science 347:1367–1671

Hocine S, Raymond P, Zenklusen D, Chao JA, Singer RH (2013) Single-molecule analysis of gene expression using two-color RNA labeling in live yeast. Nat Methods 10:119

Horvathova I, Voigt F, Kotrys AV, Zhan Y, Artus-Revel CG, Eglinger J, Stadler MB, Giorgetti L, Chao JA (2017) The dynamics of mRNA turnover revealed by single-molecule imaging in single cells. Mol Cell 68:615–625.e9

Hoskins AA, Friedman LJ, Gallagher SS, Crawford DJ, Anderson EG, Wombacher R, Ramirez N, Cornish VW, Gelles J, Moore MJ (2011) Ordered and dynamic assembly of single spliceosomes. Science 331:1289–1295

Hüttelmaier S, Zenklusen D, Lederer M, Dictenberg J, Lorenz M, Meng X, Bassell GJ, Condeelis J, Singer RH (2005) Spatial regulation of β-actin translation by Src-dependent phosphorylation of ZBP1. Nature 438:512–515

Imataka H, Gradi A, Sonenberg N (1998) A newly identified N-terminal amino acid sequence of human eIF4G binds poly(A)-binding protein and functions in poly(A)-dependent translation. EMBO J 17:7480–7489

Jakt L, Moriwaki S, Nishikawa S (2013) A continuum of transcriptional identities visualized by combinatorial fluorescent in situ hybridization. Development 140:216–225

Jaqaman K, Loerke D, Mettlen M, Kuwata H, Grinstein S, Schmid SL, Danuser G (2008) Robust single-particle tracking in live-cell time-lapse sequences. Nat Methods 5. https://doi.org/10.1038/nmeth.1237

Jeffery WR, Tomlinson CR, Brodeur RD (1983) Localization of actin messenger RNA during early ascidian development. Dev Biol 99:408–417

Johnston D, Beuchle D, Nüsslein-Volhard C (1991) Staufen, a gene required to localize maternal RNAs in the drosophila egg. Cell 66:51–63

Jonkers I, Kwak H, Lis JT (2014) Genome-wide dynamics of pol II elongation and its interplay with promoter proximal pausing, chromatin, and exons. Elife 3:e02407

Jung H, Gkogkas CG, Sonenberg N, Holt CE (2014) Remote control of gene function by local translation. Cell 157:26–40

Katz ZB, English BP, Lionnet T, Yoon YJ, Monnier N, Ovryn B, Bathe M, Singer RH (2016) Mapping translation "hot-spots" in live cells by tracking single molecules of mRNA and ribosomes. Elife 5:e10415

Kellems RE, Allison VF, Butow RA (1975) Cytoplasmic type 80S ribosomes associated with yeast mitochondria. IV. Attachment of ribosomes to the outer membrane of isolated mitochondria. J Cell Biol 65(1):1–14
Keppler A, Gendreizig S, Gronemeyer T, Pick H, Vogel H, Johnsson K (2002) A general method for the covalent labeling of fusion proteins with small molecules in vivo. Nat Biotechnol 21:86–89
Khong A, Parker R (2018) mRNP architecture in translating and stress conditions reveals an ordered pathway of mRNP compaction. J Cell Biol 217. https://doi.org/10.1083/jcb.201806183
Khong A, Matheny T, Jain S, Mitchell SF, Wheeler JR, Parker R (2017) The stress granule transcriptome reveals principles of mRNA accumulation in stress granules. Mol Cell 68:808–820.e5
Kim-Ha J, Webster P, Smith J, Macdonald P (1993) Multiple RNA regulatory elements mediate distinct steps in localization of oskar mRNA. Dev Camb Engl 119:169–178
Kislauskis E, Zhu X, Singer R (1994) Sequences responsible for intracellular localization of beta-actin messenger RNA also affect cell phenotype. J Cell Biol 127:441–451
Kotovic KM, Lockshon D, Boric L, Neugebauer KM (2003) Cotranscriptional recruitment of the U1 snRNP to intron-containing genes in yeast. Mol Cell Biol 23:5768–5779
Kramer S (2016) Simultaneous detection of mRNA transcription and decay intermediates by dual colour single mRNA FISH on subcellular resolution. Nucleic Acids Res 45. https://doi.org/10.1093/nar/gkw1245
Kühlbrandt W (2014) Cryo-EM enters a new era. Elife 3:e03678
Kwak H, Fuda NJ, Core LJ, Lis JT (2013) Precise maps of RNA polymerase reveal how promoters direct initiation and pausing. Science 339:950–953
Kyburz A, Friedlein A, Langen H, Keller W (2006) Direct interactions between subunits of CPSF and the U2 snRNP contribute to the coupling of pre-mRNA 3′ end processing and splicing. Mol Cell 23:195–205
Larson DR, Zenklusen D, Wu B, Chao JA, Singer RH (2011) Real-time observation of transcription initiation and elongation on an endogenous yeast gene. Science 332:475–478
Larsson C, Grundberg I, Söderberg O, Nilsson M (2010) In situ detection and genotyping of individual mRNA molecules. Nat Methods 7:395
Lawrence J, Singer R (1986) Intracellular localization of messenger RNAs for cytoskeletal proteins. Cell 45:407–415
Lécuyer E, Yoshida H, Parthasarathy N, Alm C, Babak T, Cerovina T, Hughes TR, Tomancak P, Krause HM (2007) Global analysis of mRNA localization reveals a prominent role in organizing cellular architecture and function. Cell 131:174–187
Lee BH, Bae S-W, Shim JJ, Sung YP, Park HY (2016) Imaging single-mRNA localization and translation in live neurons. Mol Cells 39(12):841–846
Lenstra TL, Rodriguez J, Chen H, Larson DR (2016) Transcription dynamics in living cells. Annu Rev Biophys 45:1–23
Levsky JM, Shenoy SM, Pezo RC, Singer RH (2002) Single-cell gene expression profiling. Science 297:836–840
Lin C, Chang J (1975) Electron microscopy of albumin synthesis. Science 190:465–467
Lionnet T, Czaplinski K, Darzacq X, Shav-Tal Y, Wells AL, Chao JA, Park H, de Turris V, Lopez-Jones M, Singer RH (2011) A transgenic mouse for in vivo detection of endogenous labeled mRNA. Nat Methods 8:165
Long RM, Singer RH, Meng X, Gonzalez I, Nasmyth K, Jansen R-P (1997) Mating type switching in yeast controlled by asymmetric localization of ASH1 mRNA. Science 277:383–387
Los GV, Encell LP, McDougall MG, Hartzell DD, Karassina N, Zimprich C, Wood MG, Learish R, Ohana R, Urh M et al (2008) HaloTag: A novel protein labeling technology for cell imaging and protein analysis. ACS Chem Biol 3:373–382
Lubeck E, Cai L (2012) Single-cell systems biology by super-resolution imaging and combinatorial labeling. Nat Methods 9:743

Lubeck E, Coskun AF, Zhiyentayev T, Ahmad M, Cai L (2014) Single-cell in situ RNA profiling by sequential hybridization. Nat Methods 11. https://doi.org/10.1038/nmeth.2892
Lyon KR, Aguilera LU, Morisaki T, Munsky B, Stasevich TJ (2019) Live-cell single RNA imaging reveals bursts of translational frameshifting. Mol Cell 75:172–183
Macdonald PM, Struhl G (1988) Cis- acting sequences responsible for anterior localization of bicoid mRNA in drosophila embryos. Nature 336. https://doi.org/10.1038/336595a0
Mahamid J, Pfeffer S, Schaffer M, Villa E, Danev R, Cuellar L, Förster F, Hyman AA, Plitzko JM, Baumeister W (2016) Visualizing the molecular sociology at the HeLa cell nuclear periphery. Science 351:969–972
Maiuri P, Knezevich A, Marco A, Mazza D, Kula A, McNally JG, Marcello A (2011) Fast transcription rates of RNA polymerase II in human cells. EMBO Rep 12:1280–1285
Martin KC, Ephrussi A (2009) mRNA localization: gene expression in the spatial dimension. Cell 136:719–730
Martin RM, Rino J, Carvalho C, Kirchhausen T, Carmo-Fonseca M (2013) Live-cell visualization of pre-mRNA splicing with single-molecule sensitivity. Cell Rep 4:1144–1155
Martins S, Rino J, Carvalho T, Carvalho C, Yoshida M, Klose J, de Almeida S, Carmo-Fonseca M (2011) Spliceosome assembly is coupled to RNA polymerase II dynamics at the 3′ end of human genes. Nat Struct Mol Biol 18:1115
Mehlin H, Daneholt B, Skoglund U (1992) Translocation of a specific premessenger ribonucleoprotein particle through the nuclear pore studied with electron microscope tomography. Cell 69:605–613
Melton D (1987) Translocation of a localized maternal mRNA to the vegetal pole of Xenopus oocytes. Nature 328. https://doi.org/10.1038/328080a0
Metkar M, Ozadam H, Lajoie BR, Imakaev M, Mirny LA, Dekker J, Moore MJ (2018) Higher-order organization principles of pre-translational mRNPs. Mol Cell 72(4):715–726
Miralles F, Öfverstedt L-G, Sabri N, Aissouni Y, Hellman U, Skoglund U, Visa N (2000) Electron tomography reveals posttranscriptional binding of pre-Mrnps to specific fibers in the nucleoplasm. J Cell Biol 148:271–282
Misteli T, Cáceres JF, Clement JQ, Krainer AR, Wilkinson MF, Spector DL (1998) Serine phosphorylation of SR proteins is required for their recruitment to sites of transcription in vivo. J Cell Biol 143:297–307
Molenaar C, Marras S, Slats J, Truffert J-C, Lemaître M, Raap A, Dirks R, Tanke H (2001) Linear 2′ O-methyl RNA probes for the visualization of RNA in living cells. Nucleic Acids Res 29: e89–e89
Molenaar C, Abdulle A, Gena A, Tanke HJ, Dirks RW (2004) Poly(A)+ RNAs roam the cell nucleus and pass through speckle domains in transcriptionally active and inactive cells. J Cell Biol 165:191–202
Moon SL, Morisaki T, Khong A, Lyon K, Parker R, Stasevich TJ (2019) Multicolour single-molecule tracking of mRNA interactions with RNP granules. Nat Cell Biol 21:162–168
Mor A, Suliman S, Ben-Yishay R, Yunger S, Brody Y, Shav-Tal Y (2010) Dynamics of single mRNP nucleocytoplasmic transport and export through the nuclear pore in living cells. Nat Cell Biol 12:543
Morisaki T, Lyon K, DeLuca KF, DeLuca JG, English BP, Zhang Z, Lavis LD, Grimm JB, Viswanathan S, Looger LL et al (2016) Real-time quantification of single RNA translation dynamics in living cells. Science 352:1425–1429
Mowry K, Melton D (1992) Vegetal messenger RNA localization directed by a 340-nt RNA sequence element in Xenopus oocytes. Science 255:991–994
Mueller F, Senecal A, Tantale K, Marie-Nelly H, Ly N, Collin O, Basyuk E, Bertrand E, Darzacq X, Zimmer C (2013) FISH-quant: automatic counting of transcripts in 3D FISH images. Nat Methods 10:277
Muramoto T, Cannon D, Gierliński M, Corrigan A, Barton GJ, Chubb JR (2012) Live imaging of nascent RNA dynamics reveals distinct types of transcriptional pulse regulation. Proc Natl Acad Sci U S A 109:7350–7355

Nelles DA, Fang MY, O'Connell MR, Xu JL, Markmiller SJ, Doudna JA, Yeo GW (2016) Programmable RNA tracking in live cells with CRISPR/Cas9. Cell 165:488–496

Nicolas D, Zoller B, Suter DM, Naef F (2018) Modulation of transcriptional burst frequency by histone acetylation. Proc Natl Acad Sci U S A 115:201722330

Niewidok B, Igaev M, da Graca A, Strassner A, Lenzen C, Richter CP, Piehler J, Kurre R, Brandt R (2018) Single-molecule imaging reveals dynamic biphasic partition of RNA-binding proteins in stress granules. J Cell Biol 217. https://doi.org/10.1083/jcb.201709007

Niwa M, Berget S (1991) Mutation of the AAUAAA polyadenylation signal depresses in vitro splicing of proximal but not distal introns. Genes Dev 5:2086–2095

Ochiai H, Sugawara T, Sakuma T, Yamamoto T (2014) Stochastic promoter activation affects Nanog expression variability in mouse embryonic stem cells. Sci Rep-UK 4:7125

Oeffinger M, Zenklusen D (2012) To the pore and through the pore: a story of mRNA export kinetics. Biochim Biophys Acta 1819:494–506

Osheim YN, Miller J, Beyer AL (1985) RNP particles at splice junction sequences on Drosophila chorion transcripts. Cell 43:143–151

Padovan-Merhar O, Nair GP, Biaesch AG, Mayer A, Scarfone S, Foley SW, Wu AR, Churchman SL, Singh A, Raj A (2015) Single mammalian cells compensate for differences in cellular volume and DNA copy number through independent global transcriptional mechanisms. Mol Cell 58:339–352

Paige JS, Wu KY, Jaffrey SR (2011) RNA mimics of green fluorescent protein. Science 333:642–646

Paré A, Lemons D, Kosman D, Beaver W, Freund Y, McGinnis W (2009) Visualization of individual Scr mRNAs during drosophila embryogenesis yields evidence for transcriptional bursting. Curr Biol 19:2037–2042

Pichon X, Bastide A, Safieddine A, Chouaib R, Samacoits A, Basyuk E, Peter M, Mueller F, Bertrand E (2016) Visualization of single endogenous polysomes reveals the dynamics of translation in live human cells. J Cell Biol 214:769–781

Pierron G, Weil D (2018) Re-viewing the 3D organization of mRNPs. Mol Cell 72:603–605

Politz JC, Browne ES, Wolf DE, Pederson T (1998) Intranuclear diffusion and hybridization state of oligonucleotides measured by fluorescence correlation spectroscopy in living cells. Proc Natl Acad Sci U S A 95:6043–6048

Politz JC, Tuft RA, Pederson T, Singer RH (1999) Movement of nuclear poly(A) RNA throughout the interchromatin space in living cells. Curr Biol 9:285–291

Powrie EA, Zenklusen D, Singer RH (2011) A nucleoporin, Nup60p, affects the nuclear and cytoplasmic localization of ASH1 mRNA in *S. cerevisiae*. RNA 17:134–144

Prasanth KV, Prasanth SG, Xuan Z, Hearn S, Freier SM, Bennett FC, Zhang MQ, Spector DL (2005) Regulating gene expression through RNA nuclear retention. Cell 123:249–263

Raj A, van Oudenaarden A (2008) Nature, nurture, or chance: stochastic gene expression and its consequences. Cell 135:216–226

Raj A, Peskin CS, Tranchina D, Vargas DY, Tyagi S (2006) Stochastic mRNA synthesis in mammalian cells. PLoS Biol 4:e309

Raj A, van den Bogaard P, Rifkin SA, van Oudenaarden A, Tyagi S (2008) Imaging individual mRNA molecules using multiple singly labeled probes. Nat Methods 5. https://doi.org/10.1038/nmeth.1253

Rodriguez AJ, Shenoy SM, Singer RH, Condeelis J (2006) Visualization of mRNA translation in living cells. J Cell Biol 175:67–76

Rouhanifard SH, Mellis IA, Dunagin M, Bayatpour S, Jiang CL, Dardani I, Symmons O, Emert B, Torre E, Cote A et al (2019) ClampFISH detects individual nucleic acid molecules using click chemistry–based amplification. Nat Biotechnol 37:102

Santangelo PJ, Nix B, Tsourkas A, Bao G (2004) Dual FRET molecular beacons for mRNA detection in living cells. Nucleic Acids Res 32:e57–e57

Saroufim M-A, Bensidoun P, Raymond P, Rahman S, Krause MR, Oeffinger M, Zenklusen D (2015) The nuclear basket mediates perinuclear mRNA scanning in budding yeast. J Cell Biol 211:1131–1140
Schmidt U, Basyuk E, Robert M-C, Yoshida M, Villemin J-P, Auboeuf D, Aitken S, Bertrand E (2011) Real-time imaging of cotranscriptional splicing reveals a kinetic model that reduces noise: implications for alternative splicing regulation. J Cell Biol 193:819–829
Schwartz DC, Parker R (1999) Mutations in translation initiation factors Lead to increased rates of deadenylation and decapping of mRNAs in *Saccharomyces cerevisiae*. Mol Cell Biol 19:5247–5256
Schwartz DC, Parker R (2000) mRNA decapping in yeast requires dissociation of the cap binding protein, eukaryotic translation initiation factor 4E. Mol Cell Biol 20:7933–7942
Senecal A, Munsky B, Proux F, Ly N, Braye FE, Zimmer C, Mueller F, Darzacq X (2014) Transcription factors modulate c-Fos transcriptional bursts. Cell Rep 8:75–83
Shah S, Takei Y, Zhou W, Lubeck E, Yun J, Eng C-H, Koulena N, Cronin C, Karp C, Liaw EJ et al (2018) Dynamics and spatial genomics of the nascent transcriptome by intron seqFISH. Cell 174:363–376.e16
Shav-Tal Y, Darzacq X, Shenoy SM, Fusco D, Janicki SM, Spector DL, Singer RH (2004) Dynamics of single mRNPs in nuclei of living cells. Science 304:1797–1800
Siebrasse J, Veith R, Dobay A, Leonhardt H, Daneholt B, Kubitscheck U (2008) Discontinuous movement of mRNP particles in nucleoplasmic regions devoid of chromatin. Proc Natl Acad Sci U S A 105:20291–20296
Siebrasse J, Kaminski T, Kubitscheck U (2012) Nuclear export of single native mRNA molecules observed by light sheet fluorescence microscopy. Proc Natl Acad Sci U S A 109:9426–9431
Singh O, Björkroth B, Masich S, Wieslander L, Daneholt B (1999) The intranuclear movement of Balbiani ring premessenger ribonucleoprotein particles. Exp Cell Res 251:135–146
Singh G, Pratt G, Yeo GW, Moore MJ (2015) The clothes make the mRNA: past and present trends in mRNP fashion. Annu Rev Biochem 84:1–30
Sinnamon JR, Czaplinski K (2014) RNA detection in situ with FISH-STICs. RNA 20:260–266
Skoglund U, Andersson K, Strandberg B, Daneholt B (1986) Three-dimensional structure of a specific pre-messenger RNP particle established by electron microscope tomography. Nature 319. https://doi.org/10.1038/319560a0
Smith C, Lari A, Derrer C, Ouwehand A, Rossouw A, Huisman M, Dange T, Hopman M, Joseph A, Zenklusen D et al (2015) In vivo single-particle imaging of nuclear mRNA export in budding yeast demonstrates an essential role for Mex67p. J Cell Biol 211:1121–1130
Spille J-H, Hecht M, Grube V, Cho W, Lee C, Cissé II (2019) A CRISPR/Cas9 platform for MS2-labelling of single mRNA in live stem cells. Methods 153:35–45
Staněk D, Fox A (2017) Nuclear bodies: news insights into structure and function. Curr Opin Cell Biol 46:94–101
Steward O, Levy W (1982) Preferential localization of polyribosomes under the base of dendritic spines in granule cells of the dentate gyrus. J Neurosci 2:284–291
Suter DM, Molina N, Gatfield D, Schneider K, Schibler U, Naef F (2011) Mammalian genes are transcribed with widely different bursting kinetics. Science 332:472–474
Takizawa PA, Sil A, Swedlow JR, Herskowitz I, Vale RD (1997) Actin-dependent localization of an RNA encoding a cell-fate determinant in yeast. Nature 389:90–93
Tanenbaum ME, Gilbert LA, Qi LS, Weissman JS, Vale RD (2014) A protein-tagging system for signal amplification in gene expression and fluorescence imaging. Cell 159:635–646
Tarun S, Sachs A (1996) Association of the yeast poly(A) tail binding protein with translation initiation factor eIF-4G. EMBO J 15:7168–7177
Tarun SZ, Wells SE, Deardorff JA, Sachs AB (1997) Translation initiation factor eIF4G mediates in vitro poly(A) tail-dependent translation. Proc Natl Acad Sci U S A 94:9046–9051
Teixeira D, Sheth U, Valencia-Sanchez MA, Brengues M, Parker R (2005) Processing bodies require RNA for assembly and contain nontranslating mRNAs. RNA 11:371–382

Thompson RE, Larson DR, Webb WW (2002) Precise nanometer localization analysis for individual fluorescent probes. Biophys J 82:2775–2783

Tinevez J-Y, Perry N, Schindelin J, Hoopes GM, Reynolds GD, Laplantine E, Bednarek SY, Shorte SL, Eliceiri KW (2017) TrackMate: an open and extensible platform for single-particle tracking. Methods 115:80–90

Trcek T, Larson DR, Moldón A, Query CC, Singer RH (2011) Single-molecule mRNA decay measurements reveal promoter- regulated mRNA stability in yeast. Cell 147:1484–1497

Tsanov N, Samacoits A, Chouaib R, Traboulsi A-M, Gostan T, Weber C, Zimmer C, Zibara K, Walter T, Peter M et al (2016) smiFISH and FISH-quant – a flexible single RNA detection approach with super-resolution capability. Nucleic Acids Res 44:e165–e165

Tutucci E, Vera M, Biswas J, Garcia J, Parker R, Singer RH (2017) An improved MS2 system for accurate reporting of the mRNA life cycle. Nat Methods 15:81

Tyagi S, Kramer F (1996) Molecular beacons: probes that fluoresce upon hybridization. Nat Biotechnol 14:303

Urbanek MO, Galka-Marciniak P, Olejniczak M, Krzyzosiak WJ (2014) RNA imaging in living cells - methods and applications. RNA Biol 11:1083–1095

Vargas DY, Raj A, Marras SA, Kramer F, Tyagi S (2005) Mechanism of mRNA transport in the nucleus. Proc Natl Acad Sci U S A 102:17008–17013

Vargas DY, Shah K, Batish M, Levandoski M, Sinha S, Marras S, Schedl P, Tyagi S (2011) Single-molecule imaging of transcriptionally coupled and uncoupled splicing. Cell 147:1054–1065

Vicens Q, Kieft JS, Rissland OS (2018) Revisiting the closed-loop model and the nature of mRNA 5′–3′ communication. Mol Cell 72:805–812

Viswanathan S, Williams ME, Bloss EB, Stasevich TJ, Speer CM, Nern A, Pfeiffer BD, Hooks BM, Li W-P, English BP et al (2015) High-performance probes for light and electron microscopy. Nat Methods 12:568–576

Wada Y, Ohta Y, Xu M, Tsutsumi S, Minami T, Inoue K, Komura D, Kitakami J, Oshida N, Papantonis A et al (2009) A wave of nascent transcription on activated human genes. Proc Natl Acad Sci U S A 106:18357–18361

Wakiyama M, Imataka H, Sonenberg N (2000) Interaction of eIF4G with poly(A)-binding protein stimulates translation and is critical for Xenopus oocyte maturation. Curr Biol 10:1147–1150

Waks Z, Klein AM, Silver PA (2011) Cell-to-cell variability of alternative RNA splicing. Mol Syst Biol 7:506–506

Wang F, Flanagan J, Su N, Wang L-C, Bui S, Nielson A, Wu X, Vo H-T, Ma X-J, Luo Y (2012) RNAscope A novel in situ RNA analysis platform for formalin-fixed, paraffin-embedded tissues. J Mol Diagn 14:22–29

Wang C, Han B, Zhou R, Zhuang X (2016) Real-time imaging of translation on single mRNA transcripts in live cells. Cell 165:990–1001

Wang G, Moffitt JR, Zhuang X (2018) Multiplexed imaging of high-density libraries of RNAs with MERFISH and expansion microscopy. Sci Rep-UK 8:4847

Wegener M, Müller-McNicoll M (2018) Nuclear retention of mRNAs – quality control, gene regulation and human disease. Semin Cell Dev Biol 79:131–142

Weirich CS, Erzberger JP, Flick JS, Berger JM, Thorner J, Weis K (2006) Activation of the DExD/H-box protein Dbp5 by the nuclear-pore protein Gle1 and its coactivator InsP6 is required for mRNA export. Nat Cell Biol 8. https://doi.org/10.1038/ncb1424

Wells SE, Hillner PE, Vale RD, Sachs AB (1998) Circularization of mRNA by eukaryotic translation initiation factors. Mol Cell 2:135–140

Wetterberg I, Baurén G, Wieslander L (1996) The intranuclear site of excision of each intron in Balbiani ring 3 pre-mRNA is influenced by the time remaining to transcription termination and different excision efficiencies for the various introns. RNA (New York, NY) 2:641–651

Wetterberg I, Zhao J, Masich S, Wieslander L, Skoglund U (2001) In situ transcription and splicing in the Balbiani ring 3 gene. EMBO J 20:2564–2574

Wilbertz JH, Voigt F, Horvathova I, Roth G, Zhan Y, Chao JA (2019) Single-molecule imaging of mRNA localization and regulation during the integrated stress response. Mol Cell 73:946–958.e7

Wu B, Chao JA, Singer RH (2012) Fluorescence fluctuation spectroscopy enables quantitative imaging of single mRNAs in living cells. Biophys J 102:2936–2944

Wu B, Miskolci V, Sato H, Tutucci E, Kenworthy CA, Donnelly SK, Yoon YJ, Cox D, Singer RH, Hodgson L (2015a) Synonymous modification results in high-fidelity gene expression of repetitive protein and nucleotide sequences. Genes Dev 29:876–886

Wu B, Buxbaum AR, Katz ZB, Yoon YJ, Singer RH (2015b) Quantifying protein-mRNA interactions in single live cells. Cell 162:211–220

Wu B, Eliscovich C, Yoon YJ, Singer RH (2016) Translation dynamics of single mRNAs in live cells and neurons. Science 352:1430–1435

Yan X, Hoek TA, Vale RD, Tanenbaum ME (2016) Dynamics of translation of single mRNA molecules in vivo. Cell 165:976–989

Yisraeli JK, Melton D (1988) The maternal mRNA Vg1 is correctly localized following injection into Xenopus oocytes. Nature 336:592

Yoon YJ, Wu B, Buxbaum AR, Das S, Tsai A, English BP, Grimm JB, Lavis LD, Singer RH (2016) Glutamate-induced RNA localization and translation in neurons. Proc Natl Acad Sci U S A 113: E6877–E6886

Yunger S, Rosenfeld L, Garini Y, Shav-Tal Y (2010) Single-allele analysis of transcription kinetics in living mammalian cells. Nat Methods 7:631

Zeitlinger J, Stark A, Kellis M, Hong J-W, Nechaev S, Adelman K, Levine M, Young RA (2007) RNA polymerase stalling at developmental control genes in the *Drosophila melanogaster* embryo. Nat Genet 39. https://doi.org/10.1038/ng.2007.26

Zenklusen D, Larson DR, Singer RH (2008) Single-RNA counting reveals alternative modes of gene expression in yeast. Nat Struct Mol Biol 15. https://doi.org/10.1038/nsmb.1514

Zhao N, Kamijo K, Fox P, Oda H, Morisaki T, Sato Y, Kimura H, Stasevich TJ (2019) A genetically encoded probe for imaging nascent and mature HA-tagged proteins in vivo. Nat Commun 10:2947

# Chapter 10
# Diverging RNPs: Toward Understanding lncRNA-Protein Interactions and Functions

**Martin Sauvageau**

**Abstract** RNA-protein interactions are essential to a variety of biological processes. The realization that mammalian genomes are pervasively transcribed brought a tidal wave of tens of thousands of newly identified long noncoding RNAs (lncRNAs) and raised questions about their purpose in cells. The vast majority of lncRNAs have yet to be studied, and it remains to be determined to how many of these transcripts a function can be ascribed. However, results gleaned from studying a handful of these macromolecules have started to reveal common themes of biological function and mechanism of action involving intricate RNA-protein interactions. Some lncRNAs were shown to regulate the chromatin and transcription of distant and neighboring genes in the nucleus, while others regulate the translation or localization of proteins in the cytoplasm. Some lncRNAs were found to be crucial during development, while mutations and aberrant expression of others have been associated with several types of cancer and a plethora of diseases. Over the last few years, the establishment of new technologies has been key in providing the tools to decode the rules governing lncRNA-protein interactions and functions. This chapter will highlight the general characteristics of lncRNAs, their function, and their mode of action, with a special focus on protein interactions. It will also describe the methods at the disposition of scientists to help them cross this next frontier in our understanding of lncRNA biology.

**Keywords** Long noncoding RNAs · RNA-protein interactions · RNA biology · Ribonucleoprotein complexes · Functional RNAs

---

M. Sauvageau (✉)
Montreal Clinical Research Institute (IRCM), Montréal, QC, Canada

Department of Biochemistry and Molecular Medicine, Université de Montréal, Montréal, QC, Canada
e-mail: martin.sauvageau@ircm.qc.ca



M. Oeffinger, D. Zenklusen (eds.), *The Biology of mRNA: Structure and Function*, Advances in Experimental Medicine and Biology 1203,
https://doi.org/10.1007/978-3-030-31434-7_10

## 10.1 Introduction

A vast proportion of the mammalian genome is transcribed and generates tens of thousands of long noncoding RNA molecules (lncRNAs) which can range from 200 nucleotides (nt) to several kilobases (kb) in length. The biogenesis of these noncoding transcripts is similar to messenger RNAs (mRNAs), i.e., they are transcribed by RNA polymerase II (Pol II); they are generally multi-exonic and spliced; they canonically have a 5′ cap and are for the most part polyadenylated (Cabili et al. 2011; Guttman and Rinn 2012; Guttman et al. 2010; Ni et al. 2013). However, apart from a few transcripts annotated as lncRNAs that were shown to encode small peptides (Anderson et al. 2015; Nelson et al. 2016), most are not translated and likely function at the RNA level. It is still unclear how many lncRNAs, out of the thousands identified, play an active role in the cell, but several of them were shown to be functional and affect a variety of biological processes and physiological functions, ranging from development to immune response (Li and Chang 2014; Rinn and Chang 2012; Ulitsky and Bartel 2013). The expression of lncRNAs is also frequently dysregulated or their locus disrupted by deletions, amplifications, or chromosomal translocations in various cancers and pathologies, suggesting they may contribute to the development of diseases (Grote and Herrmann 2013; Hon et al. 2017; Kotzin et al. 2016; Maass et al. 2012; Sauvageau et al. 2013; Yan et al. 2015). The added level of complexity lncRNAs bring to the regulation of biological processes has sparked a growing interest in understanding their mechanisms of action.

lncRNAs are found across all tissues, but their expression is generally more tissue specific when compared to mRNAs (Cabili et al. 2011; Derrien et al. 2012; Guttman et al. 2010). Although some lncRNAs are transcribed at high levels, most are expressed an order of magnitude lower than mRNAs, with some present at only a few copies per cell (Clark et al. 2015). This makes the study of their function using biochemical approaches particularly challenging. The cellular localization of lncRNAs also varies greatly, with some showing diffuse localization patterns, either in the nucleus, the cytoplasm, or both (Cabili et al. 2011; Derrien et al. 2012; Guttman et al. 2010), while others being localized only to specific subcellular compartments such as paraspeckles (Mao et al. 2010; Souquere et al. 2010), nucleolus, mitochondria, or the endoplasmic reticulum (Kaewsapsak et al. 2017; Leucci et al. 2016; Mercer et al. 2011; Noh et al. 2016; Vendramin et al. 2018). Similar to proteins, the subcellular localization of a lncRNA is likely to determine its interaction network with other macromolecules. Thus, when studying lncRNA function, determining its localization can provide valuable information on potential mechanisms.

However, one of the challenges in studying lncRNAs is that we currently do not have a clear understanding of the relationship between sequence, structure, and function. Unlike mRNAs which have identifiable open reading frames (ORFs) that generate proteins with modular domains, our knowledge of lncRNA functional elements or motifs is too limited to infer function or to effectively group them based on shared sequence features, as it is commonly done with protein domains. This is further complicated by the fact that lncRNA primary sequences are generally weakly conserved during evolution, making it harder to identify functional orthologs or conserved features across species (Chen et al. 2016b; Hezroni et al. 2015). Due to the lack of identifiable sequence features, current genome annotation consortia have therefore mostly used descriptive attributes, such as their genomic positional relationship with protein-coding genes, to divide lncRNAs into different biotypes (Chen et al. 2016b; Hezroni et al. 2015). Based on this, lncRNAs can be classified as (1) antisense, transcripts that overlap the genomic span of a protein-coding locus on the opposite strand; (2) sense-overlapping, transcripts that contain a coding gene in their intron on the same strand; (3) sense intronic, transcripts derived from introns of a coding gene on the same strand and not overlapping any exons; (4) intergenic (lincRNA), transcripts that lie at a distance between two genes; and (5) bidirectional promoter, transcripts that originate from within the promoter region of a protein-coding gene, with the transcription occurring on the opposite strand (Fig. 10.1). Although this can be useful for a first classification of lncRNAs when analyzing large datasets, it does not convey information on potential (shared) functions or mechanism of action.

Another challenge when studying a lncRNA locus is to determine whether its potential function is actually mediated by the RNA transcript or rather by other regulatory modalities, such as harboring DNA regulatory elements, or by inducing changes to chromatin structure and protein accessibility by the simple act of transcription of the locus (Bassett et al. 2014; Goff and Rinn 2015; Kopp and Mendell 2018). For example, recent findings have shown that some lncRNA loci, such as *Lockd*, regulate neighboring gene expression in *cis* through DNA enhancer elements independently of the noncoding transcript (Engreitz et al. 2016; Groff et al. 2016; Paralkar et al. 2016). Alternatively, multiple approaches were used to show that imprinting at the *Igf2r/Airn* locus is regulated by the local act of transcription across the lncRNA *Airn* locus and not by the RNA transcript itself (Latos et al. 2012;

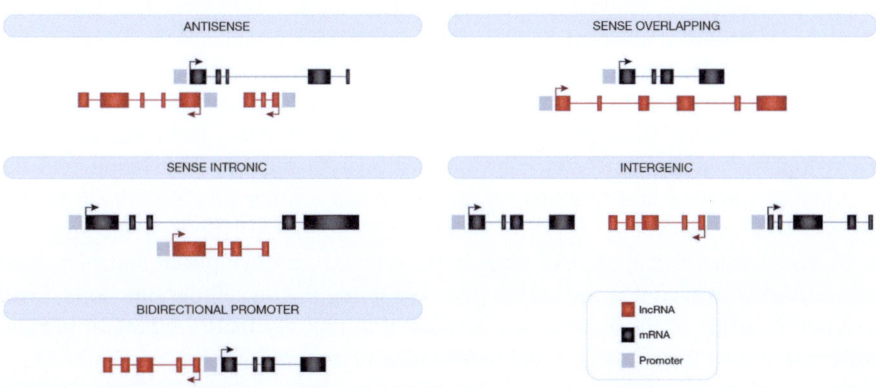


**Fig. 10.1** lncRNA biotypes. Schematic representation of distinct lncRNA biotypes based on their genomic positional relationship with protein-coding genes

Santoro et al. 2013). Thus, contrary to protein-coding genes, where frameshift mutations, domain deletions, or changes in specific residues can affect their function, our limited ability to predict the molecular mode of lncRNAs or their specific function based on sequences complicates efforts to characterize them with a single and straightforward genetic approach commonly used to study mRNAs. Techniques such as genomic deletion, insertion of a premature transcription termination signal, antisense oligos or CRISPR-Cas13-mediated post-transcriptional degradation, as well as functional rescue experiments are available to dissect the function of lncRNAs, but each comes with its own limitations that need to be considered when interpreting results. For example, in addition to affecting the sequence-specific function of a lncRNA transcript, genomic deletion of the lncRNA locus could remove embedded DNA regulatory elements affecting nearby genes, whereas insertion of a polyadenylation transcription termination signal may also impair any function mediated by the act of transcription (Bassett et al. 2014; Goff and Rinn 2015; Kopp and Mendell 2018; West et al. 2004). Similarly, CRISPR-based approaches to inhibit or activate the expression of a lncRNA (CRISPRa/i) induce changes in histone modifications which can spread over a few kb from the target site, making it difficult to distinguish whether the resulting effect is due to changes in chromatin states and potential regulatory elements, the act of transcription, or the RNA (Gilbert et al. 2014; Liu et al. 2017; Qi et al. 2013; Zalatan et al. 2015). Therefore, determining unambiguously whether a lncRNA has an RNA-based function requires using more than one of these complementary approaches (Bassett et al. 2014; Goff and Rinn 2015; Kopp and Mendell 2018). For more details on the advantages and disadvantages of using each technique to dissect lncRNA function, I will refer readers to recent reviews that discuss extensively this subject (Bassett et al. 2014; Goff and Rinn 2015; Kopp and Mendell 2018).

To date, only a handful of lncRNAs with RNA-based functions have been investigated in detail. These studies analyzing well-known lncRNAs such as *Xist*, *Malat1*, *Norad*, and a few others allowed us to make important progress in our understanding of their functions and revealed key aspects of their mechanisms. One of the most common features shared by all these lncRNAs is that their molecular function is closely connected to their interaction with specific RNA-binding proteins. In the nucleus, lncRNAs have been shown to act as transcriptional and post-transcriptional regulators as well as affect higher-order chromatin structure and nuclear organization through interactions with proteins, DNA, and other RNAs. In the cytoplasm, some lncRNAs have been shown to regulate translation and the stability of mRNAs and proteins, to affect the subcellular localization of protein targets, and to modulate signaling pathways. Here, I will discuss the different RNA-based roles that lncRNAs play in the nucleus and cytoplasm and how their interactions with specific proteins are involved in regulating cellular functions. I will also review some of the approaches developed to identify lncRNA-protein interactions and dissect their function and mechanism of action.

## 10.2 Nuclear Functions of lncRNAs

Despite similar biogenesis, lncRNAs tend to be more frequently localized in the nucleus than mRNAs (Cabili et al. 2011; Derrien et al. 2012). Searching for sequence features that mediate their preferential nuclear localization, recent studies using cell fractionation and massively parallel reporter assays found that *Alu* repeats, derived from short interspersed nuclear elements (SINEs), as well as C-rich sequences are able to promote nuclear localization of at least some lncRNAs (Carlevaro-Fita et al. 2019; Lubelsky and Ulitsky 2018; Shukla et al. 2018). Attaching these short lncRNA-derived sequences to the 3′end of an mRNA that does not normally contain such elements was found to increase its nuclear localization. Interestingly, the nuclear accumulation of transcripts is associated with binding of HNRNPK, suggesting this protein is implicated in a nuclear retention mechanism potentially distinct from its role in pre-mRNA processing (Lubelsky and Ulitsky 2018). Similarly, the nuclear matrix factor HNNRNPU was found to bind a 156 nt repeat sequence within the lncRNA *Firre* and regulate its nuclear localization (Hacisuleyman et al. 2014). Thus, even though the molecular features underlying the nuclear enrichment of some lncRNAs are not very well understood, these results suggest it likely involves specific sequence features within lncRNAs that promote interactions with proteins that help retain them in the nucleus to perform their function.

### *10.2.1 Transcriptional Regulation*

One of the best described functions of lncRNAs is to regulate transcription. Several lncRNAs have been shown to activate or repress transcription by acting either locally in proximity of their sites of transcription (in *cis*) or distally on target genes located on other chromosomes (in *trans*). Current models propose that lncRNAs can regulate transcription by interacting with transcription factors or chromatin-modifying complexes and help their recruitment to specific target loci. This could be achieved either through direct interaction with DNA and the formation of DNA: RNA triplex (Hoogsteen hydrogen bonding) and R-loops structures (Balk et al. 2013; Blank-Giwojna et al. 2019; Gibbons et al. 2018; Mondal et al. 2015; Pfeiffer et al. 2013; Postepska-Igielska et al. 2015; Zhao et al. 2018) or by mediating protein-protein interactions at target loci (Fig. 10.2a).

#### 10.2.1.1 Regulating Transcription *in cis*

Discovered more than 20 years ago, *Xist* is one of the best studied examples of a *cis*-acting lncRNA. *Xist* is a key factor regulating dosage compensation in females by inactivating one of the two X chromosomes (Lee and Jaenisch 1997; Penny et al.

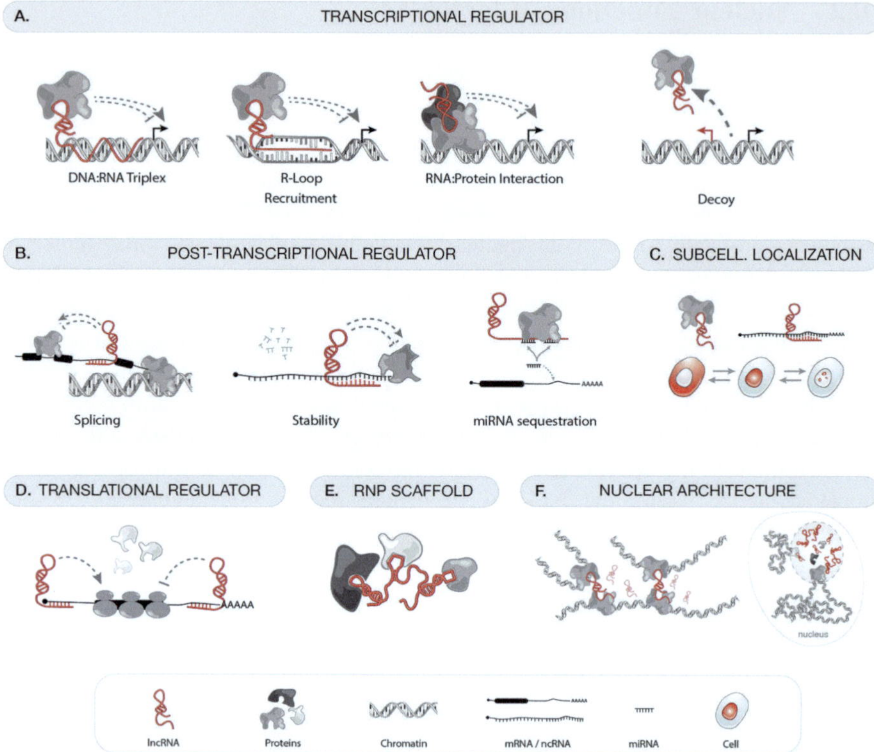


**Fig. 10.2** Models of lncRNA function. Schematic representation of the different molecular functions of lncRNAs. (**a**) lncRNAs can act as transcriptional regulators by localizing interacting proteins to specific chromatin loci directly through DNA:RNA triplex formation or R-loop formation or indirectly by acting as an adaptor between proteins. This can result in activation or repression of target genes in cis (neighboring genes) or in trans (distant located genes). lncRNAs can also act as decoy and prevent proteins to bind their target loci through binding and titering factors away as they are transcribed. (**b**) lncRNAs can regulate other RNAs post-transcriptionally by affecting their splicing, their stability or by acting as "sponges" by interacting with miRNAs and competing for binding other targets. (**c**) lncRNAs can regulate the localization or sequester proteins and other RNAs in different subcellular compartments. (**d**) Some lncRNAs can inhibit or promote translation of mRNAs. (**e**) lncRNAs can serve as scaffolds and assist in the assembly of protein complexes or potentially bring proteins in the same pathway together. (**f**) lncRNAs can regulate nuclear dynamics by affecting intra- and inter-chromosomal interactions and regulate the formation of subnuclear compartments (e.g., paraspeckles, nuclear bodies, etc.)

1996). This 17 kb long RNA is localized exclusively in the nucleus where it spreads in *cis* across the entire X chromosome from which it is transcribed. It contains six discrete repeat sequences, some of which interact with and recruit repressive chromatin-modifying complexes leading to the silencing of one of the X chromosomes (Avner and Heard 2001; Beletskii et al. 2001; Clemson et al. 1996; Engreitz et al. 2013; McHugh et al. 2015; Plath et al. 2002, 2003; Silva et al. 2003). One of these repeats, the 1.6 kb repeat A (repA), forms stem loops that directly interact with the protein SHARP to mediate epigenetic silencing of the X chromosome through recruitment of the SMRT-HDAC3 co-repressor complex (Chu et al. 2015; McHugh et al. 2015; Monfort et al. 2015). *Xist* repA was also shown to interact with the PRC2 Polycomb repressor complex, which participates in X chromosome inactivation by trimethylating histones at H3K27 (Wutz et al. 2002; Zhao et al. 2008). However, the specific deletion of repA from *Xist* strongly impairs silencing (Wutz et al. 2002) but does not affect recruitment of the PRC2 complex, raising doubts whether this portion of *Xist* really interacts directly with the complex (da Rocha et al. 2014; Plath et al. 2003). On the other hand, the repeat B does interact with HNRNPK and is necessary for *Xist* spreading along the X chromosome as well as recruitment of the PRC1 and PRC2 complexes (Colognori et al. 2019). Thus, *Xist* acts as a modular RNA scaffold (Fig. 10.2e) that coordinates the association of several chromatin-modifying complexes to silence a whole chromosome.

Antisense lncRNAs are another class of lncRNAs shown to modulate transcription. Mechanistic studies of a small number of antisense lncRNAs suggest that some are able to form direct interactions with DNA and recruit transcription factors to regulate target genes. For example, the 10 kb antisense lncRNA *Khps1* was shown to activate in *cis* the expression of its neighbor protein-coding gene *Sphk1* by forming a DNA:RNA triple helix at a *Sphk1* enhancer through a homopurine-rich sequence upstream of *Sphk1* and recruiting the transcription factors E2F1 and p300 (Blank-Giwojna et al. 2019; Postepska-Igielska et al. 2015). Interestingly, replacing the triplex forming region of *Khps1* with that of *Meg3*, another DNA:RNA triple helix-forming lncRNA (Mondal et al. 2015), tethers *Khps1* to a *Meg3* target gene (Blank-Giwojna et al. 2019), suggesting these sequences can be modular and provide target specificity. Similarly, the lncRNA *PAPAS*, which is transcribed antisense to pre-rRNA, forms DNA:RNA triplex structures with a purine-rich sequence located in the enhancer region. RNA-protein interaction and structure studies revealed that a single stranded A-rich sequence within a loop structure of *PAPAS* binds the NuRD complex subunit CHD4, which helps trigger silencing of rDNA (Zhao et al. 2018). *PARTICLE* is another example of an antisense lncRNA capable of forming a DNA: RNA triplex to repress the expression of the MAT2A gene by helping recruit the G9a and PRC2 repressor complexes at its promoter (O'Leary et al. 2015).

The telomeric repeat-containing RNA *TERRA* was also shown to form DNA: RNA hybrid structures, termed R-loops (Fig. 10.2a), at short telomeres and promote alternative lengthening of telomeres (ALT) by homology-directed repair (Balk et al. 2013). Identification of *TERRA* interactors revealed 134 proteins, among them shelterin and ALT proteins, which bind telomeres (Chu et al. 2017). Another protein identified to directly interact with *TERRA* is the chromatin remodeler ATRX, known to suppress ALT (Doksani and de Lange 2014; Lovejoy et al. 2012). Interestingly, localization of *TERRA* at telomeres appears to compete with ATRX binding as depletion of *TERRA* leads to a relocalization of ATRX to telomeres (Chu et al. 2017). But these are unlikely to be the only examples, as recent studies have begun to uncover thousands of noncoding RNAs forming DNA:RNA triplex and R-loop structures genome-wide (Sanz et al. 2016; Sentürk Cetin et al. 2019), suggesting structure as a more general mechanism for lncRNAs to recognize target genes for regulation. Further investigation of these RNAs will greatly improve our understanding of the circuitry of genome regulation.

#### 10.2.1.2 Transcription Regulation *in trans*

lncRNAs have also been shown to regulate gene expression in the nucleus by acting in *trans*. One classic example is the lncRNA *Hotair*, which is transcribed from the developmental *HoxC* locus in mammals but represses the *HoxD* gene cluster located on another chromosome (Rinn et al. 2007). Consistent with a function in regulating transcription, silencing of *Hotair* leads to a decrease in H3K27me3 repressive histone marks and a corresponding activation of *HoxD* genes. Using RNA immunoprecipitation (RIP) and pulldown assays, *Hotair* was one of the first lncRNAs reported to bind components of the PRC2 repressive complex (Rinn et al. 2007). Using similar approaches, further studies showed that PRC2 can interact with hundreds of different lncRNAs. This resulted in a model suggesting that many lncRNAs function to repress transcription by recruiting PRC2 to target genes (Khalil et al. 2009). However, tethering of *Hotair* to a luciferase reporter was shown to repress expression independent of the PRC2 complex, and overexpression of *Hotair* in breast cancer cells resulted in PRC2-independent repression of a subset of target genes (Portoso et al. 2017), suggesting that interaction and recruitment of PRC2 may not be the pathway by which *Hotair* and many other lncRNAs act. Moreover, recent studies showed promiscuous binding of PRC2 to RNAs and, in addition, that RNA binding inhibits its H3K27 methyltransferase activity (Davidovich et al. 2013; Zhang et al. 2019). Together with the here exemplified caveat that immunoprecipitation methods can lead to the identification of false positives that may not reflect in vivo interactions (see below), it remains unclear exactly how *Hotair* is able to repress transcription in *trans*.

Another lncRNA found to act in *trans* is *Ttc39aos1* (*lincRNA-Eps*), which is involved in regulating apoptosis and inflammation. This 2.5 kb RNA, transcribed from mouse chromosome 4, was first shown to suppress the expression of the pro-apoptotic gene *Pycard,* located on chromosome 7, in erythrocytes (Hu et al. 2011). Genomic deletion of *Ttc39aos1* in mice does not affect neighboring genes but leads to differential expression of multiple immune response genes in macrophages and enhances inflammation in vivo upon stimulation with lipopolysaccharides (LPS) (Atianand et al. 2016). Importantly, ectopic re-expression of *Ttc39aos1* using a retroviral vector in LPS-stimulated macrophages rescues the levels of differentially expressed genes to near wild-type levels. *Ttc39aos1* associates with chromatin particularly at the promoters of immune response genes, where it helps to maintain an epigenetically repressed state. The protein HNRNPL appears to interact with a CANACA motifs 3′ of *Ttc39aos1*, and disruption of this interaction by knockdown of HNRNPL leads to increased expression of several *Ttc39aos1* target genes and increased H3K4me3 at their promoters (Atianand et al. 2016). Thus, *Ttc39aos1* is a lncRNA that restrains inflammation by transcriptionally repressing immune response genes.

Apart from stimulating the recruitment of factors at specific loci across the genome, lncRNAs can also prevent the binding of factors by acting as decoys (Fig. 10.2a). For example, the lncRNA *Gas5* is located both in the nucleus and the cytoplasm. However, in the presence of the glucocorticoid receptor agonist dexamethasone, a fraction of *Gas5* translocate to the nucleus where it interacts with the DNA binding domain of the glucocorticoid receptor (GR) through a sequence located between nucleotides 400–598. There, *Gas5* acts as a mimic of glucocorticoid receptor elements and competes with the binding of the glucocorticoid receptor to DNA to inhibit the transcriptional activation of target genes (Kino et al. 2010). Similarly, the lncRNA *PANDA* regulates p53-mediated apoptosis by binding and sequestering the transcription factor NF-YA, thereby preventing it from activating key apoptosis target genes (Hung et al. 2011). Together, these examples illustrate that through interactions with different proteins, lncRNAs use various strategies to regulate gene expression in *trans*.

### *10.2.2 Chromatin Architecture and Nuclear Organization*

The nucleus is a highly organized and dynamic environment that allows precise transcriptional regulation and where subnuclear regions are dedicated to specific functions. For example, the nucleolus is a membrane-less compartment within the nucleus where ribosomal RNA biogenesis and processing occurs (Mélèse and Xue 1995). Other such membrane-less compartments are the nuclear speckles, which are enriched in RNA splicing factors, and paraspeckles, sites for RNA editing and nuclear retention. Some well-characterized lncRNAs have been shown to actively participate in nuclear organization. First among them is *Xist* which, in addition to being essential for the repression and compaction of one of the X chromosomes in females, is also required for the localization of the inactive X chromosome to the nuclear periphery, where it forms a domain called the Barr body. *Xist* was found to interact with an arginine-serine motif of the Lamin B receptor (LBR), a transmembrane protein that associates with the nuclear lamina (Chen et al. 2016b; Hezroni et al. 2015). Disruption of the *Xist*-LBR interaction abolishes X chromosome silencing and recruitment to the nuclear periphery (Chen et al. 2016a). The effect of *Xist* on nuclear organization is restricted to a single chromosome; however, another nuclear localized lncRNA transcribed from the X chromosome named *Firre* is involved in the formation of inter-chromosomal interactions (Hacisuleyman et al. 2014). While RNA in situ hybridization showed that *Firre* accumulates at its site of transcription, mapping of its target sites genome-wide using RNA antisense purification (RAP, see below) revealed that *Firre* interacts with several other gene loci located on different chromosomes. Interestingly, these loci are found in close proximity to the *Firre* locus in the nucleus, and this co-localization is abolished by deletion of *Firre* or knockdown of its interacting partner HNRNPU. This suggests that *Firre* is able to modulate nuclear architecture by mediating inter-chromosomal contacts (Fig. 10.2f) (Hacisuleyman et al. 2014). The exact mechanism by which

*Firre* mediates these contacts remains unknown, but similar *to Xist*, these results indicate that some lncRNAs can actively shape chromatin structure and nuclear organization.

### *10.2.3 Post-transcriptional Regulation: Splicing and RNA Editing*

Nuclear lncRNAs can be essential for the formation or the normal function of membrane-less subcompartments, such as nuclear speckles and paraspeckles (Fig. 10.2f). Nuclear speckles are often located in close proximity to highly expressed genes and are thought to act as a reservoir through which SR splicing proteins shuttle back and forth to target genes depending on their phosphorylation status (Misteli et al. 1998). The lncRNA *Malat1* is a highly conserved and abundantly expressed 7.5 kb transcript which is recruited to the periphery of nuclear speckles through direct interactions with several SR splicing factors (Tripathi et al. 2010). Knockdown of *Malat1* affects the alternative splicing of a set of pre-mRNAs and impairs the phosphorylation of SR proteins but does not affect the formation of speckles (Arun et al. 2016; Tripathi et al. 2010). *Malat1* was also found to bind actively transcribed genes (Engreitz et al. 2014; West et al. 2014), which led to the postulation that *Malat1* may guide the position of speckles near active gene loci. Apart from *Malat1* and nuclear speckles, other lncRNAs were also shown to regulate splicing (Fig. 10.2b).

*Pnky* is an evolutionarily conserved *trans*-acting lncRNA involved in cortical neurogenesis in a cell-autonomous manner in vivo. It is divergently transcribed from the neighboring proneural transcription factor *Pou3f2* and was shown to interact with PTBP1, a splicing factor that regulates neurogenesis (Andersen et al. 2019). In situ hybridization shows that *Pnky* is located in several puncta exclusively in the nucleus. Both genomic deletion of the *Pnky* locus and shRNA-mediated knockdown of *Pnky* RNA increase neuronal differentiation and deplete the neural stem pool without affecting the expression of its neighboring genes, indicating that *Pnky* does not work in *cis* (Andersen et al. 2019; Ramos et al. 2015). Instead, loss of *Pnky* expression leads to differential expression and differential splicing of multiple protein-coding genes located on different chromosomes and related to neuronal differentiation. Importantly, ectopic re-expression of *Pnky* in knockout cells is able to rescue the neural differentiation phenotype as well as the differential expression and splicing (Ramos et al. 2015). These changes are highly similar to knockdown of PTBP1, suggesting that *Pnky* and PTBP1 act together to regulate a set of protein-coding genes important for neuronal differentiation.

However, lncRNAs have also been described as a component of paraspeckles, nuclear ribonucleoprotein bodies that regulate gene expression by sequestering specific mRNAs and RNA-binding proteins involved in transcription and RNA processing. The well-conserved and highly expressed lncRNA *Neat1* is essential for paraspeckle formation and organization (Clemson et al. 2009; Mao et al. 2010; Sasaki et al. 2009). Following alternative processing of its 3′end, *Neat1* is separated into two distinct isoforms, a short 3.7 kb *Neat1_1* transcript and a long 22.7 kb *Neat1_2* transcript, both of which are important for proper paraspeckle formation (Hutchinson et al. 2007; Naganuma et al. 2012; Yamazaki et al. 2018). *Neat1* transcripts form highly organized spheres within paraspeckles in which both the 5′ and 3′ ends of the long *Neat1_2* isoform are located at the periphery and its body in the center, whereas the short *Neat1_1* isoform is located only at the periphery (Souquere et al. 2010; West et al. 2016). Within these structures, *Neat1* interacts with more than 40 proteins many of which are RNA-binding proteins involved in mRNA processing, such as SFPQ, NONO, PSPC1, TDP43, and FUS (Naganuma et al. 2012; Yamazaki and Hirose 2015). Some of these proteins, such SFPQ and TDP43, are sequestered within paraspeckles through interacting with *Neat1*, thereby preventing them from binding their mRNA target(s) (Hirose et al. 2014; Imamura et al. 2014). *Neat1* and paraspeckles have been suggested to regulate specific mRNAs expression by retaining them within paraspeckles and interfering with their export to the cytoplasm. This was shown for mRNAs containing inverted *Alu* repeats or SINEs in their 3′ untranslated region (UTR), which form double-strand RNA duplexes (Chen and Carmichael 2009; Prasanth et al. 2005). The RNA deaminase ADAR recognizes these structures and converts adenosines into inosines (A-to-I editing). This results in mRNAs with extensive A-to-I edited inverted *Alu* repeats, which are retained in paraspeckles through an interaction with the inosine-specific RNA-binding protein NONO (Chen et al. 2008; Prasanth et al. 2005; Zhang and Carmichael 2001). Interestingly, sequestration of SFPQ in paraspeckles prevents it from activating the RNA deaminase ADARB2 at its promoter (Hirose et al. 2014), suggesting that *Neat1* paraspeckles also regulates A-to-I editing of mRNA at the transcriptional level. Taken together, this demonstrates that nuclear lncRNAs can regulate the expression of mRNAs post-transcriptionally at multiple levels in combination with defined protein partners.

## 10.3 Cytoplasmic Functions of lncRNAs

In addition to the large number of nuclear transcripts, a significant fraction of lncRNAs is found in the cytoplasm, although their functions are not as well studied, or understood, as those in the nucleus. However, there are a few examples where lncRNAs were shown to be actively involved in regulating protein localization, translation, and mRNA stability as well as modulating specific signaling pathways (Fig. 10.2c, d).

### 10.3.1 Protein Localization

Similar to the decoy or sequestering function of some lncRNAs in the nucleus, cytoplasmic lncRNAs can also modulate protein location and sequester proteins away (Fig. 10.2c). The best characterized example is *NORAD*, a highly conserved 5.3 kb lncRNA required to maintain genome stability (Lee et al. 2016; Tichon et al. 2016). *NORAD* was found to interact with PUMILIO RNA-binding proteins, which inhibit gene expression via binding of the 3′UTR of target mRNAs, inducing deadenylation and decapping (Miller and Olivas 2011). Analysis of the *NORAD* sequence revealed a conserved 8 nt sequence repeated 18 times along the transcript. This motif (UGUANAUA or UGUANAUN) perfectly matches the PUMILIO response elements (PRE) (Lee et al. 2016; Tichon et al. 2016). *NORAD* is estimated to be expressed at several hundred copies per cell; thus, with multiple PREs within its sequence, this lncRNA has the capacity to bind multiple PUMILIO proteins per RNA molecule. Loss of *NORAD* leads to PUMILIO hyperactivity and downregulation of PUMILIO target mRNAs, many of which are involved in maintenance of genome stability. Accordingly, *NORAD* depleted cells show increased chromosomal instability. Thus, this lncRNA protects the genome integrity by restricting PUMILIO from binding its target mRNAs for degradation (Lee et al. 2016; Tichon et al. 2016). Recently, *NORAD* was also found to bind RBMX, another RNA-binding protein involved in DNA damage; however, the consequence of this interaction is still unclear (Elguindy et al. 2019; Munschauer et al. 2018).

Another lncRNA, the 2.7 kb-long *NRON,* was shown to regulate subcellular protein localization. In T cells, *NRON* associates with the transcription factor nuclear factor of activated T cells (NFAT) in the cytoplasm and sequesters it in an inactive form within a complex that includes the nuclear import factor KPNB1, the calmodulin-binding protein IQGAP1, and the inhibitory kinase LRRK2. Following T-cell activation, increased Ca2+ levels lead to a disassembly of the *NRON* scaffold complex which allows NFAT to translocate in the nucleus where it can regulate cytokine target genes (Sharma et al. 2011; Willingham et al. 2005). Thus, by tittering proteins away or controlling their activation state, cytoplasmic lncRNAs have evolved different strategies to regulate protein function.

### 10.3.2 Translation Regulation

Besides ribosomal RNA, other lncRNAs were shown to regulate translation (Fig. 10.2d). One such lncRNA is *UCHL1-AS1*, a 1.2 kb antisense transcript that partially overlaps (73 nt) with the 5′ UTR of *UCHL1*, a gene that encodes a ubiquitin hydrolase. Under normal conditions, *UCHL1-AS1* is predominantly found in the nucleus. However, when cells are treated with the mTOR inhibitor rapamycin, *UCHL1-AS1* translocates to the cytoplasm where it binds the 5′UTR of the *UCHL1* mRNA through its 73 nt complementary sequence (Carrieri et al. 2012).

This interaction helps target the *UCHL1* mRNA to polysomes to increase translation. The non-overlapping 3′ end of *UCHL1-AS1* also contains an inverted SINE B2 transposable element sequence, which is required to increase translation of *UCHL1* (Carrieri et al. 2012). The exact mechanism of this translational regulation is currently unknown, but it likely involves the modulation of specific protein interactions. This arrangement—where an antisense transcript shares a complementary sequence (binding domain) in a 5′ head-to-head fashion with an mRNA followed by an inverted SINE B2 repeat (effector domain) further downstream—is common to multiple lncRNAs organized in a sense/antisense pair with an mRNA (Carrieri et al. 2012; Zucchelli et al. 2015). These lncRNAs were termed SINEUPs. Importantly, it was shown that SINEUP sequence organization is modular and can be engineered to create synthetic RNAs that increase the translation of target mRNAs (Zucchelli et al. 2015). These results not only show that lncRNA-mRNA interactions are regulatory but also highlight the importance of repeat sequences in the function of lncRNAs.

### *10.3.3 Regulating RNA Stability*

Staufen1-mediated RNA decay (SMD) is a process whereby Staufen1 (STAU1), an RNA-binding protein, binds specifically to double-strand RNA structures to promote degradation with the help of the helicase UPF1. The double-strand RNA structures recognized by STAU1 can be formed by the intramolecular base pairing of sequences in the 3′UTR of an mRNA or by intermolecular base pairing of 3′UTR sequences of an mRNA with a partially complementary *Alu* sequence of another RNA (Gong and Maquat 2011; Kim et al. 2005). The lncRNA *½-sbsRNA1* is a cytoplasmic transcript that contains an *Alu* sequence which was specifically shown to pair with an *Alu* sequence in the 3′UTR of the *SERPINE1* mRNA. Interaction between the lncRNA *½-sbsRNA1* and *SERPINE1* is necessary for STAU1 binding and degradation of *SERPINE1* through SMD, and knockdown of *½-sbsRNA1* leads to an increase in target mRNA expression (Gong and Maquat 2011). In addition to *½-sbsRNA1*, three other *Alu* element-containing lncRNAs (named *½-sbsRNA2–4*) were found to regulate specific mRNA targets containing partially complementary sequences in their 3′UTR (Gong and Maquat 2011). Curiously, STAU1 can also bind inverted *Alu* repeats within 3′UTR of mRNAs, and this interaction seems to compete with NONO binding in paraspeckles. The resulting interaction leads to export of mRNAs to the cytoplasm and can enhance their translation (Capshew et al. 2012; Elbarbary et al. 2013). It is currently not known why interaction of STAU1 with inverted *Alu* repeats within an mRNA does not trigger SMD, similar to intermolecular and partial complementary interactions between mRNA and cytoplasmic *½-sbsRNAs Alu* sequences. In any case, with thousands of lncRNAs containing *Alu* elements, it is predicted that many more cytoplasmic lncRNAs function through STAU1-mediated decay, potentially creating a complex regulatory network of lncRNA-mRNA pairs.

### 10.3.4 Protein Signaling and Activation

Binding of a lncRNA to a specific region of a protein can potentially mask a site or compete for the interaction with another molecule. As such, lncRNAs have been shown to regulate specific protein-protein interactions. For example, the lncRNA *NKILA* is upregulated by the transcription factor NFkB and also interacts with NFkB/IkB. This interaction was found to mask the phosphorylation motif of IkB, leading to the activation of NFkB (Liu et al. 2015). Similarly, the lncRNA *lnc-DC* regulates the differentiation of dendritic cells by directly binding to cytoplasmic STAT3 protein. Interaction with *lnc-DC* promotes the phosphorylation of STAT3 on tyrosine 705 by blocking the binding of SHP1, a phosphatase known to inhibit STAT3 (Wang et al. 2014). Another example of a lncRNA involved in signal transduction is *LINK-A*, a 1.5 kb transcript frequently upregulated in breast cancer. Nucleotides 481–540 and 781–840 of LINK-A were shown to interact with the SH3 and C-terminal regions of the tyrosine kinase domain of BRK (Lin et al. 2016). Through its interaction with BRK, *LINK-A* was found to facilitate the recruitment of BRK to an EGFR-GPNMB complex and induce a conformational change leading to its activation. On the contrary, knockdown of *LINK-A* abolishes the recruitment of BRK to EGFR-GPNMB and its activation (Lin et al. 2016). Modulating interactions between different proteins therefore seems to be an important function of different lncRNAs.

## 10.4 Characterizing lncRNA-Protein Interactions

Characterizing RNA-protein interactions has been a focus in studying different aspects of gene regulation for decades, and many methods have been developed to study RNA-protein interactions. However, identifying lncRNA-binding proteins, as well as dissecting the rules of how protein interactions modulate lncRNA function, is turning out to be a challenging endeavor. One complication in the study of lncRNA/lncRNPs is that contrary to mRNAs, which have identifiable ORFs that code for well-described protein domains, we have a poor understanding of how sequence features of lncRNAs are organized, and we are far from being able to predict the function of a lncRNA based on its primary sequence. The poor conservation level of lncRNA primary sequence, the versatility of RNA to generate similar secondary structures from different sequences, and the difficulties of predicting RNA structures all limit our ability to identify common domains and motifs or modular structures that might drive lncRNA function. Advances in techniques to experimentally determine RNA accessibility and structure will greatly help on that front. Interestingly, as shown for different lncRNAs described above, repetitive elements such as *Alu*, SINEs, and other repeat sequences seem to be important features in conferring functions to RNAs. Interestingly, whereas *Alu* sequences can be found in lncRNAs as well as UTRs of mRNAs, local repeats, which are defined as sequences that repeat within one given genomic locus, are more abundant in lncRNAs than mRNAs (Hacisuleyman et al. 2016). One of these local repeats, termed RRD and found in the lncRNA *Firre*, mediates interaction with hnRNPU and is essential to keep the lncRNA in the nucleus (Hacisuleyman et al. 2014). Certain families of transposable elements are also enriched in lncRNAs compared to mRNAs (Johnson and Guigó 2014; Kapusta et al. 2013; Kelley and Rinn 2012) and the presence of transposable elements in RNA sequences was found in several instances to be associated with protein binding (Kelley et al. 2014). Given that a large fraction of lncRNAs contain transposable elements and repeat sequences, a more detailed and systematic analysis of their roles as potential RNA domains for protein interactions offers an encouraging path forward in identifying potential common sequence elements to better understand lncRNA functions.

RNA and proteins have been living alongside each other for millennia. The thousands of lncRNAs located in the nucleus, cytoplasm, and other subcellular compartments are likely to be part of ribonucleoprotein complexes. Different candidate centric studies, as well as systematic characterization of RBP-targeted RNAs such as performed by the ENCODE consortium, revealed that many of the 1500+ RNA-binding proteins (RBP) encoded by mammalian genomes (Gerstberger et al. 2014) interact with lncRNAs (Hendrickson et al. 2016; Sundararaman et al. 2016). Therefore, in-depth characterization as well as further identification of lncRNA-RBP interactions using protein-centric approaches will generate important information that will shed light on the function of many lncRNAs, possibly in a similar way to what the characterization of histone marks has done for epigenetics. Significant effort has been made to develop biochemical approaches to identify proteins that interact with target lncRNAs, some of them adapted from approaches previously used to study other types of RNA-proteins interactions. While most methods to study RNA-protein interactions were protein-centric until recently, including *RNA immunoprecipitation* (RIP) and *cross-linking and immunoprecipitation* (CLIP) (Ule et al. 2003), the recent development of RNA-centric approaches, such as the *chromatin isolation by RNA purification* (ChIRP), *capture hybridization analysis of RNA targets* (CHART), and *RNA antisense purification* (RAP) (Chu et al. 2015; McHugh et al. 2015; Simon et al. 2011), has already proven to be important tools to further advance our understanding of the roles noncoding RNAs play in cells. One thing to note, however, is that the low expression level of many lncRNAs can make the use of biochemical purification and characterization challenging. Here, I will describe some of the methods that have been developed or adapted to study lncRNA-protein interactions.

## *10.4.1 Protein-Centric Approaches*

### 10.4.1.1 RNA Immunoprecipitation (RIP)

RNA immunoprecipitation is suited to detect interactions of individual proteins with specific RNA species in vitro or in vivo *and* has been applied to characterize

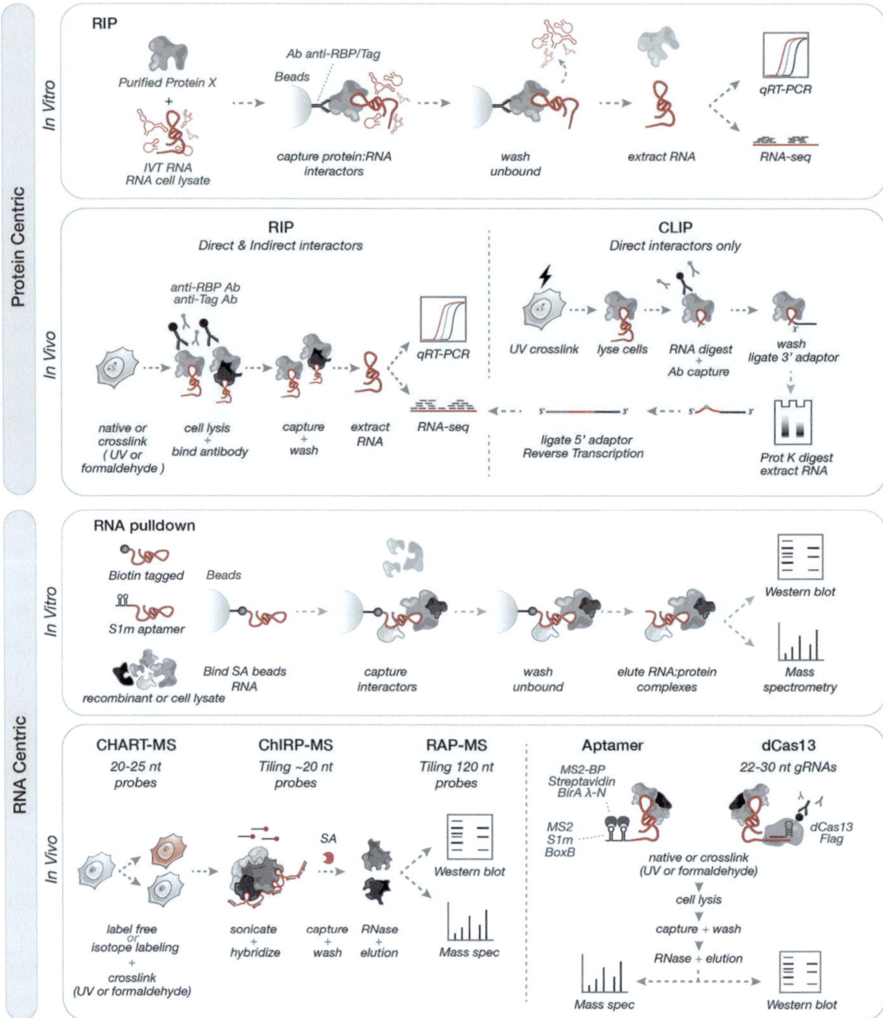


**Fig. 10.3** Methods to identify lncRNA-protein interactions in vitro and in vivo. Protein-centric approaches (*Top panels*). The various techniques and steps to identify RNAs interacting with a specific protein in vitro (RIP) and in vivo (RIP and CLIP) using anti-protein or anti-Tag antibodies are schematized. RNA-centric approaches (*Lower panels*). The various techniques and steps to identify RNA-protein interactions in vitro (RNA pulldowns) and in vivo (RAP-MS, ChIRP-MS, CHART, aptamer, dCas13-Tag) using RNA as bait are schematized

RNA-protein interactions for many years (Fig. 10.3). For in vitro RIP, a purified protein of interest is first mixed with an in vitro-transcribed RNA or RNAs from a heterogeneous cell lysate, and an antibody coupled to magnetic beads is subsequently used to capture the protein. RNAs directly and stably interacting with the protein will co-precipitate, whereas non-associated RNAs will be eliminated by rigorous washes. Interacting RNAs are then extracted and can be measured using real-time PCR and Northern blotting or identified by RNA sequencing (Zhao et al. 2010). For in vivo RIP, RNA-protein interactions are isolated directly from a cell lysate following lysis. Prior to performing in vivo RIP, one important consideration is whether to cross-link cells in order to capture potentially transient or unstable interactions or keep them in a native state. Native RIP is more suitable to identify RNAs bound directly and stably to the protein of interest, whereas cross-linking RIP can identify not only less stably bound RNAs but also RNAs bound indirectly through interactions with other proteins within a complex. To date, RIP has been one of the most commonly used methods to study lncRNA-protein interactions. However, interpretation of in vivo RIP results requires careful controls, since it was shown that some RNA-protein interactions are not necessarily present in cells in vivo but instead form in solution post cell lysis (Mili and Steitz 2004). Therefore, it is advisable to confirm RNA-protein interactions using a second method and complementary experiments.

#### 10.4.1.2 Cross-Linking Immunoprecipitation (CLIP)

CLIP overcomes some of the drawbacks of RIP by cross-linking RNA-protein complexes with ultraviolet light (UV), which only cross-links RNA to proteins in close proximity as opposed to formaldehyde which cross-links more readily proteins. It also allows the identification of protein binding site on the interacting RNA. Post UV cross-linking, and following cell lysis, samples are treated with RNase in a controlled manner to generate short RNA fragments, and only the RNA regions interacting with a protein will be protected. The covalently cross-linked RNA-protein partners are then immunoprecipitated using an antibody against the protein of interest, and stringent washes of the isolates are performed. Adaptors are subsequently ligated to the 3′ end of the RNA, proteins digested with proteinase K, and RNAs extracted for subsequent 5′ end adaptor ligation, library preparation, and deep sequencing (Fig. 10.3). To identify more readily the exact nucleotides interacting with a protein, several variants of CLIP have been developed. For example, PAR-CLIP introduces photoactivatable ribonucleosides, such as 4-thiouridine or 6-thioguanosine, into nascent transcripts to induce strong and efficient cross-linking by UV. Following cDNA synthesis of the purified RNA, this ribonucleoside is changed to a cytosine, indicating the site of interaction between the RNA and the protein (Hafner et al. 2010). Individual nucleotide resolution CLIP (iCLIP) was developed to identify more precisely RNA-binding site without cDNA mutation. Similar to standard CLIP, adapters are ligated at the 3′ end of RNA fragments following RNase treatment. However, instead of ligating an adaptor at the 5′ end, the cDNA is circularized and subsequently processed for sequencing. The bond created by UV cross-linking between RNA and proteins leaves a peptide adduct on the RNA interaction site, and the reverse transcriptase is often stalled at these adducts. By circularizing the cDNA, these sites can be precisely mapped leading to the identification of RNA-protein contacts at nucleotide resolution (König et al. 2010). CLIP and similar methods have been highly successful in identifying precise RNA-protein interactions in the past decade. However, these techniques are technically challenging and require large amounts of material, which poses a problem for the study of low abundance lncRNAs.

## *10.4.2 RNA-Centric Approaches*

Whereas protein-centric approaches are useful in obtaining maps of RBPs that associate with lncRNAs and to determine binding sites on specific mRNAs, they are not suited to obtain a complete interactome for a specific lncRNAs. To achieve this, different RNA centric methods that isolate lncRNPs from cells have been developed.

### 10.4.2.1 RNA Pulldown

RNA pulldown is an in vitro assay that allows to identify proteins that interact with an RNA of interest (Fig. 10.3). In this technique, in vitro-transcribed RNA is labeled with biotin or synthesized with an S1m aptamer at the 5′ or 3′ end. A recombinant protein or a cell lysate is then mixed with the labeled RNA, and complexes are affinity-purified using streptavidin-conjugated magnetic beads. Following more or less stringent washing, interacting proteins (direct and indirect) are identified by Western blotting or mass spectrometry. This technique is relatively simple to implement and is often used as a secondary assay to confirm whether an interaction between a lncRNA and a protein is direct. However, when using protein lysates, both direct and indirect interactions are generally detected.

### 10.4.2.2 Biotinylated Oligo Approaches (ChIRP-MS, CHART-MS, RAP-MAS)

With an increasing need for RNA-centric biochemical purification methods, various new methods to systematically map lncRNA-protein interactions in vivo have been developed and applied to characterize the proteome of specific candidates. Among the most frequently used methods is *C*ompre*h*ensive *I*dentification of *R*NA-binding *P*roteins by *M*ass *S*pectrometry (ChIRP-MS), a method that uses 20 nt biotinylated oligonucleotides complementary to and tiled across a lncRNA of interest, as a handle to pull down interacting proteins. Cultured cells are first cross-linked with formaldehyde and cells are lysed and sonicated to solubilize the lysate. Biotinylated oligos are then added to the cell extract allowing to hybridize with the target lncRNA. RNA-protein complexes are then captured on streptavidin-conjugated magnetic beads followed by washes to eliminate non-interacting proteins. After elution, the isolated interacting proteins are subjected to liquid chromatography-tandem mass spectrometry for identification (Chu et al. 2015).

*C*apture *H*ybridization *A*nalysis of *R*NA *T*argets *M*ass *S*pectrometry (CHART-MS) is conceptually and technically similar to ChIRP. However, whereas ChIRP uses short oligos across the entire lncRNA without a priori knowledge of any functional RNA domains, CHART empirically determines which probes to use by mapping good candidate hybridization regions using an RNase H assay that degrades DNA:RNA hybrids (Simon et al. 2011). Proteins interacting with NEAT1 and MALAT1 in paraspeckles and nuclear speckles, respectively, were identified successfully with CHART-MS (West et al. 2014).

Similar to ChIRP and CHART, *R*NA *A*ffinity *P*urification Mass Spectrometry (RAP) uses antisense biotinylated oligos which hybridize to a target lncRNA and pulldown interacting proteins. Compared to ChIRP-MS and CHART-MS, the most distinctive feature of RAP-MS is that it uses longer but fewer biotinylated tiling probes (>60 nucleotides), which form very stable RNA-DNA hybrids (McHugh et al. 2014). This allows more stringent hybridization and wash conditions and hence reduces nonspecific RNA-protein interactions albeit at a higher cost due to the length of the probes. Both formaldehyde or UV have been used as cross-linking agents for RAP, and the approach has been combined with *S*table *I*sotope *L*abeling by *A*mino acid *C*ell culture (SILAC) to quantitatively measure proteins associated with *Xist*, such as SHARP, LBR, and PTBP1 (McHugh et al. 2015).

#### 10.4.2.3 Aptamer-Based Approaches

Aptamers are functional RNA sequences with the ability to specifically bind to proteins or small molecules with high affinity (Ellington and Szostak 1990). They can, therefore, be used as handles for RNA pulldown experiments by linking them to the 5′ or 3′ end of a target RNA and introduced in vivo through either a vector-based ectopic expression system or endogenous tagging by homology-directed repair. The S1m aptamer is particularly suitable since it has a high affinity for streptavidin and hence allows target RNAs to be purified directly using streptavidin-conjugated magnetic beads (Leppek and Stoecklin 2014; Srisawat and Engelke 2001). This approach was successfully used with full-length and truncated versions of the cardiac-enriched lncRNA *Chaer* to identify a 524 nt region necessary for its interaction with EZH2 (Wang et al. 2016). The MS2 aptamer is another widely used 19 nt hairpin derived from the MS2 RNA phage that interacts with the MS2 coat protein with high affinity and specificity. The incorporation of ≥4 repeats of the MS2 aptamer into the target RNA, 5′ or 3′, allows the isolation of RNA-protein complexes on amylose resin using immobilized maltose-binding protein fused to MS2-binding coat protein (Parrott et al. 2000; Tsai et al. 2011) or by co-expressing a FLAG-tagged MS2 coat protein together with the RNA-MS2 aptamer fusion (Gong and Maquat 2014; Gong et al. 2012). Alternatively, a BoxB aptamer can be linked to an RNA and used in combination with a BirA-λN fusion protein that can bind the BoxB aptamer to biotinylate neighboring interacting proteins, a technique termed RNA-Protein

Interaction Detection (RaPID). Interacting proteins can subsequently be purified using streptavidin beads and identified by Western blotting or mass spectrometry. This was successfully used to identify host proteins interacting with the Zika virus RNA (Ramanathan et al. 2018).

There's been a great leap in the development of methods allowing us to probe RNA-protein interactions by using either molecules as handles, but many challenges still remain. UV and formaldehyde cross-linking methods are generally inefficient and require large amounts of cells. Improvements in cross-linking methods would thus greatly help capturing RNA-protein interactions with fewer cells or in rare subpopulations. Similarly, many of these methods are not well suited to probing interactions of low abundance transcripts. The discovery of CRISPR-Cas13, which is easily programmable and can interact specifically with RNA, can potentially be adapted to serve as a handle to purify endogenous RNA-protein interactions directly from cells. Combining this with proximity labeling approaches such as BioID or APEX can potentially be a powerful approach to identify endogenous lncRNPs. Of course, orthogonal approaches (both RNA and protein-centric) are always necessary to validate interactions.

## 10.5 Concluding Remarks

Messenger ribonucleoprotein complexes are involved in regulating mRNA metabolism from their transcriptional birth to their degradation. Long noncoding RNAs, through their interaction with proteins (lncRNPs), can also participate in regulating mRNA metabolism, but also exhibit a wide range of properties and functions to actively regulate cellular processes at almost all levels. We have only just begun to scratch the surface of their importance in modulating cellular pathways. While at this point, it is still hard to predict how many out of the thousands lncRNAs identified are actually functional at the RNA level, many are likely to be found to regulate key cellular processes. Nevertheless, for this to happen, thorough and detailed approaches are needed in order to (1) distinguish whether the act of lncRNA transcription or the RNA transcript itself performs a function in *cis* or in *trans*; (2) determine how lncRNAs integrate into known signaling pathways by identifying their protein interaction partners; (3) decipher lncRNA sequence-structure-function relationship; and (4) determine the impact of lncRNA function on normal physiology, disease, and progression. The task ahead is colossal and will require painstaking and sustained work. Fortunately, the last few years have seen the development of many novel technologies as well as robust genetic and biochemical methods which will greatly improve our ability to tackle these questions and allow us to explore the vast world of lncRNAs. Exiting times are ahead in exploring the lncRNA world.

**Acknowledgments** I would like to thank Drs Marlene Oeffinger and Daniel Zenklusen for insightful discussions and critically reading the manuscript as well as members of my laboratory for their input.

**Funding** M.S. is a Junior Research Scholar of the Fonds de Recherche du Québéc Santé (FRQS). This work is also supported by a Canadian Institutes of Health Research (CIHR) Project Grant, a National Sciences and Engineering Research Council of Canada (NSERC) Discovery Grant, and a Canadian Foundation for Innovation (CFI) John R. Evans Leaders Fund Grant to M.S.

## References

Andersen RE, Hong S, Lim JJ, Cui M, Harpur BA, Hwang E, Delgado RN, Ramos AD, Liu S, Blencowe BJ et al (2019) The long noncoding RNA Pnky is a trans-acting regulator of cortical development *In Vivo*. Dev cell 49:632–642.e7

Anderson DM, Anderson KM, Chang C-L, Makarewich CA, Nelson BR, McAnally JR, Kasaragod P, Shelton JM, Liou J, Bassel-Duby R et al (2015) A micropeptide encoded by a putative long noncoding RNA regulates muscle performance. Cell 160:595–606

Arun G, Diermeier S, Akerman M, Chang K-C, Wilkinson EJ, Hearn S, Kim Y, MacLeod RA, Krainer AR, Norton L et al (2016) Differentiation of mammary tumors and reduction in metastasis upon Malat1 lncRNA loss. Genes Dev 30:34–51

Atianand MK, Hu W, Satpathy AT, Shen Y, Ricci EP, Alvarez-Dominguez JR, Bhatta A, Schattgen SA, McGowan JD, Blin J et al (2016) A long noncoding RNA lincRNA-EPS acts as a transcriptional brake to restrain inflammation. Cell 165:1672–1685

Avner P, Heard E (2001) X-chromosome inactivation: counting, choice and initiation. Nat Rev Genet 2:59

Balk B, Maicher A, Dees M, Klermund J, Luke-Glaser S, Bender K, Luke B (2013) Telomeric RNA-DNA hybrids affect telomere-length dynamics and senescence. Nat Struct Mol Biol 20:1199–1205

Bassett AR, Akhtar A, Barlow DP, Bird AP, Brockdorff N, Duboule D, Ephrussi A, Ferguson-Smith AC, Gingeras TR, Haerty W et al (2014) Considerations when investigating lncRNA function in vivo. Elife 3:e03058

Beletskii A, Hong Y-K, Pehrson J, Egholm M, Strauss WM (2001) PNA interference mapping demonstrates functional domains in the noncoding RNA Xist. Proc Natl Acad Sci U S A 98:9215–9220

Blank-Giwojna A, Postepska-Igielska A, Grummt I (2019) lncRNA KHPS1 activates a poised enhancer by triplex-dependent recruitment of epigenomic regulators. Cell Rep 26:2904–2915.e4

Cabili MN, Trapnell C, Goff L, Koziol M, Tazon-Vega B, Regev A, Rinn JL (2011) Integrative annotation of human large intergenic noncoding RNAs reveals global properties and specific subclasses. Genes Dev 25:1915–1927

Capshew CR, Dusenbury KL, Hundley HA (2012) Inverted Alu dsRNA structures do not affect localization but can alter translation efficiency of human mRNAs independent of RNA editing. Nucleic Acids Res 40:8637–8645

Carlevaro-Fita J, Polidori T, Das M, Navarro C, Zoller TI, Johnson R (2019) Ancient exapted transposable elements promote nuclear enrichment of human long noncoding RNAs. Genome Res 29:208–222

Carrieri C, Cimatti L, Biagioli M, Beugnet A, Zucchelli S, Fedele S, Pesce E, Ferrer I, Collavin L, Santoro C et al (2012) Long non-coding antisense RNA controls Uchl1 translation through an embedded SINEB2 repeat. Nature 491:454

Chen L-L, Carmichael GG (2009) Altered nuclear retention of mRNAs containing inverted repeats in human embryonic stem cells: functional role of a nuclear noncoding RNA. Mol Cell 35:467–478

Chen L, DeCerbo JN, Carmichael GG (2008) Alu element-mediated gene silencing. EMBO J 27:1694–1705

Chen C-K, Blanco M, Jackson C, Aznauryan E, Ollikainen N, Surka C, Chow A, Cerase A, McDonel P, Guttman M (2016a) Xist recruits the X chromosome to the nuclear lamina to enable chromosome-wide silencing. Science 354:468–472

Chen J, Shishkin AA, Zhu X, Kadri S, Maza I, Guttman M, Hanna JH, Regev A, Garber M (2016b) Evolutionary analysis across mammals reveals distinct classes of long non-coding RNAs. Genome Biol 17:19. https://doi.org/10.1186/s13059-016-0880-9

Chu C, Zhang Q, da Rocha S, Flynn RA, Bharadwaj M, Calabrese MJ, Magnuson T, Heard E, Chang HY (2015) Systematic discovery of Xist RNA binding proteins. Cell 161:404–416

Chu H-P, Cifuentes-Rojas C, Kesner B, Aeby E, Lee H, Wei C, Oh H, Boukhali M, Haas W, Lee JT (2017) TERRA RNA antagonizes ATRX and protects telomeres. Cell 170:86–101.e16

Clark MB, Mercer TR, Bussotti G, Leonardi T, Haynes KR, Crawford J, Brunck ME, Cao K-A, Thomas GP, Chen WY et al (2015) Quantitative gene profiling of long noncoding RNAs with targeted RNA sequencing. Nat Methods 12(4):339–342. https://doi.org/10.1038/nmeth.3321

Clemson C, McNeil J, Willard H, Lawrence J (1996) XIST RNA paints the inactive X chromosome at interphase: evidence for a novel RNA involved in nuclear/chromosome structure. J Cell Biol 132:259–275

Clemson CM, Hutchinson JN, Sara SA, Ensminger AW, Fox AH, Chess A, Lawrence JB (2009) An architectural role for a nuclear noncoding RNA: NEAT1 RNA is essential for the structure of Paraspeckles. Mol Cell 33:717–726

Colognori D, Sunwoo H, Kriz AJ, Wang C-Y, Lee JT (2019) Xist Deletional analysis reveals an interdependency between Xist RNA and Polycomb complexes for spreading along the inactive X. Mol Cell 74:101–117.e10

da Rocha S, Boeva V, Escamilla-Del-Arenal M, Ancelin K, Granier C, Matias N, Sanulli S, Chow J, Schulz E, Picard C et al (2014) Jarid2 is implicated in the initial Xist-induced targeting of PRC2 to the inactive X chromosome. Mol Cell 53:301–316

Davidovich C, Zheng L, Goodrich KJ, Cech TR (2013) Promiscuous RNA binding by Polycomb repressive complex 2. Nat Struct Mol Biol 20:1250–1257

Derrien T, Johnson R, Bussotti G, Tanzer A, Djebali S, Tilgner H, Guernec G, Martin D, Merkel A, Knowles DG et al (2012) The GENCODE v7 catalog of human long noncoding RNAs: analysis of their gene structure, evolution, and expression. Genome Res 22:1775–1789

Doksani Y, de Lange T (2014) The role of double-strand break repair pathways at functional and dysfunctional telomeres. CSH Perspect Biol 6:a016576

Elbarbary RA, Li W, Tian B, Maquat LE (2013) STAU1 binding 3′ UTR IRAlus complements nuclear retention to protect cells from PKR-mediated translational shutdown. Genes Dev 27:1495–1510

Elguindy MM, Kopp F, Goodarzi M, Rehfeld F, Thomas A, Chang T-C, Mendell JT (2019) PUMILIO, but not RBMX, binding is required for regulation of genomic stability by noncoding RNA NORAD. bioRxiv. https://doi.org/10.1101/645960

Ellington AD, Szostak JW (1990) In vitro selection of RNA molecules that bind specific ligands. Nature 346:346818a0

Engreitz JM, Pandya-Jones A, McDonel P, Shishkin A, Sirokman K, Surka C, Kadri S, Xing J, Goren A, Lander ES et al (2013) The Xist lncRNA exploits three-dimensional genome architecture to spread across the X chromosome. Science 341:1237973

Engreitz JM, Sirokman K, McDonel P, Shishkin AA, Surka C, Russell P, Grossman SR, Chow AY, Guttman M, Lander ES (2014) RNA-RNA interactions enable specific targeting of noncoding RNAs to nascent pre-mRNAs and chromatin sites. Cell 159:188–199

Engreitz JM, Haines JE, Perez EM, Munson G, Chen J, Kane M, McDonel PE, Guttman M, Lander ES (2016) Local regulation of gene expression by lncRNA promoters, transcription and splicing. Nature 539:452

Gerstberger S, Hafner M, Tuschl T (2014) A census of human RNA-binding proteins. Nat Rev Genet 15:829–845

Gibbons HR, Shaginurova G, Kim LC, Chapman N, Spurlock CF, Aune TM (2018) Divergent lncRNA GATA3-AS1 regulates GATA3 transcription in T-helper 2 cells. Front Immunol 9:2512

Gilbert LA, Horlbeck MA, Adamson B, Villalta JE, Chen Y, Whitehead EH, Guimaraes C, Panning B, Ploegh HL, Bassik MC et al (2014) Genome-scale CRISPR-mediated control of gene repression and activation. Cell 159:647–661

Goff LA, Rinn JL (2015) Linking RNA biology to lncRNAs. Genome Res 25:1456–1465

Gong C, Maquat LE (2011) lncRNAs transactivate STAU1-mediated mRNA decay by duplexing with 3′ UTRs via Alu elements. Nature 470:284

Gong C, Maquat LE (2014) Regulatory non-coding RNAs, methods and protocols. Methods Mol Biol 1206:81–86

Gong C, Popp M, Maquat LE (2012) Biochemical analysis of long non-coding RNA-containing ribonucleoprotein complexes. Methods 58:88–93

Groff AF, Sanchez-Gomez DB, Soruco M, Gerhardinger C, Barutcu RA, Li E, Elcavage L, Plana O, Sanchez LV, Lee JC et al (2016) In vivo characterization of Linc-p21 reveals functional *cis*-regulatory DNA elements. Cell Rep 16:2178–2186

Grote P, Herrmann BG (2013) The long non-coding RNA Fendrr links epigenetic control mechanisms to gene regulatory networks in mammalian embryogenesis. RNA Biol 10:1579–1585

Guttman M, Rinn JL (2012) Modular regulatory principles of large non-coding RNAs. Nature 482:339

Guttman M, Garber M, Levin JZ, Donaghey J, Robinson J, Adiconis X, Fan L, Koziol MJ, Gnirke A, Nusbaum C et al (2010) Ab initio reconstruction of cell type–specific transcriptomes in mouse reveals the conserved multi-exonic structure of lincRNAs. Nat Biotechnol 28:503

Hacisuleyman E, Goff LA, Trapnell C, Williams A, Henao-Mejia J, Sun L, McClanahan P, Hendrickson DG, Sauvageau M, Kelley DR et al (2014) Topological organization of multichromosomal regions by the long intergenic noncoding RNA Firre. Nat Struct Mol Biol 21:198–206

Hacisuleyman E, Shukla CJ, Weiner CL, Rinn JL (2016) Function and evolution of local repeats in the firre locus. Nat Commun 7:11021

Hafner M, Landthaler M, Burger L, Khorshid M, Hausser J, Berninger P, Rothballer A, Ascano M, Jungkamp A-C, Munschauer M et al (2010) Transcriptome-wide identification of RNA-binding protein and MicroRNA target sites by PAR-CLIP. Cell 141:129–141

Hendrickson DG, Kelley DR, Tenen D, Bernstein B, Rinn JL (2016) Widespread RNA binding by chromatin-associated proteins. Genome Biol 17:28

Hezroni H, Koppstein D, Schwartz MG, Avrutin A, Bartel DP, Ulitsky I (2015) Principles of long noncoding RNA evolution derived from direct comparison of transcriptomes in 17 species. Cell Rep 11:1110–1122

Hirose T, Virnicchi G, Tanigawa A, Naganuma T, Li R, Kimura H, Yokoi T, Nakagawa S, Bénard M, Fox AH et al (2014) NEAT1 long noncoding RNA regulates transcription via protein sequestration within subnuclear bodies. Mol Biol Cell 25:169–183

Hon C-C, Ramilowski JA, Harshbarger J, Bertin N, Rackham OJ, Gough J, Denisenko E, Schmeier S, Poulsen TM, Severin J et al (2017) An atlas of human long non-coding RNAs with accurate 5′ ends. Nature 543:199

Hu W, Yuan B, Flygare J, Lodish HF (2011) Long noncoding RNA-mediated anti-apoptotic activity in murine erythroid terminal differentiation. Genes Dev 25:2573–2578

Hung T, Wang Y, Lin MF, Koegel AK, Kotake Y, Grant GD, Horlings HM, Shah N, Umbricht C, Wang P et al (2011) Extensive and coordinated transcription of noncoding RNAs within cell-cycle promoters. Nat Genet 43:621

Hutchinson JN, Ensminger AW, Clemson CM, Lynch CR, Lawrence JB, Chess A (2007) A screen for nuclear transcripts identifies two linked noncoding RNAs associated with SC35 splicing domains. BMC Genomics 8:39

Imamura K, Imamachi N, Akizuki G, Kumakura M, Kawaguchi A, Nagata K, Kato A, Kawaguchi Y, Sato H, Yoneda M et al (2014) Long noncoding RNA NEAT1-dependent

SFPQ relocation from promoter region to Paraspeckle mediates IL8 expression upon immune stimuli. Mol Cell 53:393–406

Johnson R, Guigó R (2014) The RIDL hypothesis: transposable elements as functional domains of long noncoding RNAs. RNA 20:959–976

Kaewsapsak P, Shechner D, Mallard W, Rinn JL, Ting AY (2017) Live-cell mapping of organelle-associated RNAs via proximity biotinylation combined with protein-RNA crosslinking. Elife 6: e29224

Kapusta A, Kronenberg Z, Lynch VJ, Zhuo X, Ramsay L, Bourque G, Yandell M, Feschotte C (2013) Transposable elements are major contributors to the origin, diversification, and regulation of vertebrate long noncoding RNAs. PLoS Genet 9:e1003470

Kelley D, Rinn J (2012) Transposable elements reveal a stem cell-specific class of long noncoding RNAs. Genome Biol 13:R107

Kelley DR, Hendrickson DG, Tenen D, Rinn JL (2014) Transposable elements modulate human RNA abundance and splicing via specific RNA-protein interactions. Genome Biol 15:537

Khalil AM, Guttman M, Huarte M, Garber M, Raj A, Morales D, Thomas K, Presser A, Bernstein BE, van Oudenaarden A et al (2009) Many human large intergenic noncoding RNAs associate with chromatin-modifying complexes and affect gene expression. Proc Natl Acad Sci U S A 106:11667–11672

Kim Y, Furic L, DesGroseillers L, Maquat LE (2005) Mammalian Staufen1 recruits Upf1 to specific mRNA 3′UTRs so as to elicit mRNA decay. Cell 120:195–208

Kino T, Hurt DE, Ichijo T, Nader N, Chrousos GP (2010) Noncoding RNA Gas5 is a growth arrest–and starvation-associated repressor of the glucocorticoid receptor. Sci Signal 3:ra8

König J, Zarnack K, Rot G, Curk T, Kayikci M, Zupan B, Turner DJ, Luscombe NM, Ule J (2010) iCLIP reveals the function of hnRNP particles in splicing at individual nucleotide resolution. Nat Struct Mol Biol 17:909

Kopp F, Mendell JT (2018) Functional classification and experimental dissection of long noncoding RNAs. Cell 172:393–407

Kotzin JJ, Spencer SP, McCright SJ, Kumar DB, Collet MA, Mowel WK, Elliott EN, Uyar A, Makiya MA, Dunagin MC et al (2016) The long non-coding RNA Morrbid regulates Bim and short-lived myeloid cell lifespan. Nature 537:239

Latos PA, Pauler FM, Koerner MV, Şenergin BH, Hudson QJ, Stocsits RR, Allhoff W, Stricker SH, Klement RM, Warczok KE et al (2012) *Airn* transcriptional overlap, but not its lncRNA products, induces imprinted *Igf2r* silencing. Science 338:1469–1472

Lee JT, Jaenisch R (1997) Long-range cis effects of ectopic X-inactivation centres on a mouse autosome. Nature 386:386275a0

Lee S, Kopp F, Chang T-C, Sataluri A, Chen B, Sivakumar S, Yu H, Xie Y, Mendell JT (2016) Noncoding RNA NORAD regulates genomic stability by sequestering PUMILIO proteins. Cell 164:69–80

Leppek K, Stoecklin G (2014) An optimized streptavidin-binding RNA aptamer for purification of ribonucleoprotein complexes identifies novel ARE-binding proteins. Nucleic Acids Res 42: e13–e13

Leucci E, Vendramin R, Spinazzi M, Laurette P, Fiers M, Wouters J, Radaelli E, Eyckerman S, Leonelli C, Vanderheyden K et al (2016) Melanoma addiction to the long non-coding RNA SAMMSON. Nature 531:518

Li L, Chang HY (2014) Physiological roles of long noncoding RNAs: insight from knockout mice. Trends Cell Biol 24:594–602

Lin A, Li C, Xing Z, Hu Q, Liang K, Han L, Wang C, Hawke DH, Wang S, Zhang Y et al (2016) The LINK-A lncRNA activates normoxic HIF1α signalling in triple-negative breast cancer. Nat Cell Biol 18(2):213–224. https://doi.org/10.1038/ncb3295

Liu B, Sun L, Liu Q, Gong C, Yao Y, Lv X, Lin L, Yao H, Su F, Li D et al (2015) A cytoplasmic NF-κB interacting long noncoding RNA blocks IκB phosphorylation and suppresses breast Cancer metastasis. Cancer Cell 27:370–381

Liu JS, Horlbeck MA, Cho S, Birk HS, Malatesta M, He D, Attenello FJ, Villalta JE, Cho MY, Chen Y et al (2017) CRISPRi-based genome-scale identification of functional long noncoding RNA loci in human cells. Science 355:eaah7111

Lovejoy CA, Li W, Reisenweber S, Thongthip S, Bruno J, de Lange T, De S, Petrini JH, Sung PA, Jasin M et al (2012) Loss of ATRX, genome instability, and an altered DNA damage response are hallmarks of the alternative lengthening of telomeres pathway. PLoS Genet 8:e1002772

Lubelsky Y, Ulitsky I (2018) Sequences enriched in Alu repeats drive nuclear localization of long RNAs in human cells. Nature 555:107

Maass PG, Rump A, Schulz H, Stricker S, Schulze L, Platzer K, Aydin A, Tinschert S, Goldring MB, Luft FC et al (2012) A misplaced lncRNA causes brachydactyly in humans. J Clin Invest 122:3990–4002

Mao YS, Sunwoo H, Zhang B, Spector DL (2010) Direct visualization of the co-transcriptional assembly of a nuclear body by noncoding RNAs. Nat Cell Biol 13:95

McHugh CA, Russell P, Guttman M (2014) Methods for comprehensive experimental identification of RNA-protein interactions. Genome Biol 15:203

McHugh CA, Chen C-K, Chow A, Surka CF, Tran C, McDonel P, Pandya-Jones A, Blanco M, Burghard C, Moradian A et al (2015) The Xist lncRNA interacts directly with SHARP to silence transcription through HDAC3. Nature 521:232

Mélèse T, Xue Z (1995) The nucleolus: an organelle formed by the act of building a ribosome. Curr Opin Cell Biol 7:319–324

Mercer TR, Neph S, Dinger ME, Crawford J, Smith MA, Shearwood A-MJ, Haugen E, Bracken CP, Rackham O, Stamatoyannopoulos JA et al (2011) The human mitochondrial transcriptome. Cell 146:645–658

Mili S, Steitz JA (2004) Evidence for reassociation of RNA-binding proteins after cell lysis: implications for the interpretation of immunoprecipitation analyses. RNA 10:1692–1694

Miller MA, Olivas WM (2011) Roles of Puf proteins in mRNA degradation and translation. Wiley Interdiscip Rev RNA 2:471–492

Misteli T, Cáceres JF, Clement JQ, Krainer AR, Wilkinson MF, Spector DL (1998) Serine phosphorylation of SR proteins is required for their recruitment to sites of transcription in vivo. J Cell Biol 143:297–307

Mondal T, Subhash S, Vaid R, Enroth S, Uday S, Reinius B, Mitra S, Mohammed A, James A, Hoberg E et al (2015) MEG3 long noncoding RNA regulates the TGF-β pathway genes through formation of RNA–DNA triplex structures. Nat Commun 6:7743

Monfort A, Di Minin G, Postlmayr A, Freimann R, Arieti F, Thore S, Wutz A (2015) Identification of Spen as a crucial factor for Xist function through forward genetic screening in haploid embryonic stem cells. Cell Rep 12:554–561

Munschauer M, Nguyen CT, Sirokman K, Hartigan CR, Hogstrom L, Engreitz JM, Ulirsch JC, Fulco CP, Subramanian V, Chen J et al (2018) The NORAD lncRNA assembles a topoisomerase complex critical for genome stability. Nature 561:132–136

Naganuma T, Nakagawa S, Tanigawa A, Sasaki YF, Goshima N, Hirose T (2012) Alternative 3′-end processing of long noncoding RNA initiates construction of nuclear paraspeckles. EMBO J 31:4020–4034

Nelson BR, Makarewich CA, Anderson DM, Winders BR, Troupes CD, Wu F, Reese AL, McAnally JR, Chen X, Kavalali ET et al (2016) A peptide encoded by a transcript annotated as long noncoding RNA enhances SERCA activity in muscle. Science 351:271–275

Ni T, Yang Y, Hafez D, Yang W, Kiesewetter K, Wakabayashi Y, Ohler U, Peng W, Zhu J (2013) Distinct polyadenylation landscapes of diverse human tissues revealed by a modified PA-seq strategy. BMC Genomics 14:615

Noh J, Kim K, Abdelmohsen K, Yoon J-H, Panda AC, Munk R, Kim J, Curtis J, Moad CA, Wohler CM et al (2016) HuR and GRSF1 modulate the nuclear export and mitochondrial localization of the lncRNA RMRP. Genes Dev 30:1224–1239

O'Leary V, Ovsepian S, Carrascosa L, Buske F, Radulovic V, Niyazi M, Moertl S, Trau M, Atkinson M, Anastasov N (2015) PARTICLE, a triplex-forming long ncRNA, regulates locus-specific methylation in response to low-dose irradiation. Cell Rep 11:474–485

Paralkar VR, Taborda CC, Huang P, Yao Y, Kossenkov AV, Prasad R, Luan J, Davies J, Hughes JR, Hardison RC et al (2016) Unlinking an lncRNA from its associated cis element. Mol Cell 62:104–110

Parrott AM, Lago H, Adams CJ, Ashcroft AE, Stonehouse NJ, Stockley PG (2000) RNA aptamers for the MS2 bacteriophage coat protein and the wild-type RNA operator have similar solution behaviour. Nucleic Acids Res 28:489–497

Penny GD, Kay GF, Sheardown SA, Rastan S, Brockdorff N (1996) Requirement for Xist in X chromosome inactivation. Nature 379. https://doi.org/10.1038/379131a0

Pfeiffer V, Crittin J, Grolimund L, Lingner J (2013) The THO complex component Thp2 counteracts telomeric R-loops and telomere shortening. EMBO J 32:2861–2871

Plath K, Mlynarczyk-Evans S, Nusinow DA, Panning B (2002) Xist RNAand the mechanism of X chromosome inactivation. Annu Rev Genet 36:233–278

Plath K, Fang J, Mlynarczyk-Evans SK, Cao R, Worringer KA, Wang H, de la Cruz CC, Otte AP, Panning B, Zhang Y (2003) Role of histone H3 lysine 27 methylation in X inactivation. Science 300:131–135

Portoso M, Ragazzini R, Brenčič Ž, Moiani A, Michaud A, Vassilev I, Wassef M, Servant N, Sargueil B, Margueron R (2017) PRC2 is dispensable for HOTAIR-mediated transcriptional repression. EMBO J 36:981–994

Postepska-Igielska A, Giwojna A, Gasri-Plotnitsky L, Schmitt N, Dold A, Ginsberg D, Grummt I (2015) LncRNA Khps1 regulates expression of the proto-oncogene SPHK1 via triplex-mediated changes in chromatin structure. Mol Cell 60:626–636

Prasanth KV, Prasanth SG, Xuan Z, Hearn S, Freier SM, Bennett FC, Zhang MQ, Spector DL (2005) Regulating gene expression through RNA nuclear retention. Cell 123:249–263

Qi LS, Larson MH, Gilbert LA, Doudna JA, Weissman JS, Arkin AP, Lim WA (2013) Repurposing CRISPR as an RNA-guided platform for sequence-specific control of gene expression. Cell 152:1173–1183

Ramanathan M, Majzoub K, Rao DS, Neela PH, Zarnegar BJ, Mondal S, Roth JG, Gai H, Kovalski JR, Siprashvili Z et al (2018) RNA–protein interaction detection in living cells. Nat Methods 15:207

Ramos AD, Andersen RE, Liu S, Nowakowski T, Hong S, Gertz CC, Salinas RD, Zarabi H, Kriegstein AR, Lim DA (2015) The long noncoding RNA Pnky regulates neuronal differentiation of embryonic and postnatal neural stem cells. Cell Stem Cell 16:439–447

Rinn JL, Chang HY (2012) Genome regulation by long noncoding RNAs. Annu Rev Biochem 81:145–166

Rinn JL, Kertesz M, Wang JK, Squazzo SL, Xu X, Brugmann SA, Goodnough HL, Helms JA, Farnham PJ, Segal E et al (2007) Functional demarcation of active and silent chromatin domains in human HOX loci by noncoding RNAs. Cell 129:1311–1323

Santoro F, Mayer D, Klement RM, Warczok KE, Stukalov A, Barlow DP, Pauler FM (2013) Imprinted Igf2r silencing depends on continuous Airn lncRNA expression and is not restricted to a developmental window. Development 140:1184–1195

Sanz LA, Hartono SR, Lim Y, Steyaert S, Rajpurkar A, Ginno PA, Xu X, Chédin F (2016) Prevalent, dynamic, and conserved R-loop structures associate with specific epigenomic signatures in mammals. Mol Cell 63:167–178

Sasaki YT, Ideue T, Sano M, Mituyama T, Hirose T (2009) MENε/β noncoding RNAs are essential for structural integrity of nuclear paraspeckles. Proc Natl Acad Sci U S A 106:2525–2530

Sauvageau M, Goff LA, Lodato S, Bonev B, Groff AF, Gerhardinger C, Sanchez-Gomez DB, Hacisuleyman E, Li E, Spence M et al (2013) Multiple knockout mouse models reveal lincRNAs are required for life and brain development. Elife 2:e01749

Sentürk Cetin N, Kuo C-C, Ribarska T, Li R, Costa IG, Grummt I (2019) Isolation and genome-wide characterization of cellular DNA:RNA triplex structures. Nucleic Acids Res 47. https://doi.org/10.1093/nar/gky1305

Sharma S, Findlay GM, Bandukwala HS, Oberdoerffer S, Baust B, Li Z, Schmidt V, Hogan PG, Sacks DB, Rao A (2011) Dephosphorylation of the nuclear factor of activated T cells (NFAT) transcription factor is regulated by an RNA-protein scaffold complex. Proc Natl Acad Sci U S A 108:11381–11386

Shukla CJ, McCorkindale AL, Gerhardinger C, Korthauer KD, Cabili MN, Shechner D, Irizarry RA, Maass PG, Rinn JL (2018) High-throughput identification of RNA nuclear enrichment sequences. EMBO J 37:e98452

Silva J, Mak W, Zvetkova I, Appanah R, Nesterova TB, Webster Z, Peters A, Jenuwein T, Otte AP, Brockdorff N (2003) Establishment of histone H3 methylation on the inactive X chromosome requires transient recruitment of Eed-Enx1 Polycomb group complexes. Dev Cell 4:481–495

Simon MD, Wang CI, Kharchenko PV, West JA, Chapman BA, Alekseyenko AA, Borowsky ML, Kuroda MI, Kingston RE (2011) The genomic binding sites of a noncoding RNA. Proc Natl Acad Sci U S A 108:20497–20502

Souquere S, Beauclair G, Harper F, Fox A, Pierron G (2010) Highly ordered spatial organization of the structural long noncoding NEAT1 RNAs within Paraspeckle nuclear bodies. Mol Biol Cell 21:4020–4027

Srisawat C, Engelke DR (2001) Streptavidin aptamers: affinity tags for the study of RNAs and ribonucleoproteins. RNA 7:632–641

Sundararaman B, Zhan L, Blue SM, Stanton R, Elkins K, Olson S, Wei X, Van Nostrand EL, Pratt GA, Huelga SC et al (2016) Resources for the comprehensive discovery of functional RNA elements. Mol Cell 61:903–913

Tichon A, Gil N, Lubelsky Y, Solomon T, Lemze D, Itzkovitz S, Stern-Ginossar N, Ulitsky I (2016) A conserved abundant cytoplasmic long noncoding RNA modulates repression by Pumilio proteins in human cells. Nat Commun 7:12209

Tripathi V, Ellis JD, Shen Z, Song DY, Pan Q, Watt AT, Freier SM, Bennett FC, Sharma A, Bubulya PA et al (2010) The nuclear-retained noncoding RNA MALAT1 regulates alternative splicing by modulating SR splicing factor phosphorylation. Mol Cell 39:925–938

Tsai B, Wang X, Huang L, Waterman ML (2011) Quantitative profiling of in vivo-assembled RNA-protein complexes using a novel integrated proteomic approach. Mol Cell Proteomics 10: M110.007385

Ule J, Jensen KB, Ruggiu M, Mele A, Ule A, Darnell RB (2003) CLIP identifies Nova-regulated RNA networks in the brain. Science 302:1212–1215

Ulitsky I, Bartel DP (2013) lincRNAs: genomics, evolution, and mechanisms. Cell 154:26–46

Vendramin R, Verheyden Y, Ishikawa H, Goedert L, Nicolas E, Saraf K, Armaos A, Ponti R, Izumikawa K, Mestdagh P et al (2018) SAMMSON fosters cancer cell fitness by concertedly enhancing mitochondrial and cytosolic translation. Nat Struct Mol Biol 25:1035–1046

Wang P, Xue Y, Han Y, Lin L, Wu C, Xu S, Jiang Z, Xu J, Liu Q, Cao X (2014) The STAT3-binding long noncoding RNA lnc-DC controls human dendritic cell differentiation. Science 344:310–313

Wang Z, Zhang X-J, Ji Y-X, Zhang P, Deng K-Q, Gong J, Ren S, Wang X, Chen I, Wang H et al (2016) The long noncoding RNA Chaer defines an epigenetic checkpoint in cardiac hypertrophy. Nat Med 22:1131–1139

West S, Gromak N, Proudfoot NJ (2004) Human $5' \rightarrow 3'$ exonuclease Xrn2 promotes transcription termination at co-transcriptional cleavage sites. Nature 432:522

West JA, Davis CP, Sunwoo H, Simon MD, Sadreyev RI, Wang PI, Tolstorukov MY, Kingston RE (2014) The long noncoding RNAs NEAT1 and MALAT1 bind active chromatin sites. Mol Cell 55:791–802

West JA, Mito M, Kurosaka S, Takumi T, Tanegashima C, Chujo T, Yanaka K, Kingston RE, Hirose T, Bond C et al (2016) Structural, super-resolution microscopy analysis of paraspeckle nuclear body organization. J Cell Biol 214:817–830

Willingham A, Orth A, Batalov S, Peters E, Wen B, Aza-Blanc P, Hogenesch J, Schultz P (2005) A strategy for probing the function of noncoding RNAs finds a repressor of NFAT. Science 309:1570–1573

Wutz A, Rasmussen TP, Jaenisch R (2002) Chromosomal silencing and localization are mediated by different domains of Xist RNA. Nat Genet 30:167–174

Yamazaki T, Hirose T (2015) The building process of the functional paraspeckle with long non-coding RNAs. Front Biosci Elite Ed 7:1–41

Yamazaki T, Souquere S, Chujo T, Kobelke S, Chong Y, Fox AH, Bond CS, Nakagawa S, Pierron G, Hirose T (2018) Functional domains of NEAT1 architectural lncRNA induce Paraspeckle assembly through phase separation. Mol Cell 70:1038–1053.e7

Yan X, Hu Z, Feng Y, Hu X, Yuan J, Zhao SD, Zhang Y, Yang L, Shan W, He Q et al (2015) Comprehensive genomic characterization of long non-coding RNAs across human cancers. Cancer Cell 28:529–540

Zalatan JG, Lee ME, Almeida R, Gilbert LA, Whitehead EH, La Russa M, Tsai JC, Weissman JS, Dueber JE, Qi LS et al (2015) Engineering complex synthetic transcriptional programs with CRISPR RNA scaffolds. Cell 160:339–350

Zhang Z, Carmichael GG (2001) The fate of dsRNA in the nucleus A p54nrb-containing complex mediates the nuclear retention of promiscuously A-to-I edited RNAs. Cell 106:465–476

Zhang Q, McKenzie NJ, Warneford-Thomson R, Gail EH, Flanigan SF, Owen BM, Lauman R, Levina V, Garcia BA, Schittenhelm RB et al (2019) RNA exploits an exposed regulatory site to inhibit the enzymatic activity of PRC2. Nat Struct Mol Biol 26:237–247

Zhao J, Sun BK, Erwin JA, Song J-J, Lee JT (2008) Polycomb proteins targeted by a short repeat RNA to the mouse X chromosome. Science 322:750–756

Zhao J, Ohsumi TK, Kung JT, Ogawa Y, Grau DJ, Sarma K, Song J, Kingston RE, Borowsky M, Lee JT (2010) Genome-wide identification of Polycomb-associated RNAs by RIP-seq. Mol Cell 40:939–953

Zhao Z, Sentürk N, Song C, Grummt I (2018) lncRNA PAPAS tethered to the rDNA enhancer recruits hypophosphorylated CHD4/NuRD to repress rRNA synthesis at elevated temperatures. Genes Dev 32:836–848

Zucchelli S, Fasolo F, Russo R, Cimatti L, Patrucco L, Takahashi H, Jones MH, Santoro C, Sblattero D, Cotella D et al (2015) SINEUPs are modular antisense long non-coding RNAs that increase synthesis of target proteins in cells. Front Cell Neurosci 9:174

MIX
Papier aus verantwortungsvollen Quellen
Paper from responsible sources
FSC® C105338
FSC
www.fsc.org

If you have any concerns about our products,
you can contact us on
**ProductSafety@springernature.com**

In case Publisher is established outside the EU,
the EU authorized representative is:
**Springer Nature Customer Service Center GmbH**
**Europaplatz 3, 69115 Heidelberg, Germany**

Printed by Libri Plureos GmbH
in Hamburg, Germany